Efficient Technology for the Pretreatment of Biomass

Efficient Technology for the Pretreatment of Biomass

Editors

Ivet Ferrer
Helene Carrere
Aline Carvalho Da Costa
Cigdem Eskicioglu

MDPI • Basel • Beijing • Wuhan • Barcelona • Belgrade • Manchester • Tokyo • Cluj • Tianjin

Editors
Ivet Ferrer
Department of Civil and Environmental Engineering
Universitat Politècnica de Catalunya·BarcelonaTech
Barcelona
Spain

Helene Carrere
INRAE, LBE, Laboratory of Environmental Biotechnology
Montpellier University
Narbonne
France

Aline Carvalho Da Costa
School of Chemical Engineering
University of Campinas (UNICAMP)
Campinas
Brazil

Cigdem Eskicioglu
School of Engineering
University of British Columbia, Okanagan Campus
Kelowna
Canada

Editorial Office
MDPI
St. Alban-Anlage 66
4052 Basel, Switzerland

This is a reprint of articles from the Special Issue published online in the open access journal *Molecules* (ISSN 1420-3049) (available at: www.mdpi.com/journal/molecules/special_issues/pretreatment).

For citation purposes, cite each article independently as indicated on the article page online and as indicated below:

LastName, A.A.; LastName, B.B.; LastName, C.C. Article Title. *Journal Name* **Year**, *Volume Number*, Page Range.

ISBN 978-3-0365-2391-0 (Hbk)
ISBN 978-3-0365-2390-3 (PDF)

Contents

About the Editors

Ivet Ferrer

Dr. Ivet Ferrer is a Full Professor at the Department of Civil and Environmental Engineering of the Universitat Politècnica de Catalunya·BarcelonaTech. Currently, she leads the Research Group of Environmental Engineering and Microbiology (GEMMA-UPC). Her main research topic is the optimization of biomass anaerobic digestion by applying pretreatments and co-digestion. She has also addressed the implementation and assessment of low-tech digesters for rural areas. Her research is now focused on the recovery of resources from waste streams, including high-value bioproducts and biogas.

Helene Carrere

Dr. Hélène Carrère is an INRAE Research Director. She joined INRAE in 1994 to work on separation processes in Microbiology and Food Process Engineering unit. She moved to the Laboratory of Environmental Biotechnology in 2001 where her research activities focus on the study and development of pretreatments to improve the performance of anaerobic digestion and fermentation for the recovery of various feedstocks (e.g., sewage sludge, lignocellulosic biomass, manure, food waste, etc.).

Aline Carvalho Da Costa

Dr. Aline Carvalho da Costa is an Associate Professor at the School of Chemical Engineering of the University of Campinas and does research in the production of bioproducts and biofuels from renewable sources within the biorefinery concept. Among her main research areas are bioprocess modeling and optimization, as well as first and second generation ethanol production with an emphasis on pre-treatment technologies, enzymatic hydrolysis, and fermentation.

Cigdem Eskicioglu

Dr. Cigdem Eskicioglu is a Full Professor and NSERC/Metro Vancouver Senior Industrial Research Chair in the area of Resource Recovery from Wastewater in the School of Engineering at University of British Columbia (UBC)'s Okanagan Campus. She is also the Founder/Leader of the Bioreactor Technology Group (BTG), which focuses on advanced wastewater treatment processes for more efficient contaminant removal, energy conservation and production, resource recovery and mitigation of trace contaminants of emerging concern.

Preface to "Efficient Technology for the Pretreatment of Biomass"

Biomass corresponds to organic matter of animal, vegetable, microbial, or algal origin. Biomass use as feedstock for biomaterial, chemicals, platform molecules, biofuel or bioenergy are the most reliable alternatives to limit fossil fuel consumption and to reduce greenhouse gas emissions. Resource recovery from different kinds of waste, such as sludge, food waste, municipal solid waste, and animal waste (manure and slaughterhouse waste) is particularly interesting from an environmental point of view, as it also reduces environmental pollution. In addition, lignocellulosic biomass and algae, which do not compete for food production, represent an important source of renewable resources (i.e., energy and other value-added products). However, a pretreatment step is generally required before biomass (bio)-conversion into valuable products in order to increase the process yield and/or productivity.

Pretreatments are applied upstream of various conversion processes of biomass into biofuel or biomaterial with valuable end products such as bioethanol, biohydrogen, biomethane, biomolecules or biomaterials. Pretreatments cover a wide range of processes that include mechanical, thermal, chemical and biological techniques. This step is recognized as crucial and cost intensive for the development of biorefineries. Thus, more research is necessary to identify the most effective and economical pretreatment options for different biomass sources.

This Special Issue aims to gather research papers on recent developments of biomass pretreatments for biomaterial, chemicals, biofuel or bioenergy production, in the fields of Chemistry Sciences, Process Engineering, Chemical Engineering, Modeling and Control, Energy and Fuels, and Bioprocesses.

Ivet Ferrer, Helene Carrere, Aline Carvalho Da Costa, Cigdem Eskicioglu
Editors

molecules

Article

Strategies to Optimize Microalgae Conversion to Biogas: Co-Digestion, Pretreatment and Hydraulic Retention Time

Maria Solé-Bundó [1], Humbert Salvadó [2], Fabiana Passos [3], Marianna Garfí [1] and Ivet Ferrer [1,*]

1 GEMMA—Group of Environmental Engineering and Microbiology, Department of Civil and Environmental Engineering, Universitat Politècnica de Catalunya·BarcelonaTech, c/Jordi Girona 1-3, Building D1, E-08034 Barcelona, Spain; maria.sole-bundo@upc.edu (M.S.-B.); marianna.garfi@upc.edu (M.G.)
2 Department of Evolutionary Biology, Ecology and Environmental Sciences, Faculty of Biology, Universitat de Barcelona, Av. Diagonal 643, 08007 Barcelona, Spain; hsalvado@ub.edu
3 Department of Sanitary and Environmental Engineering, Federal University of Minas Gerais, Antonio Carlos Avenue 6627, 31270-090 Belo Horizonte, Brazil; fabiana@desa.ufmg.br
* Correspondence: ivet.ferrer@upc.edu; Tel.: +34-93-401-6463

Received: 13 July 2018; Accepted: 16 August 2018; Published: 21 August 2018

Abstract: This study aims at optimizing the anaerobic digestion (AD) of biomass in microalgal-based wastewater treatment systems. It comprises the co-digestion of microalgae with primary sludge, the thermal pretreatment (75 °C for 10 h) of microalgae and the role of the hydraulic retention time (HRT) in anaerobic digesters. Initially, a batch test comparing different microalgae (untreated and pretreated) and primary sludge proportions showed how the co-digestion improved the AD kinetics. The highest methane yield was observed by adding 75% of primary sludge to pretreated microalgae (339 mL CH_4/g VS). This condition was then investigated in mesophilic lab-scale reactors. The average methane yield was 0.46 L CH_4/g VS, which represented a 2.9-fold increase compared to pretreated microalgae mono-digestion. Conversely, microalgae showed a low methane yield despite the thermal pretreatment (0.16 L CH_4/g VS). Indeed, microscopic analysis confirmed the presence of microalgae species with resistant cell walls (i.e., *Stigioclonium* sp. and diatoms). In order to improve their anaerobic biodegradability, the HRT was increased from 20 to 30 days, which led to a 50% methane yield increase. Overall, microalgae AD was substantially improved by the co-digestion with primary sludge, even without pretreatment, and increasing the HRT enhanced the AD of microalgae with resistant cell walls.

Keywords: anaerobic digestion; bioenergy; co-digestion; hydraulic retention time; microalgal biomass; primary sludge; thermal pretreatment

1. Introduction

Algal biofuels call for low-cost technologies to be competitive with fossil fuels. In this context, microalgae cultivation in wastewater reduces freshwater and nutrient consumption while providing sanitation. Microalgal-based wastewater treatment systems consist of open ponds (e.g., high rate algal ponds (HRAPs)) capable of removing organic matter without aeration in the biological reactor, as with conventional activated sludge systems. Indeed, heterotrophic bacteria use the oxygen released through microalgae photosynthesis. The biomass grown in the ponds is then harvested to obtain a clarified effluent. Harvested biomass can be valorized as an organic fertilizer [1] or to produce bioenergy, with anaerobic digestion (AD) being the most straightforward technology for this purpose [2,3].

However, microalgae AD is limited by their resistant cell wall, which hampers the conversion into methane [4]. Thus, the application of pretreatment methods to damage or weaken the microalgae cell wall increases the bioavailability of intracellular contents to anaerobic microorganisms [5,6]. Even so, some pretreatments might result in higher costs (e.g., chemicals or biological products) or energy requirements (e.g., thermal or mechanical techniques) than the benefits obtained by implementing the pretreatment step (energy gain). This is a relevant aspect when choosing the most appropriate pretreatment for each substrate [7]. In this sense, microalgae thermal pretreatment at low temperature (<100 °C) has shown a promising energy balance [8].

In addition, the high nitrogen content (i.e., low C/N ratio) of microalgae can lead to methanogen inhibition due to ammonia toxicity during the AD process [9,10]. To overcome this issue, possible solutions include the reduction of protein levels in microalgae biomass by culturing them in low nitrogen media or the use of ammonia-tolerant anaerobic inoculum [11,12]. More commonly, the co-digestion (i.e., the simultaneous digestion of two or more substrates) of microalgae with other carbon-rich biomass has been proposed to reduce the ammonia concentration levels in the reactors while increasing the organic loading rate (OLR) [6,13]. In such a case, co-substrates obtained near or at the same treatment plant are preferred to avoid transport costs [14]. This strategy could be easily implemented in microalgal-based wastewater treatment plants (WWTPs), where harvested microalgal biomass could be co-digested with primary sludge from primary settlers. Indeed, primary sludge is more readily digestible and has less protein content than microalgae [15], so it could enhance microalgae biodegradability while increasing the OLR. To the best of our knowledge, only a few studies have evaluated the co-digestion of microalgae with primary sludge and always in batch tests [15,16]. Given that some benefits were pointed out (e.g., methane yield increase), these results should be validated better in continuous reactors.

The aim of this study is to optimize the AD process in WWTPs based on HRAP. Thus, the co-digestion of primary sludge from primary settlers and harvested microalgal biomass from HRAP (hereafter called microalgae) was investigated in both batch and continuous reactors. Moreover, a thermal pretreatment at 75 °C for 10 h was applied to microalgae, and the HRT of anaerobic digesters was increased to evaluate their effect on the microalgae methane yield. Microscopic analyses were used to help in understanding how microalgae were degraded during the pretreatment and AD process. Finally, an energy assessment of each studied scenario was calculated to attest the viability of full-scale application.

2. Results

The co-digestion of microalgae and primary sludge at different proportions was initially studied by means of biochemical methane potential (BMP) tests (Section 2.1.1). Subsequently, two continuous lab-scale anaerobic reactors were run in parallel (Table 1). During the first period, the co-digestion of pretreated microalgae with primary sludge was investigated (Section 2.1.2). During the second one, microalgae mono-digestion (with and without pretreatment) at longer HRT was compared (Section 2.2.1), including a microscopic analysis (Section 2.2.2).

Table 1. Experimental conditions during the mesophilic anaerobic digestion (AD) in lab-scale reactors. HRT: hydraulic retention time; VS: volatile solids.

	Period I (HRT = 20 Days)	**Period II (HRT = 30 Days)**
Digester 1	25% VS pretreated [1] microalgae + 75% VS primary sludge	Untreated microalgae
Digester 2	Pretreated [1] microalgae	Pretreated [1] microalgae

[1] 75 °C for 10 h.

2.1. Improving Microalgae Anaerobic Digestion by Co-Digestion with Primary Sludge and Thermal Pretreatment

2.1.1. Anaerobic Co-Digestion of Microalgae and Primary Sludge in Batch Tests

The co-digestion of microalgae with primary sludge was evaluated at different proportions (25%, 50% and 75% of microalgae, on a volatile solids (VS) basis) (Table 2). Additionally, in some trials, microalgae were pretreated at 75 °C for 10 h in order to solubilize the biomass and enhance the anaerobic digestion rate and extent [8]. Indeed, the microalgae methane yield was increased by 62% (from 90 to 146 mL CH_4/g VS) and the first-order kinetics constant (k) by 128% (from 0.07 to 0.16 day^{-1}) after the pretreatment (Table 2). However, primary sludge showed the highest methane yield (380 mL CH_4/g VS) and faster kinetics (k = 0.24 day^{-1}) as compared to untreated and pretreated microalgae. This is due to the nature of primary sludge, which is more readily digestible than microalgae.

Table 2. Ultimate methane yield (mean values ± standard deviation) and first-order kinetics constant (k) (error variance (s^2) represented in brackets) obtained in the biochemical methane potential (BMP) test.

Trial	Methane Yield ($mL\ CH_4/g\ VS$)		First-Order Kinetics (k) (Day^{-1})	
	Experimental Values [1]	Calculated Values [2]	Experimental Values [1]	Calculated Values [3]
Microalgae (M)	90 ± 2	-	0.07 (≤30)	-
75% M + 25% PS [4]	133 ± 6	162	0.27 (≤74)	0.16 (70)
50% M + 50% PS [4]	216 ± 1	234	0.28 (≤80)	0.20 (88)
25% M + 75% PS [4]	291 ± 9	306	0.27 (≤108)	0.23 (113)
Pretreated Microalgae (Mp)	146 ± 6	-	0.16 (≤75)	-
75% Mp + 25% PS [4]	183 ± 2	204	0.25 (≤85)	0.20 (72)
50% Mp + 50% PS [4]	249 ± 17	262	0.28 (≤99)	0.22 (82)
25% Mp + 75% PS [4]	339 ± 2	320	0.25 (≤150)	0.23 (107)
Primary Sludge (PS)	378 ± 4	-	0.24 (≤162)	-

[1] Experimental data from BMP tests; [2] Theoretical values calculated as the sum of the ultimate methane yield of each substrate mono-digestion times their proportion in the trial; [3] Values obtained from the curves that represent the theoretical values calculated as the sum of the ultimate methane yield of each substrate mono-digestion times their proportion in the trial over time; [4] volatile solids basis.

However, the co-digestion of microalgae with primary sludge substantially improved the anaerobic digestion kinetics (k = 0.25–0.28 day^{-1}) as compared to mono-digestion trials. Also, when comparing the experimental values of kinetics from co-digestion trials with those values calculated from the theoretical curves obtained as the sum of mono-digestion experimental values (Table 2), the experimental k value was always higher than the theoretical one. This means that mixing both substrates accelerated the AD process, as already observed in other cases [17,18]. This could contribute to reducing costs by decreasing the digesters' hydraulic retention time (HRT) and thus their volume. Still regarding the kinetics, no differences were observed between pretreated and untreated trials, since microalgae and primary sludge co-digestion without pretreatment already improved by far the anaerobic digestion rate. On the contrary, the pretreatment itself had already accelerated the kinetics of the process, so the effects of the co-digestion results were less discernible than for untreated substrates [19,20].

Otherwise, the higher the proportion of primary sludge, the higher the methane yield (Figure 1), with 339 mL CH_4/g VS being the highest methane yield achieved with the co-digestion of 75% primary sludge and 25% pretreated microalgae. These findings suggest that there was no synergic effect with respect to the ultimate methane production when co-digesting both substrates.

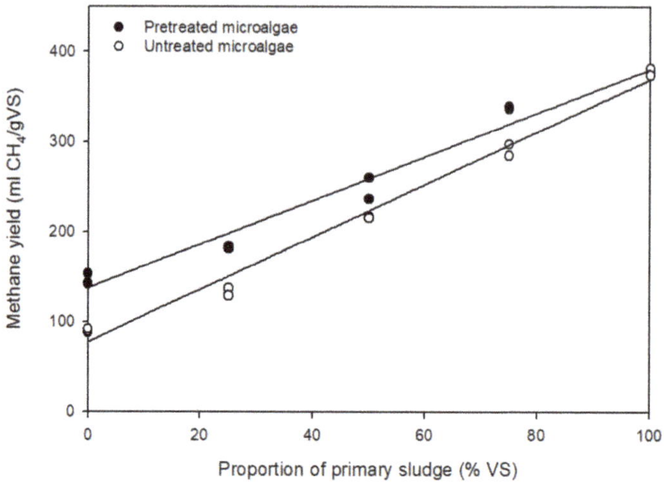

Figure 1. Correlation between the methane yield and the primary sludge proportion added to untreated and pretreated microalgae.

2.1.2. Anaerobic Co-Digestion of Microalgae and Primary Sludge in Lab-Scale Reactors

The best co-digestion condition (25–75% VS of thermally pretreated microalgae and primary sludge) from BMP tests was thereafter compared to the mono-digestion of thermally pretreated microalgae in lab-scale reactors (Table 3, Figure 2). During the whole experimental period, both reactors were operated with an OLR of around 1.2 g VS/(L·day), given the concentration of VS in microalgae harvested and thickened by gravity (around 4% TS and 2.5% VS) and the HRT (20 days).

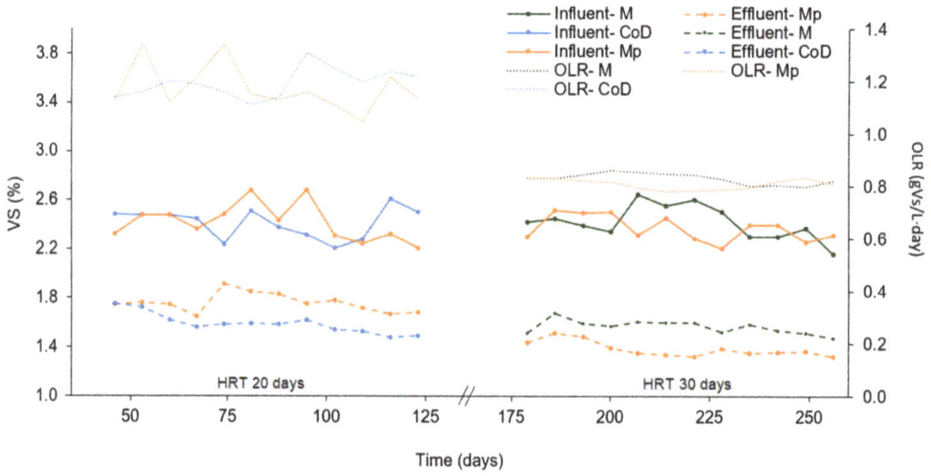

Figure 2. Influent and effluent volatile solids of untreated microalgae (M), thermally pretreated microalgae (Mp) and in co-digestion with primary sludge (CoD) for the studied periods: Period I at a HRT of 20 days and Period II at a HRT of 30 days.

Table 3. Biogas production, solids removal, influent (substrate) and effluent (digestate) characteristics from untreated or thermally pretreated microalgae AD and co-digestion with primary sludge in lab-scale reactors. Mean ± standard deviation. OLR: organic loading rate.

		Period I		Period II	
		Microalgae,p	**Co-Digestion**	**Microalgae**	**Microalgae,p**
Operational Conditions	HRT (days)	20	20	30	30
	OLR (g VS/L·day)	1.21 ± 0.06	1.17 ± 0.09	0.85 ± 0.01	0.81 ± 0.02
Biogas Production	Methane production rate (L CH_4/L·day)	0.20 ± 0.05	0.53 ± 0.29 [a]	0.12 ± 0.08	0.19 ± 0.07 [b]
	Methane yield (L CH_4/g VS)	0.16 ± 0.05	0.46 ± 0.27 [a]	0.14 ± 0.07	0.24 ± 0.07 [b]
	Methane content in biogas (% CH_4)	66.2 ± 2.62	71.7 ± 0.9 [a]	67.6 ± 1.6	69.5 ± 1.7
Removal Efficiency	TS removal (%)	16.6 ± 4.1	19.0 ± 1.7 [a]	18.6 ± 1.7	26.2 ± 3.7 [b]
	VS removal (%)	27.9 ± 1.9	34.3 ± 2.4 [a]	36.2 ± 2.5	39.5 ± 3.7 [b]
Influent Characteristics	TS [% (*w*/*w*)]	3.87 ± 0.28	4.13 ± 0.29	3.63 ± 0.48	3.42 ± 0.28
	VS [% (*w*/*w*)]	2.47 ± 0.17	2.38 ± 0.15	2.42 ± 0.14	2.37 ± 0.10
	VS/TS (%)	64 ± 3 [a]	58 ± 3	56 ± 2	55 ± 2
	COD (g O_2/L)	42.0 ± 6.7	42.9 ± 7.7	26.6 ± 1.6	25.2 ± 1.8
	TKN (g/L)	n.a.	n.a.	2.4 ± 0.1	2.3 ± 0.1
	N-NH_4 (g/L)	0.16 ± 0.07	0.13 ± 0.06	0.06 ± 0.01	0.26 ± 0.06 [b]
Effluent Characteristics	pH	7.55 ± 0.15 [a]	7.30 ± 0.08	7.35 ± 0.11	7.55 ± 0.08 [b]
	TS [% (*w*/*w*)]	3.49 ± 0.34	3.53 ± 0.18	2.87 ± 0.16	2.67 ± 0.27
	VS [% (*w*/*w*)]	1.77 ± 0.09 [a]	1.62 ± 0.11	1.58 ± 0.06	1.45 ± 0.11
	VS/TS (%)	51 ± 3 [a]	46 ± 2	56 ± 2	55 ± 2
	COD (g/L)	30.9 ± 2.1	29.0 ± 3.0	26.6 ± 1.6	25.2 ± 2.1
	N-NH_4 (g/L)	1.1 ± 0.2 [a]	0.6 ± 0.1	0.7 ± 0.1	0.8 ± 0.1
	VFA (mg COD/L)	124 (<756 [1])	44 (<757 [1])	0 (<0 [1])	130 (<596 [1])
	CST (s)	982 ± 61 [a]	290 ± 11	795 ± 71	919 ± 21 [b]

[1] Maximum value achieved. p = pretreated; TKN = total Kjeldahl nitrogen; VFA = volatile fatty acids; CST = capillarity suction time. [a,b] Stand for significantly higher values between paired columns ("a" for period I and "b" for period II) ($\alpha = 0.05$).

In the co-digestion reactor, the average methane yield was 0.46 L CH_4/g VS, which represented a 2.9-fold increase as compared to pretreated microalgae mono-digestion (0.16 L CH_4/g VS). In addition, the methane production rate increased from 0.20 to 0.53 L CH_4/(L·day). Despite this important increase in methane yield and methane production rate, the average VS removal was not much different (34.3% for co-digestion vs. 27.9% for mono-digestion). A possible reason for this is that primary sludge had a higher lipid content than microalgae, which are mainly composed of proteins. Indeed, our previous study quantified the content of lipids (45% and 24% VS) and proteins (29% and 58% VS) for primary sludge and microalgae, respectively [13]. Comparing the methane potential of both macromolecules, lipids can achieve 1.014 L CH_4/g VS and proteins only 0.851 L CH_4/g VS [21]. Therefore, the conversion potential of primary sludge to methane is higher than microalgae, as already observed in the BMP tests. The methane yield of the co-digestion reactor was higher than that obtained co-digesting sewage sludge with *Spirulina maxima* (50% VS each) at 20 days of HRT (0.36 L CH_4/g VS) [22], and similar to that obtained co-digesting *Scenedesmus* sp. or native microalgal biomass (25% VS) with sewage sludge (75% VS) at 15 days of HRT (0.39 and 0.51 L CH_4/g VS, respectively) [23].

Concerning the stability of digesters, pH values were stable during the whole period, ranging from 7.35 to 7.55 (Table 3). Regarding the ammonium concentration, the highest value was observed in the mono-digestion reactor with pretreated microalgae (1.1 g N-NH_4/L) due to a higher protein release during the AD process. This value is close to the threshold which resulted in AD inhibition [24]. Therefore, if reactors had been operated at higher OLRs, the inhibition of ammonia toxicity may have occurred. Conversely, co-digestion with primary sludge reduced the ammonium concentration in the digester to 0.6 g N-NH_4/L. In this case, the OLR could have been increased without approaching the ammonia inhibition threshold. VFA concentrations were also very low in both reactors (Table 3) Finally, an important aspect for the digestate management and final disposal is its dewaterability. While the digestate from thermally pretreated microalgae digestion presented a poor dewaterability (CST value of 982 s), the results were consistently improved by the co-digestion with primary sludge (CST value of 290 s). In this sense, the co-digestion substantially improved the effluent dewaterability since primary sludge has less affinity for water than microalgae.

2.2. *Effect of the Thermal Pretreatment on Microalgae Anaerobic Digestion*

2.2.1. Anaerobic Digestion of Thermally Pretreated Microalgae in Lab-Scale Reactors

As previously discussed, microalgae showed a low methane yield despite the thermal pretreatment (0.16 L CH_4/g VS). In order to improve their anaerobic biodegradability, the digester HRT was increased from 20 to 30 days. In parallel, another digester with untreated microalgae was operated as control. During this period, the methane production rate of pretreated microalgae increased by 58% (from 0.12 to 0.19 L CH_4/(L·day)) and the methane yield by 71% (from 0.14 to 0.24 L CH_4/g VS) as compared to control (Table 3). Accordingly, the VS removal also increased from 36.2 to 39.5% (Table 3).

Regarding the ammonium concentration, it was higher in the pretreated reactor digestate than in the control (0.8 g N-NH_4/L vs. 0.7 g N-NH_4/L), suggesting a higher protein solubilization in the case of pretreatment. However, as a result of increasing the HRT, the OLR decreased from 1.2 to 0.8 g VS/(L·day). Consequently, the N-NH_4 concentration in the reactor was reduced in comparison with the previous period at 20 days of HRT (0.8 vs. 1.1 g N-NH_4/L).

The methane yield increase observed in this study is in agreement with the results obtained by Passos and Ferrer [8], who reported an increase of 70% after applying a thermal pretreatment at 95 °C for 10 h to similar microalgae species. However, different conclusions regarding the effect of the thermal pretreatment on microalgae can be found in the literature. For instance, no significant effect was observed after a pretreatment at 70 °C for 3 h to *Scenedesmus* sp., but the same pretreatment at 90 °C enhanced the anaerobic biodegradability of *Scenedesmus* sp. from 22 to 48% in BMP tests [25]. Other authors found no influence of the thermal pretreatment, but did find an effect of the thermochemical pretreatment, which increased methane yield by 40% in some microalgae species [26]. Indeed, the effect of the thermal pretreatment highly depends on the microalgae species and the conditions applied, and so a pilot-scale evaluation of the pretreatment performance is required before scaling-up.

In terms of digestate dewaterability, both the untreated and thermally pretreated microalgae showed a poor dewaterability, with higher CST values (795 and 919 s, respectively) than the co-digestion reactor (290 s).

2.2.2. Microscopic Analysis

Microalgae were periodically characterized by optical microscopy over the whole experimental period. Qualitative results showed how microalgal biomass was flocculated. The main green microalgae species belonged to the genus *Chlorella* and *Stigeoclonium*, along with diatoms (Figure 3a,b). These microalgae species remained predominant during the whole period, although the relative abundance varied over time, which is common in open ponds treating wastewater [27].

Figure 3. Microscopic images of microalgae before (**a**,**b**) and after (**c**,**d**) the thermal pretreatment, along with the digestates from untreated microalgae AD (**e**) and thermally pretreated microalgae AD (**f**) at a HRT of 30 days.

After the thermal pretreatment, microalgae clearly appeared to be less pigmented than fresh microalgae and most of the cells were dead (Figure 3c,d). Also, in the pretreated sample, a higher amount of amorphous material was found because of organic matter release. However, most of the cell walls were found unbroken. This was especially the case for diatoms (Figure 3c) and *Stigeoclonium* sp. (Figure 3d), which presented a higher resistance to the pretreatment. Indeed, other authors concluded that the thermal pretreatment was not able to break microalgae cell walls but it did damage or weaken them [28,29].

To further evaluate the effect of the thermal pretreatment on microalgae AD, microscopic images from the digestate of pretreated microalgae (Figure 3f) were compared to those from the digestate of untreated microalgae (Figure 3e). In this manner, it was possible to elucidate whether pretreated cells were more accessible to methanogens, even if cell walls were not lysed after the pretreatment step. A higher amount of particulate substances was observed in the untreated microalgae digestate (Figure 3e), although entire microalgae cells were found in both digestates even after 30 days of digestion.

Next, a quantitative analysis was conducted by counting the two most abundant microalgae species, *Chlorella* sp. and diatoms, in the influent and effluent (Figure 4). This analysis confirmed the qualitative results. While the number of *Chlorella* sp. individuals was reduced by the thermal pretreatment, no significant differences were observed for diatoms. Indeed, both of them present a resistant cell wall, but their characteristics and composition differ. On the one hand, *Chlorella* sp. has mainly a carbohydrate-based cell wall, and carbohydrates solubilization can be boosted by the thermal pretreatment [30]. On the other hand, diatoms have a siliceous-based cell wall, which resists the effect of temperature.

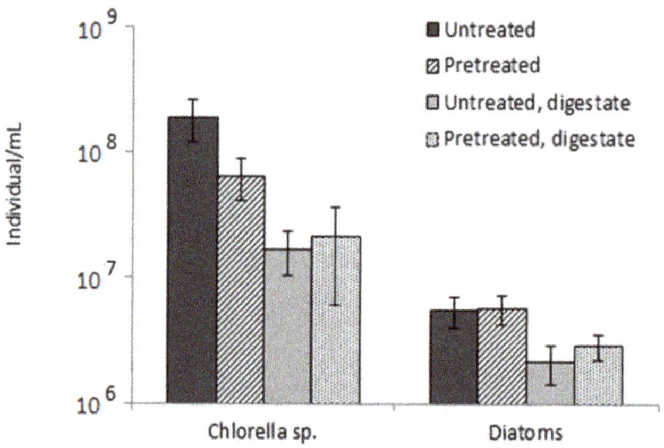

Figure 4. *Chlorella* sp. and diatoms counting in the influents (untreated; pretreated) and effluents (untreated digestate; pretreated digestate) during Period II. Mean values and standard deviation are represented.

In spite of this, both microalgae species were partially removed during the AD process according to digestate counting. While *Chlorella* showed around one logarithmic unit removal, a much lower removal efficiency was observed for diatoms, leading to a higher relative abundance of diatoms in the digestates. Comparing both *Chlorella* and diatom abundance in untreated and pretreated microalgae digesters, no significant differences were found. Even so, the pretreated microalgae digester showed a higher methane yield and VS removal. This may be because, although having same quantity of entire cells, those cells that were attacked by microorganisms were more degraded in the pretreated microalgae reactor.

2.3. *Effect of the HRT on Microalgae Anaerobic Biodegradability*

The effect of the HRT can be evaluated by comparing the results on pretreated microalgae AD obtained in both periods (at 20 and 30 days of HRT). When the HRT was increased to 30 days, the methane yield of pretreated microalgae increased by 50% (from 0.16 to 0.24 L CH_4/g VS) compared to that obtained at 20 days of HRT (Table 3, Figure 5). Indeed, the VS removal was also higher with a HRT of 30 days (39.5%) as compared to 20 days (27.9%).

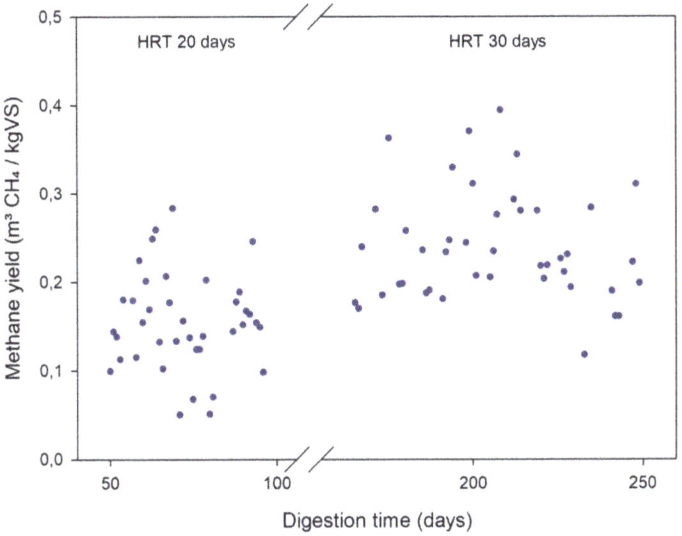

Figure 5. Daily methane yield of thermally pretreated microalgae for the two studied periods: Period I at an HRT of 20 days and Period II at an HRT of 30 days.

Although one expected benefit of applying a pretreatment is the kinetics improvement and thereby a reduction of the HRT [7], the methane yield increase reported in this study was still significant when he HRT was increased from 20 to 30 days. Thus, operating microalgae digesters at moderate HRTs seems appropriate, even if applying pretreatments. As discussed in the previous section, the thermal pretreatment weakened the microalgae cell wall but without completely lysing and releasing all intracellular material. Therefore, increasing the HRT enhanced the chance for microorganisms to access microalgae intracellular material through their weakened or damaged cell wall. These results are in agreement with previous studies. For instance, applying a thermal pretreatment to microalgae did not show any significant differences with a HRT of 15 days, but it increased the methane yield by 72% with a HRT of 20 days [8]. It has been suggested that the operation of digesters at high sludge retention times (SRT) promotes the presence of low growth-rate microorganisms and increases the hydrolytic potential of the system [31]. Comparing a thermophilic continuous stirred tank reactor working at 50 days of HRT (and SRT) with an anaerobic membrane bioreactor (AnMBR) with a SRT of 70 days, higher microbial diversity could be found in digesters working at higher HRT system [31].

3. Discussion

Results have shown how the co-digestion with primary sludge can substantially improve the microalgae mono-digestion by increasing the methane yield, decreasing the ammonia concentration, which may enable increasing the OLR, and improving the digestate dewaterability.

This study assessed different proportions of primary sludge and microalgae in batch tests and determined the best one in continuous lab-scale reactors. The truth is that in full-scale microalgal-based

WWTPs, this proportion would change over the year. Indeed, the microalgal biomass production shows a strong seasonality [27] depending on the HRAPs operation conditions, influent characteristics, etc. [32]. These factors determine not only the amount but also the microalgae species in the system [27,33], and the microalgae species also affect the anaerobic digestion rate and extent, depending especially on the characteristics of the cell wall [27]. Overall, the implementation of anaerobic digesters in HRAP plants involves working with different proportions of microalgae and primary sludge and different microalgae species over the year. All these factors should be considered when it comes to sizing an AD plant integrated to a HRAP system. For instance, if the proportion of primary sludge is expected to be high, the biogas production is also expected to be high, and the operation of the digesters should be feasible at 20 days of HRT. However, if the proportion of microalgae is expected to be high, then it is necessary to assess the most appropriate strategy to follow (increasing the HRT and/or applying a pretreatment).

In this study, the thermal pretreatment increased the microalgae methane yield, but not as much as expected due to the presence of microalgae species with hardly degradable cell walls (i.e., *Stigioclonium* sp.). However, when the reactors were operated at a longer HRT (30 days), the methane yield of pretreated microalgae increased considerably (from 0.16 to 0.24 L CH_4/g VS). When considering these alternatives, different issues should be addressed: firstly, the balance between the energy requirements in comparison the energy gain of the pretreatment step; secondly, the increase of volume, surface area and costs resulting from an increased HRT.

Consequently, an energy assessment was carried out by scaling-up the results of the lab-scale reactors during both experimental periods (I: co-digestion vs. pretreated microalgae mono-digestion at 20 days of HRT; II: pretreated vs. untreated microalgae at 30 days of HRT). Flow rates between 10–100 m^3/day were considered (Table 4). The assessment compared the energy required to apply the pretreatment (if any) and anaerobic digestion (Ei) with the energy obtained through the biogas produced in each case (Eo). In this way, when the energy ratio (Eo/Ei) is higher than 1, there is an energy gain. As can be seen in Table 4, this value was higher than 1 in all scenarios, meaning that the energy balance was always positive. However, the best results were obtained with the co-digestion of microalgae and primary sludge (energy ratio between 3.5–4). This means that the energy produced with the co-digestion is at least 3.5-fold the energy consumed. Regarding the thermal pretreatment, this also showed an energy gain in all cases. However, the energy ratio increased from 1.1–1.2 to 1.5–1.7 by increasing the HRT from 20 to 30 days. When comparing the energy gain with untreated and pretreated microalgae at the same HRT of 30 days, the results are very similar (from 1.4–1.6 to 1.5–1.7). Bearing in mind the investment and operation costs of the pretreatment, this would not be worthwhile in terms of energy production, and would only become so if other benefits such as hygenisation were considered.

Table 4. Results of the energy assessment for the co-digestion and pretreated microalgae mono-digestion at 20 days of HRT, and for the untreated and pretreated microalgae mono-digestion at 30 days of HRT, with different flow rates (Q = 10, 25 and 100 m^3/day). Ei (i.e., energy input) and Eo (i.e., energy output).

	Period I						**Period II**					
	Microalgae,p			**Co-Digestion**			**Microalgae**			**Microalgae,p**		
Q (m^3/day)	10	25	100	10	25	100	10	25	100	10	25	100
Ei (GJ/day)	1.15	2.75	10.46	0.96	2.28	8.58	0.99	2.31	8.53	1.24	2.93	11.04
Eo (GJ/day)	1.29	3.22	12.89	3.42	8.54	34.15	1.35	3.38	13.53	1.84	4.59	18.37
$\Delta E = Eo - Ei$ (GJ/day)	0.14	0.47	2.43	2.45	6.26	25.27	0.36	1.08	5.00	0.60	1.66	7.32
Eo/Ei (-)	1.1	1.2	1.2	3.5	3.7	4.0	1.4	1.5	1.6	1.5	1.6	1.7

To sum up, the most suitable option to anaerobically digest microalgae from HRAPs would be the co-digestion with primary sludge at a 20-day HRT if the proportion of sludge was high, and at 30 days if the proportion of microalgae was high. The energy gain could be used to cover the energy demand of the WWTP, moving towards energy-neutral WWTPs [32]. Thus, integrating anaerobic co-digestion in HRAPs is a good strategy to transform resources from wastewater into valuable products. It enables resource recovery from wastewater, which is a prerequisite for the technological development of a cradle-to-cradle bio-based economy [34].

4. Materials and Methods

4.1. Substrates Caracteristics

The microalgal biomass (hereafter called microalgae) used in this study consisted of a microalgae-bacteria consortium grown in a pilot raceway pond (0.5 m^3) that treated wastewater from a municipal sewer, as described by Passos et al. [27]. Microalgal biomass was harvested from secondary settlers and gravity thickened in laboratory Imhoff cones at 4 °C for 24 h. The pilot plant was located at the laboratory of the GEMMA research group (Barcelona, Spain).

Thickened primary sludge and digested sludge used as inoculum in BMP tests and continuous reactors came from a municipal WWTP near Barcelona. The inoculum was collected before the start-up of each assay, while primary sludge was periodically collected (every 3 weeks) and stored at 4 °C before use.

Thickened microalgae presented an average concentration of 3.7% TS and 2.7% VS, while primary sludge had an average concentration of 4.6% TS and 3.4% VS. In order to use the same OLR in all digesters, both substrates were diluted to achieve 2.5% VS.

4.2. Thermal Pretreatment

The thermal pretreatment of microalgae was carried out in glass bottles with a total volume of 250 mL and a liquid volume of 150 mL Bottles were placed in an incubator under continuous stirring at a constant temperature of 75 °C for 10 h. In semi-continuous experiments, microalgae were collected and pretreated once a week. Pretreated biomass was then stored at 4 °C before use.

4.3. Biochemical Methane Potential Tests

BMP tests were used to study the anaerobic biodegradability of co-digestion trials of primary sludge and microalgae, with and without thermal pretreatment. To this end, three proportion conditions were tested: (i) 25% of microalgae and 75% of primary sludge, (ii) 50% of microalgae and 50% of primary sludge and, (iii) 75% of microalgae and 25% of primary sludge on a VS basis. All conditions were conducted with untreated and pretreated microalgae.

The substrate to inoculum (S/I) ratio was 0.5 g COD/g VS, according to Arias et al. [35]. After adding the proper amount of both substrates and inoculum, serum bottles (160 mL) were filled with distilled water up to 100 mL, flushed with helium gas, sealed with butyl rubber stoppers and incubated at 35 °C until biogas production ceased. Accumulated biogas was measured with a manometer (GMH 3161 Greisinger, Regenstauf, Germany) and the methane content in biogas was periodically analyzed by gas chromatography. A blank treatment was used to quantify the amount of methane produced by the inoculum alone. Each co-digestion condition was performed in duplicate, whereas control trials (microalgae, pretreated microalgae and primary sludge) and blank were performed in triplicate.

4.4. Continuous Anaerobic Digestion

Microalgae anaerobic (co-)digestion was performed and monitored using two lab-scale reactors (2 L), with an effective volume of 1.5 L. Reactors were operated under mesophilic conditions (37 ± 1 °C) by implementing an electric heating cover (Selecta, Barcelona, Spain). Constant mixing was provided by a magnetic stirrer (Thermo Scientific, Waltham, MA, USA). Reactors were operated on a daily feeding basis, where the same volume was purged from and added to digesters using plastic syringes.

During the first experimental period, one of the digesters was fed with pretreated microalgae (i.e., control), while the second one was fed with pretreated microalgae (25% VS) and primary sludge (75% VS). Both reactors were operated at a HRT of 20 days and were considered to be under steady-state after 2.5 HRTs. Afterwards, the anaerobic digestion performance was further monitored during 2 complete HRTs (~6 weeks). During the second experimental period, the HRT was increased to 30 days. One reactor was still fed with pretreated microalgae, while the other one was fed with untreated microalgae (i.e., control). They were also considered to be under steady-state after 2.5 HRTs and anaerobic digestion performance was further monitored during 2 complete HRTs (~8.5 weeks). The total operation period of the digesters was 225 days.

Biogas production was measured by the water displacement method and the methane content in biogas was periodically analyzed by GC. The volume of biogas produced was expressed under standard temperature (0 °C) and pressure (1 atm) conditions (STP).

4.5. Microscopic Analysis

Microalgae were periodically identified over the semi-continuous reactors operation. The analysis was carried out with an optic microscope (Motic BA310E, Motic, Hong Kong, China), equipped with a camera (NiKon DS-Fi2, Nanjing, China) using the software NISElements Viewer (Prague, Czech Republic). Microalgae genus were identified from classical specific literature [36,37].

To prove the effect of the thermal pretreatment and AD on microalgae, four sampling campaigns were conducted. In each campaign the following samples were analyzed: (i) untreated microalgae; (ii) thermally pretreated microalgae; (iii) effluent (digestate) from untreated microalgae AD and (iv) effluent (digestate) from pretreated microalgae AD. From these samples, microalgae species were identified and two of the most abundant were quantified (*Chlorella* sp. and diatoms). For their quantification, each well homogenized sample was examined by bright and contrast phase microscopy using a Zeiss microscope Axioskop 40 (Goettingen, Germany). To quantify *Chlorella* sp. and diatoms, two subsamples of 20 µL, were counted at 400 magnification. In each subsample, 30 microscopic fields across the cover-slide were counted using coverslides of 20 mm side [38]. Previous to the cell counting, aggregated flocs of these unicellular species were broken down by means of an ultrasound technique [39].

4.6. Analitical Procedures

The TS and VS analysis was performed according to the standard methods [40]. The quantification of total COD concentration was performed according to the closed reflux colorimetric method outlined by the standard methods [40]. TKN was determined by titration after a mineralization step performed by a BUCHI 370-K distillator/titrator. The concentration of the ammonium nitrogen ($N\text{-}NH_4^+$) was measured according to the method by Solorzano [41]. pH was determined with a Crison Portable 506 pH-meter (Alella, Spain). Digestate dewaterability was evaluated by means of the capillary suction time (CST) test (Triton Electronics Ltd., Cambridge, UK). Volatile fatty acids (VFA) concentrations in continuous flow digesters were measured once a week by injecting 1 µL of each sample, once centrifuged (4200 rpm for 8 min) and filtered (0.2 µm), into an Agilent 7820A GC (Santa Clara, USA) after sulphuric acid and diisopropyl ether addition. The GC was equipped with an auto-sampler, flame ionization detector and a capillary column (DP-FFAB Agilent 30 m × 0.25 mm × 0.25 µm), and operated at injector and detector temperatures of 200 and 300 °C, respectively, with helium as carrier gas.

Biogas composition was determined by calculating the percentage of methane and carbon dioxide in the digesters headspace. Gases were measured by means of a GC (Thermo Finnigan, Austin, TX, USA) equipped with a thermal conductivity detector (TCD) (Hayesep packed column). The carrier gas was helium and injector/detector/oven temperatures of 150/250/35 °C, respectively. The methane content in biogas from BMP tests was measured each sampling day, while in continuous reactors it was quantified twice a week.

4.7. Statistics and Kinetic Data Analysis

The effect of independent variables during the continuous anaerobic (co-)digestion was evaluated via multi-factor analysis of variance (ANOVA) considering a 95% confidence level ($\alpha = 0.05$) using the R Statistics Software. For the BMPs and the microscopic counting, mean values and standard deviations were considered.

To evaluate the kinetics of the process from BMP tests results, experimental data was adjusted to a first-order kinetic model (Equation (1)) by the least square method.

$$B = B_0 \{1 - \exp[-k \cdot t]\} \tag{1}$$

where B_0 stands for the methane production potential (mL CH_4/g VS), k is the first order kinetic rate constant (day^{-1}), B is the accumulated methane production at time t (mL CH_4/g VS) and t is time (day).

The error variance (s^2) was estimated by the following equation (Equation (2)):

$$s^2 = \frac{\sum_1^i (y_i - \hat{y}_i)^2}{N - K} \tag{2}$$

where y_i is the experimental value, $\hat{y}_i$ is the value estimated by the model, N is the number of samples and K is the number of model parameters.

4.8. Energy Assessment

The theoretical energy balance of full-scale reactors was estimated by up-scaling experimental data to medium-size WWTP with flow rates of 10-25-100 m^3/day. Electricity and heat requirements for microalgae pretreatment and anaerobic digestion were calculated according to Passos and Ferrer [8].

Input heat was calculated as the energy required to heat influent biomass from ambient temperature (Ta) to digestion temperature (Td), according to Equation (3) The density (ρ) and specific heat (γ) of microalgae and primary sludge were assumed to be the same as those of water, 1000 kg/m^3 and 4.18 kJ/(kg·°C), respectively. Heat losses through the reactor wall were considered and the heat transfer coefficient (k) was assumed to be 1 W/(m^2·day). The reactor wall surface area was calculated from the reactor useful volume, considering a 2:1 diameter to height ratio, while the reactor bottom and top were not accounted for.

$$E_{i,heat} = \rho \cdot Q \cdot \gamma \cdot (Td - Ta) + k \cdot A \cdot (Td - Ta) \cdot 86.4 \tag{3}$$

where Ei,heat is the input heat (kJ/day); ρ is the density (kg/m^3); Q is the flow rate (m^3/day); γ is the specific heat (kJ/(kg·°C)); Td is the anaerobic digestion temperature (37 °C); Ta is the ambient temperature (20 °C); k is the heat transfer coefficient (W/(m^2·°C)); and A is the surface area of the reactor wall (m^2).

When thermal pretreatment was involved, heat recovery was considered. Input heat was calculated as the energy required to heat influent biomass from Ta to pretreatment temperature (Tp), subtracted by the heat recovered when cooling down biomass from Tp to Td (Equation (4)). Heat would be recovered by means of a heat exchanger, with an efficiency ϕ of 85%.

$$E_{i,heat} = \rho \cdot Q \cdot \gamma \cdot (Tp - Ta) - \rho \cdot Q \cdot \gamma \cdot (Tp - Td)\ \phi + k \cdot A \cdot (Td - Ta) \cdot 86.4 \tag{4}$$

where Ei,heat is the input heat (kJ/day); ρ is the density (kg/m^3); Q is the flow rate (m^3/day); γ is the specific heat (kJ/(kg·°C)); Td is the anaerobic digestion temperature (37 °C); Ta is the ambient temperature (20 °C); Tp is the pretreatment temperature (75 °C); ϕ is the heat recovery efficiency (85%); k is the heat transfer coefficient (W/(m^2·°C)); and A is the surface area of the reactor wall (m^2).

Furthermore, input electricity for anaerobic digestion was estimated as the energy required for biomass pumping and reactor mixing, which were assumed to be 1800 kJ/m^3 and 300 kJ/($m^3_{reactor}$ day), respectively (Equation (5)):

$$\text{Ei,electricity} = Q\cdot\theta + V\cdot\omega \quad (5)$$

where Ei,electricity is the input electricity (kJ/day); Q is the flow rate (m^3/day); θ is the electricity consumption for pumping (kJ/m^3); V is the useful volume (m^3); and ω is the electricity consumption for mixing (kJ/($m^3_{reactor}\cdot$day)).

The energy output of the process was calculated from the methane production rate of each reactor, according to Equation (6). The lower heating value of methane (ξ) was assumed to be 35,800 kJ/m^3 CH_4. An efficiency of 90% on energy conversion was considered (η).

$$\text{Eo} = \text{P,CH}_4\cdot\xi\cdot V\cdot\eta \quad (6)$$

where Eo is the output energy (kJ/d); P,CH_4 is the methane production rate (m^3 CH_4/($m^3_{reactor}\cdot$day)); ξ is the lower heating value of methane (kJ/m^3 CH_4); V is the useful volume (m^3); and η is the energy conversion efficiency.

Finally, results were expressed as energy balance (ΔE) and energy ratio (Eo/Ei). The energy balance was calculated as the difference between the energy output and energy input (heat and electricity) (Equation (7)), while the energy ratio was calculated as the energy output over the energy input (heat and electricity) (Equation (8)).

$$\Delta E = \text{Eo} - (\text{Ei,heat} + \text{Ei,electricity}) \quad (7)$$

$$\text{Eo/Ei} = \text{Eo}/(\text{Ei,heat} + \text{Ei,electricity}) \quad (8)$$

Author Contributions: I.F., M.G. and M.S.-B. conceived and designed the experiments; F.P. performed the BMP test assay and M.S.-B. conducted the experiments with the continuous reactors; H.S. participated with the microscopic analysis; and I.F., M.G. and M.S.-B. analyzed the data and wrote the paper.

Funding: This research was funded by the Spanish Ministry of Economy and Competitiveness (project FOTOBIOGAS CTQ2014-57293-C3-3-R). Maria Solé is grateful to the Universitat Politècnica de Catalunya·BarcelonaTech for her PhD scholarship. Marianna Garfí is grateful to the Spanish Ministry of Economy and Competitiveness (Plan Estatal de Investigación Científica y Técnica y de Innovación 2013–2016, Subprograma Ramón y Cajal (RYC) 2016).

Conflicts of Interest: The authors declare no conflict of interest.

References

1. Arashiro, L.T.; Montero, N.; Ferrer, I.; Acién, F.G.; Gómez, C.; Garfí, M. Life cycle assessment of high rate algal ponds for wastewater treatment and resource recovery. *Sci. Total Environ.* **2018**, *622–623*, 1118–1130. [CrossRef] [PubMed]
2. Uggetti, E.; Passos, F.; Solé, M.; Garfí, M.; Ferrer, I. Recent Achievements in the Production of Biogas from Microalgae. *Waste Biomass Valorization* **2017**, *8*, 129–139. [CrossRef]
3. Ward, A.J.; Lewis, D.M.; Green, F.B. Anaerobic digestion of algae biomass: A review. *Algal Res.* **2014**, *5*, 204–214. [CrossRef]
4. González-Fernández, C.; Sialve, B.; Bernet, N.; Steyer, J.-P. Impact of microalgae characteristics on their conversion to biofuel. Part II: Focus on biomethane production. *Biofuels Bioprod. Biorefin.* **2012**, *6*, 205–218. [CrossRef]
5. Passos, F.; Uggetti, E.; Carrère, H.; Ferrer, I. Pretreatment of microalgae to improve biogas production: A review. *Bioresour. Technol.* **2014**, *172*, 403–412. [CrossRef] [PubMed]
6. Jankowska, E.; Sahu, A.K.; Oleskowicz-Popiel, P. Biogas from microalgae: Review on microalgae's cultivation, harvesting and pretreatment for anaerobic digestion. *Renew. Sustain. Energy Rev.* **2017**, *75*, 692–709. [CrossRef]

7. Carrere, H.; Antonopoulou, G.; Affes, R.; Passos, F.; Battimelli, A.; Lyberatos, G.; Ferrer, I. Review of feedstock pretreatment strategies for improved anaerobic digestion: From lab-scale research to full-scale application. *Bioresour. Technol.* **2016**, *199*, 386–397. [CrossRef] [PubMed]
8. Passos, F.; Ferrer, I. Microalgae Conversion to Biogas: Thermal Pretreatment Contribution on Net Energy Production. *Environ. Sci. Technol.* **2014**, *48*, 7171–7178. [CrossRef] [PubMed]
9. Herrmann, C.; Kalita, N.; Wall, D.; Xia, A.; Murphy, J.D. Optimised biogas production from microalgae through co-digestion with carbon-rich co-substrates. *Bioresour. Technol.* **2016**, *214*, 328–337. [CrossRef] [PubMed]
10. Ehimen, E.A.; Sun, Z.F.; Carrington, C.G.; Birch, E.J.; Eaton-Rye, J.J. Anaerobic digestion of microalgae residues resulting from the biodiesel production process. *Appl. Energy* **2011**, *88*, 3454–3463. [CrossRef]
11. Magdalena, J.; Ballesteros, M.; González-Fernandez, C. Efficient Anaerobic Digestion of Microalgae Biomass: Proteins as a Key Macromolecule. *Molecules* **2018**, *23*, 1098. [CrossRef] [PubMed]
12. Mahdy, A.; Fotidis, I.A.; Mancini, E.; Ballesteros, M.; González-Fernández, C.; Angelidaki, I. Ammonia tolerant inocula provide a good base for anaerobic digestion of microalgae in third generation biogas process. *Bioresour. Technol.* **2017**, *225*, 272–278. [CrossRef] [PubMed]
13. Solé-Bundó, M.; Matamoros, V.; Garfí, M.; Ferrer, I. Co-digestion of primary sludge and microalgae from wastewater treatment systems: Effect on biogas production and emerging contaminants removal. Unpublished data. **2018**, under preparation.
14. Mata-Alvarez, J.; Dosta, J.; Romero-Güiza, M.S.; Fonoll, X.; Peces, M.; Astals, S. A critical review on anaerobic co-digestion achievements between 2010 and 2013. *Renew. Sustain. Energy Rev.* **2014**, *36*, 412–427. [CrossRef]
15. Mahdy, A.; Mendez, L.; Ballesteros, M.; González-Fernández, C. Algaculture integration in conventional wastewater treatment plants: Anaerobic digestion comparison of primary and secondary sludge with microalgae biomass. *Bioresour. Technol.* **2015**, *184*, 236–244. [CrossRef] [PubMed]
16. Hlavínek, P.; Stvríteský, L.; Pešoutová, R.; Houdková, L. Biogas Production from Algal Biomass from Municipal Wastewater Treatment. *Waste Biomass Valorization* **2016**, *7*, 747–752. [CrossRef]
17. Beltran, C.; Jeison, D.; Fermoso, F.G.; Borja, R. Batch anaerobic co-digestion of waste activated sludge and microalgae (*Chlorella sorokiniana*) at mesophilic temperature. *J. Environ. Sci. Health A Tox. Hazard Subst. Environ. Eng.* **2016**, *51*, 847–850. [CrossRef] [PubMed]
18. Neumann, P.; Torres, A.; Fermoso, F.G.; Borja, R.; Jeison, D. Anaerobic co-digestion of lipid-spent microalgae with waste activated sludge and glycerol in batch mode. *Int. Biodeterior. Biodegrad.* **2015**, *100*, 85–88. [CrossRef]
19. Astals, S.; Musenze, R.S.; Bai, X.; Tannock, S.; Tait, S.; Pratt, S.; Jensen, P.D. Anaerobic co-digestion of pig manure and algae: Impact of intracellular algal products recovery on co-digestion performance. *Bioresour. Technol.* **2015**, *181*, 97–104. [CrossRef] [PubMed]
20. Solé-Bundó, M.; Eskicioglu, C.; Garfí, M.; Carrère, H.; Ferrer, I. Anaerobic co-digestion of microalgal biomass and wheat straw with and without thermo-alkaline pretreatment. *Bioresour. Technol.* **2017**, *237*, 89–98. [CrossRef] [PubMed]
21. Sialve, B.; Bernet, N.; Bernard, O. Anaerobic digestion of microalgae as a necessary step to make microalgal biodiesel sustainable. *Biotechnol. Adv.* **2009**, *27*, 409–416. [CrossRef] [PubMed]
22. Samson, R.; LeDuy, A. Improved performance of anaerobic digestion ofSpirulinamaxima algal biomass by addition of carbon-rich wastes. *Biotechnol. Lett.* **1983**, *5*, 677–682. [CrossRef]
23. Peng, S.; Colosi, L.M. Anaerobic Digestion of Algae Biomass to Produce Energy during Wastewater Treatment. *Water Environ. Res.* **2016**, *88*, 29–39. [CrossRef] [PubMed]
24. Rajagopal, R.; Massé, D.I.; Singh, G. A critical review on inhibition of anaerobic digestion process by excess ammonia. *Bioresour. Technol.* **2013**, *143*, 632–641. [CrossRef] [PubMed]
25. González-Fernández, C.; Sialve, B.; Bernet, N.; Steyer, J.P. Thermal pretreatment to improve methane production of Scenedesmus biomass. *Biomass Bioenergy* **2012**, *40*, 105–111. [CrossRef]
26. Bohutskyi, P.; Betenbaugh, M.J.; Bouwer, E.J. The effects of alternative pretreatment strategies on anaerobic digestion and methane production from different algal strains. *Bioresour. Technol.* **2014**, *155*, 366–372. [CrossRef] [PubMed]
27. Passos, F.; Gutiérrez, R.; Brockmann, D.; Steyer, J.-P.; García, J.; Ferrer, I. Microalgae production in wastewater treatment systems, anaerobic digestion and modelling using ADM1. *Algal Res.* **2015**, *10*, 55–63. [CrossRef]

28. Passos, F.; Hernández-Mariné, M.; García, J.; Ferrer, I. Long-term anaerobic digestion of microalgae grown in HRAP for wastewater treatment. Effect of microwave pretreatment. *Water Res.* **2014**, *49*, 351–359. [CrossRef] [PubMed]
29. Ometto, F.; Quiroga, G.; Pšenička, P.; Whitton, R.; Jefferson, B.; Villa, R. Impacts of microalgae pre-treatments for improved anaerobic digestion: Thermal treatment, thermal hydrolysis, ultrasound and enzymatic hydrolysis. *Water Res.* **2014**, *65*, 350–361. [CrossRef] [PubMed]
30. Solé-Bundó, M.; Carrère, H.; Garfí, M.; Ferrer, I. Enhancement of microalgae anaerobic digestion by thermo-alkaline pretreatment with lime (CaO). *Algal Res.* **2017**, *24*, 199–206. [CrossRef]
31. Greses, S.; Zamorano-López, N.; Borrás, L.; Ferrer, J.; Seco, A.; Aguado, D. Effect of long residence time and high temperature over anaerobic biodegradation of Scenedesmus microalgae grown in wastewater. *J. Environ. Manag.* **2018**, *218*, 425–434. [CrossRef] [PubMed]
32. Passos, F.; Gutiérrez, R.; Uggetti, E.; Garfí, M.; García, J.; Ferrer, I. Towards energy neutral microalgae-based wastewater treatment plants. *Algal Res.* **2017**, *28*, 235–243. [CrossRef]
33. Gutiérrez, R.; Ferrer, I.; González-Molina, A.; Salvadó, H.; García, J.; Uggetti, E. Microalgae recycling improves biomass recovery from wastewater treatment high rate algal ponds. *Water Res.* **2016**, *106*, 539–549. [CrossRef] [PubMed]
34. Puyol, D.; Batstone, D.J.; Hülsen, T.; Astals, S.; Peces, M.; Krömer, J.O. Resource Recovery from Wastewater by Biological Technologies: Opportunities, Challenges, and Prospects. *Front. Microbiol.* **2016**, *7*, 2106. [CrossRef] [PubMed]
35. Arias, D.M.; Solé-Bundó, M.; Garfí, M.; Ferrer, I.; García, J.; Uggetti, E. Integrating microalgae tertiary treatment into activated sludge systems for energy and nutrients recovery from wastewater. *Bioresour. Technol.* **2018**. [CrossRef] [PubMed]
36. Palmer, C.M. *Algas en Los Abastecimientos de Agua. Manual Ilustrado Acerca de la Identificación, Importancia y Control de Las Algas en Los Abastecimientos de Agua*; Editorial Interamericana: México City, México, 1962.
37. Bourrelly, P. *Les Algues D'eau Douce, in: Les Algues Vertes*, 1st ed.; Societé nouvelle des éditions doubée: Paris, France, 1985.
38. Salvadó, H.; Palomo, A.; Mas, M.; Puigagut, J.; Gracia, M. Dynamics of nematodes in a high organic loading rotating biological contactors. *Water Res.* **2004**, *38*, 2571–2578. [CrossRef] [PubMed]
39. Abzazou, T.; Salvadó, H.; Bruguera-Casamada, C.; Simón, P.; Lardín, C.; Araujo, R.M. Assessment of total bacterial cells in extended aeration activated sludge plants using flow cytometry as a microbial monitoring tool. *Environ. Sci. Pollut. Res.* **2015**, *22*, 11446–11455. [CrossRef] [PubMed]
40. American Public Health Association (APHA); American Water Works Association (AWWA); Water Environment Federation (WEF). *Standard Methods for the Examination of Water and Wastewater*; APHA: Washington, DC, USA; AWWA: Singapore; WEF: Cologny, Switzerland, 2005.
41. Solorzano, L. Determination of ammonia in natural waters by the phenolhypochlorite method. *Limnol. Oceangr.* **1969**, *14*. [CrossRef]

Sample Availability: Samples of the compounds are not available from the authors.

molecules

MDPI

Article

Comparison of Different Electricity-Based Thermal Pretreatment Methods for Enhanced Bioenergy Production from Municipal Sludge

E. Hosseini Koupaie, T. Johnson and C. Eskicioglu *

UBC Bioreactor Technology Group, School of Engineering, The University of British Columbia Okanagan Campus, 1137 Alumni Avenue, Kelowna, BC V1V 1V7, Canada; ehssan.hosseini.k@gmail.com (E.H.K.); thomas.johnson@ubc.ca (T.J.)

* Correspondence: cigdem.eskicioglu@ubc.ca; Tel.: +1-250-807-8544, Fax: +1-250-807-9850

Received: 30 June 2018; Accepted: 9 August 2018; Published: 11 August 2018

check for updates

Abstract: This paper presents results for a comprehensive study that compares the performance of three electricity-based thermal pretreatment methods for improving the effectiveness of anaerobic digestion (AD) to process municipal wastewater sludge. The study compares thermal pretreatment using conventional heating (CH), microwave (MW), and radio frequency (RF) heating techniques. The effectiveness of the pretreatment methods was assessed in terms of chemical oxygen demand (COD) and biopolymers solubilization, AD bioenergy production, input electrical energy, and overall net energy production of the sequential pretreatment/AD process. The heating applicators for the bench-scale testing consisted of a custom-built pressure-sealed heating vessel for CH experiments, an off-the-shelf programmable MW oven operating at a frequency of 2.45 GHz for MW heating experiments, and a newly developed 1 kW RF heating system operating at a frequency of 13.56 MHz for RF heating experiments. Under identical thermal profiles, all three thermal pretreatment methods achieved similar sludge disintegration in terms of COD and biopolymer solubilization as well as AD bioenergy production (p-value > 0.05). According to the energy assessment results, the application of CH and MW pretreatments resulted in overall negative energy production, while positive net energy production was obtained through the sequential pretreatment/AD process utilizing RF pretreatment.

Keywords: anaerobic digestion; bioenergy; municipal sludge; solubilization; thermal pretreatment

1. Introduction

Municipalities rely on physical, chemical, and biological treatment processes to treat their municipal and industrial wastewater. As a result of these treatment processes, municipal sludge, a by-product of treatment, is generated in wastewater treatment plants (WWTP). Currently, about 0.7 million tons of dry municipal sludge are produced annually in Canada [1]. In the United States and Europe sludge volumes are even higher, and annual production ranges from 7 to 10 million [2,3]. To service the demands of growing cities and respond to the increasingly stringent wastewater regulations, existing treatment plants are expanding, resulting in increased production of municipal sludge. Therefore, the management of wastewater residual sludge has now become one of the world's largest and most critical management challenges.

Among different sludge handling/disposal methods (i.e., incineration, composting, and landfilling), AD has aroused more attention in recent years due to its potential for generating renewable energy in the form of methane gas. In addition to the bioenergy production, the cost of the moving, handling, and processing the waste sludge is minimized due to the significant volume reduction after the AD process [4]. AD is a complex biochemical process comprised of four main sequential stages:

hydrolysis, acidogenesis, acetogenesis, and methanogenesis [5]. Of the four stages, the hydrolysis stage is known as a rate-limiting stage because high molecular-weight organics are converted into low molecular-weight or soluble compounds [6]. In terms of the municipal sludge, the hydrolysis stage is particularly limited due to a high content of microbial cells and extracellular polymeric substances resisting enzymatic reactions [7].

Previous research has proven that thermal pretreatment (hydrolysis) can accelerate the digestion process by increasing the soluble fraction of organic matter before AD. The thermal pretreatment methods primarily use conventional (conductive) heating (CH) or microwave (MW) irradiation [8–11]. In CH, the heat transfer mechanism is through a thermal conduction process where energy is transferred from more energetic to less energetic particles due to the thermal gradient [12]. The thermal gradient can lead to non-uniform heating as well as a transient thermal lag throughout the load. As a way of overcoming the limitations of CH, during the last decade, more attention has been given to the application of MW heating at a frequency of 2.45 GHz for sludge hydrolysis [9,11,13–15]. Unlike CH, in MW heating, the electric field interacts directly with molecules in the load and increases kinetic energy to heat the load. The main drawbacks of MW heating is the short penetration depth of the electromagnetic waves which creates non-uniform heating throughout the load and the low energy efficiency (~60%) of high power MW generators [16].

Studies have been conducted to compare the effects of CH and MW pretreatment methods for enhanced sludge solubilization and AD performance. From these studies, some researchers concluded that MW heating is a more effective than CH because of athermal (non-thermal) effects where the electric field intensity directly damages cellular structures rather than through thermal effects [14,15,17–21]. However, other researchers have concluded that CH is a better thermal pretreatment method compared to MW heating [9,11,13,14]. In other studies, no significant differences between CH and MW pretreatments have been measured with respect to sludge solubilization or biogas production [9,17,22,23]. From these studies, it is difficult to draw a conclusion on the merits of CH and MW heating due to the contradictory observations. Additionally, most of these studies were limited to the performance evaluation of the CH and MW systems without conducting an energy assessment and this motivates further study.

In this paper, a third thermal pretreatment method using RF heating was added to a comparative study of CH and MW heating. The RF heating system was a custom heating apparatus that was specifically designed to efficiently heat municipal sludge based on the electrical properties of the load. Experiments were conducted to compare the three electricity-based thermal pretreatment methods (CH, MW, and RF). Measurements were made to quantify the disintegration of municipal sludge and determine the bioenergy production from the AD process. The electrical energy required for each thermal pretreatment process was also measured to calculate the overall energy efficiency of the thermal pretreatment process, and conclusions on energy efficiency for the three methods are summarized.

2. Results and Discussion

2.1. Comparison of CH, MW, and RF Pretreatments for Sludge Disintegration

Figure 1a compares the degree of COD solubilization for the CH and MW pretreatment systems under different temperatures and heating rates. According to Equation (1), the degree of solubilization (DS) represents the percentage of the substrate (in terms of COD, sugar, protein, and humic acid) that is converted from the particulate to soluble phase during the pretreatment.

$$\text{DS}\ (\%) = 100 \times \frac{(\text{S}_2 - \text{S}_1)}{(\text{T}_1 - \text{S}_1)} \quad (1)$$

where, S_1 and T_1 are the concentration of the soluble and total fraction before pretreatment, respectively and S_2 is the concentration of the soluble fraction after pretreatment (in mg/L).

As expected, regardless of the thermal pretreatment method applied, the concentration of soluble COD increased after the pretreatment process. According to Figure 1a, the COD solubilization increased with temperature and decreased with heating rate. For CH pretreatment, the maximum (DS = 26.3%) solubilization was measured at a temperature of 160 °C and a thermal heating rate of 3 °C/min, while minimum (DS = 5.4%) solubilization was measured at a temperature of 80 °C for a heating rate of 11 °C/min. The results of the COD solubilization using the MW and RF pretreatments under various temperatures and holding times are compared in Figure 1b. Increasing the pretreatment temperature and holding time had statistically significant effects on COD solubilization. As per Figure 1b, the maximum (DS = 17.1%) and minimum (DS = 7.0%) COD solubilization were obtained under the pretreatment temperatures of 120 °C and 60 °C and holding times of zero and 120 min, respectively. Consistent with the results of the COD solubilization tests, the solubilization of sugar, protein, and humic acid were increased by increasing the pretreatment temperature and holding time and decreasing the heating rate.

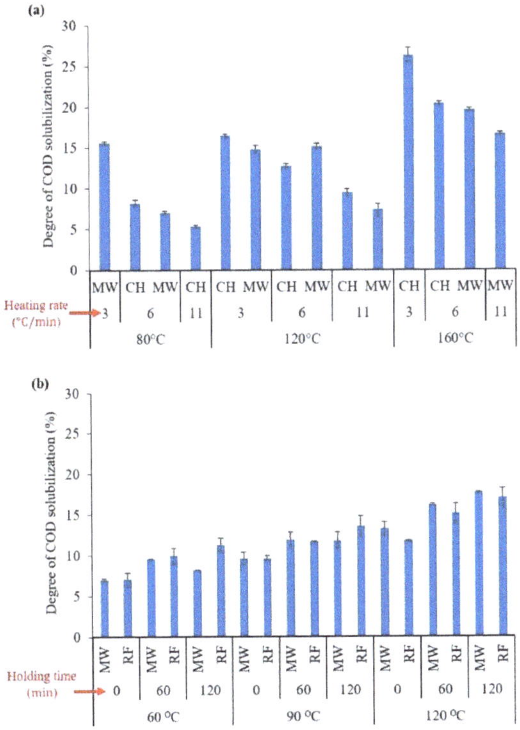

Figure 1. Comparison of CH, MW, and RF pretreatments for the solubilization of COD from; (**a**) CH vs. MW study and (**b**) MW vs. RF study.

In Table 1, the *p*-values associated with each of the experimental independent variables are shown. There was no statistically significant difference among the three pretreatment methods (CH, MW, and RF) in terms of COD and biopolymers solubilization (p-value > 0.05). The main effect plots of the COD, sugar, protein, and humic acid solubilization associated with the "CH vs. MW" and "MW vs. RF" studies are shown in Figure 2a,b respectively. As observed from Figure 2, despite the significant effects of pretreatment temperature, heating rate, and holding time on COD and biopolymers solubilization, there was no significant difference among the application of different thermal pretreatment methods (CH, MW, and RF) for sludge solubilization.

Table 1. Summary of the *p*-values obtained via an overall statistical analysis.

Variable	Levels	COD	Sugar	Protein	Humic Acid
Temperature (°C)	80, 90, 120, 160	0.000	0.000	0.000	0.000
Rate (°C/min)	3, 6, 11	0.000	0.000	0.000	0.000
Holding time (min)	0, 60, 120	0.019	0.000	0.002	0.013
Method	CH, MW, RF	0.321	0.317	0.512	0.770

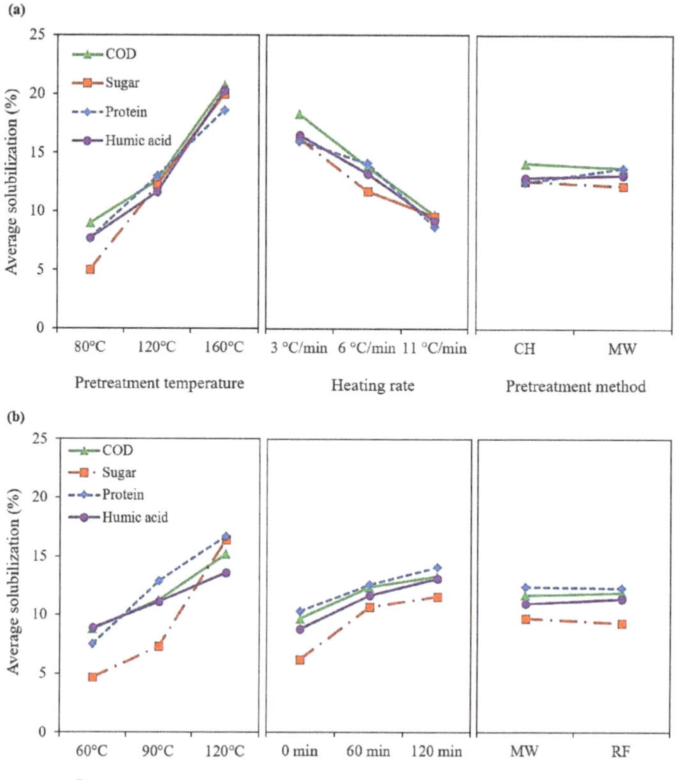

Figure 2. The main effect plot of COD, sugar, protein, and humic acid solubilization from; (**a**) CH vs. MW study; (**b**) MW vs. RF study.

It should be mentioned that the findings of other research studies evaluating the effects of pretreatment temperature, heating rate, and holding time are in agreement with those of this study [10,24–27]. However, in terms of the effect of pretreatment method (CH vs. MW), it is difficult to derive conclusions about any possible differences between CH and MW pretreatment methods due to the significant inconsistency among the results of the published research [9,11,13–15,21,28]. Because under identical thermal profile, the three different pretreatment methods compared in this research achieved the same level of sludge solubilization, it is inferred that the main reason behind the contradictory results of the previous research is the inability to maintain identical thermal profiles among thermal pretreatment methods. Considering the statistically significant effects of the final temperature, heating rate, and holding time on sludge disintegration, any comparison among thermal pretreatment methods should be conducted under identical thermal profiles. Otherwise, it may result in unreliable and contradictory conclusions as observed in the literature [9,11,13–15,17–21].

2.2. Comparison of CH, MW, and RF Pretreatments for Bioenergy Production

Following the solubilization tests, a series of mesophilic and thermophilic batch digesters were set up to compare the bioenergy production from the municipal sludge that was pretreated with CH, MW, and RF methods. According to the results obtained through the "CH vs. MW" study, except a few pretreatment scenarios which were conducted under the highest heating rate of 11 °C/min, thermal pretreatment increased the bioenergy production compared to the non-pretreated sludge samples. Consistent with the results of the solubilization tests, statistically significant effects of the pretreatment temperature and heating rate were also observed on the production of bioenergy through the mesophilic and thermophilic AD of municipal sludge (p-value < 0.05). It was also proven that both the CH and MW pretreatment methods can achieve similar bioenergy production if they are applied under identical thermal profiles (p-value > 0.05).

Figure 3a compares the bioenergy production of the digesters fed with MW-, RF-, and non-pretreated sludge in a unit of kJ/g sludge-added. The percentage improvements in the bioenergy production from the thermally-pretreated digesters (compared to the control digester) are also shown in Figure 3b. According to Figure 3b, all digesters fed with thermally-pretreated sludge produced a higher amount of bioenergy (in the form of methane) compared to the control (non-pretreated) digesters. The maximum bioenergy production (0.419 kJ/g sludge-added) was obtained using the RF pretreatment at a temperature and holding time of 120 °C and 120 min, respectively. Depending on the condition of the pretreatment applied (temperature and holding time), the output energy of the pretreated digesters was increased 5 to 21% compared to the control digester.

The statistical analysis revealed that both the pretreatment temperature and holding time had statistically significant effects on the bioenergy production (p-value < 0.05). However, no statistically significant difference was found among the digesters fed with MW- and RF-pretreated sludge (p-value > 0.05) in terms of the output energy. These results further confirm the outcomes of the solubilization study in which the three thermal pretreatment methods were proven to have similar effects on the improvement of sludge solubilization. The overall statistical analysis revealed that if the thermal profile is identical, the type of the pretreatment method used (CH, MW, and RF) is not a significant factor determining the production of bioenergy through the digestion of municipal sludge. This outcome is in contrast to that of the previous studies in which one of the thermal pretreatment methods (i.e., CH and MW) is suggested as a superior method over another for improved bioenergy production [13–15,21].

As per Figure 3, the overall trend of the thermal pretreatment effect on the bioenergy production was that the higher the pretreatment temperature or heating rate is, the higher the output energy of the digesters is. However, due to higher electrical energy consumption, the net energy (E_{net}) of the sequential pretreatment/AD system may not necessarily be higher at elevated pretreatment temperatures. Figure 4 compares the electrical energy consumption of the CH, MW, and RF pretreatment methods. As per Figure 4a, under any pretreatment condition (thermal profile) used, the MW pretreatment system consumed 56–66% more electrical energy compared to the CH system. According to Figure 1b, regardless of the thermal pretreatment condition used, the energy consumption during the MW pretreatment was significantly (229–441%) higher than that of the RF pretreatment.

In this research, the "CH vs. MW" comparison was performed under different pretreatment temperatures and heating rates and a fixed holding time (0 min). On the other hand, the "MW vs. RF" comparison was conducted under a fixed heating rate (3 °C/min) and various pretreatment temperatures and holding times. Therefore, Figure 4c compares the electrical energy consumption of the pretreatment systems during a given pretreatment condition at which all the three systems were used. Under this pretreatment condition (temperature = 120 °C, heating rate = 3 °C/min, holding time = 0 min), the CH, MW, and RF pretreatment systems consumed 2.0, 3.3, and 0.6 kJ electrical energy per gram of sludge, respectively (Figure 4c).

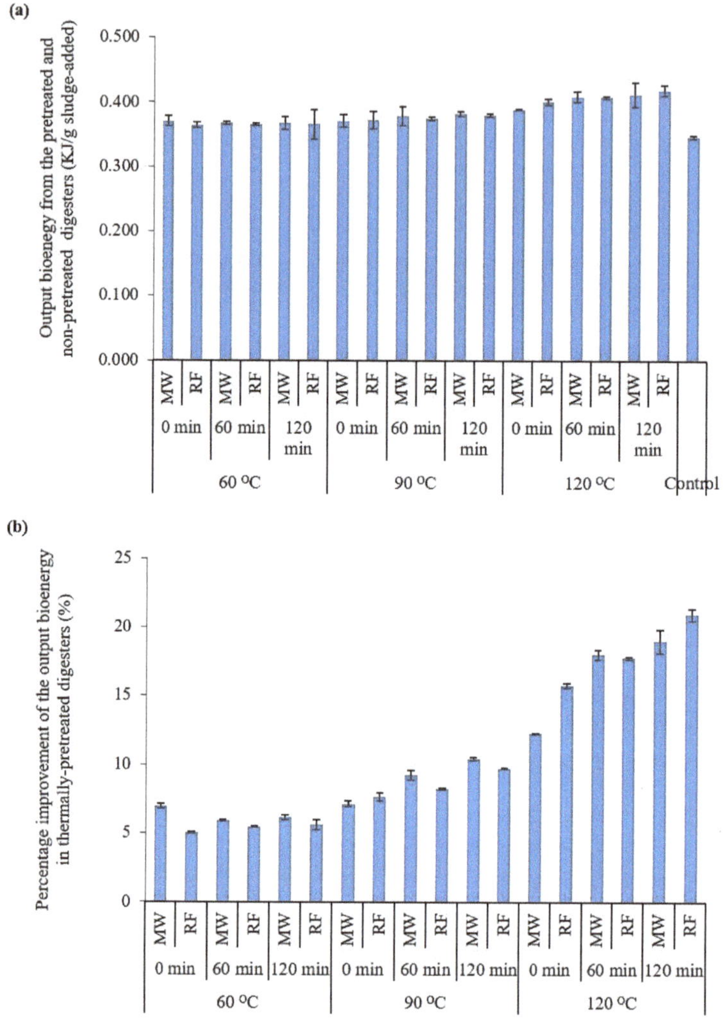

Figure 3. (**a**) Output energy from the pretreated and non-pretreated digesters; (**b**) percentage improvement (relative to the control digester) in output energy.

It has been already demonstrated that under an identical thermal profile, the bioenergy production of the digesters fed with thermally-pretreated sludge is independent of the type of the pretreatment system used. Therefore, the lower the input energy consumption of the pretreatment system is, the higher the net energy production of the sequential pretreatment/AD process will be.

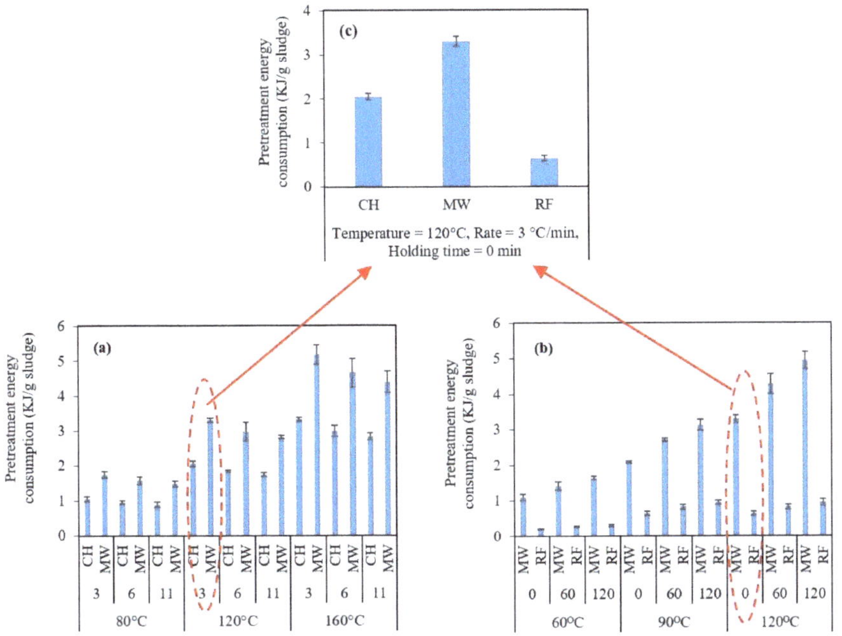

Figure 4. Electrical energy consumption (input energy) during different pretreatment condition; (**a**) CH vs. MW; (**b**) MW vs. RF; (**c**) CH vs. MW vs. RF.

The net energy of an advanced AD system calculated from Equation (2) does not include the amount of the thermal energy that can be recovered from the pretreated sludge before feeding to the digester. The recovered thermal energy can be used to preheat the sludge, increase its temperature to some extent, and therefore reduce the input energy of the system. An efficiency factor of 75–90% for the thermal energy recovery via a heat exchanger is suggested by other researchers in the field [29–32]. In this study, an efficiency factor of 80% was selected. Figure 5 compares the net energy production through the advanced AD process utilizing the MW and RF pretreatment system. As shown in Figure 5, due to high electrical energy consumption (input energy), the MW pretreatment resulted in a negative energy balance for the pretreatment temperatures of above 60 °C, but, the application of the RF pretreatment achieved a positive net energy balance under all the pretreatment conditions tested.

Despite the positive net energy production achieved via sequential pretreatment/AD process utilizing RF heating, the net energy increase via methane generation still stayed below the energy input requirement for RF heating. Therefore, the control (non-pretreated) digester had the highest net energy production. Given the secondary benefits of thermal hydrolysis of municipal sludge established in the literature such as improved pathogen destruction and faster dewaterability [33,34], the results of the current research conducted under batch flow regime warrant a more comprehensive energy analysis with data generated from larger scale continuously fed digesters (simulating full-scale AD more closely) using RF pretreatment on thickened sludge. The application of RF heating on thickened sludge at much higher solids concentrations (i.e., > 10% total solids (TS), as seen in patented thermal hydrolysis processes) will expect to achieve higher net energy than the control digesters. This outcome will be significant considering the fact that according to Cano et. al. (2015), despite the enhanced solubilization or biogas production achieved, almost all of the pretreatment technologies consuming electricity cannot satisfy their energy requirement [35]. Energy analyses from continuous-flow AD studies incorporating RF pretreatment of thickened sludge are currently underway at UBC Bioreactor Technology Group.

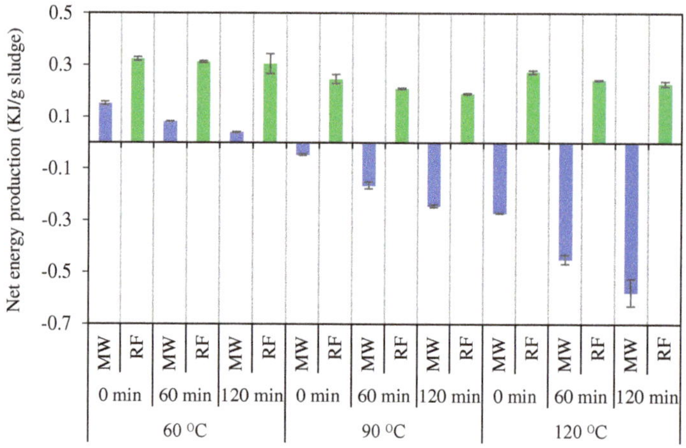

Figure 5. The net energy production of the MW- and RF-pretreated digesters.

3. Materials and Methods

3.1. Municipal Sludge Characteristics

Table 2 summarizes the main characteristics of the thickened waste-activated sludge (TWAS) and dewatered sludge cake (DWSC) which were used in this research. The sludge samples were collected from the City of Kelowna's municipal wastewater treatment plant (WWTP) located in the Okanagan Valley in the southern interior of the Province of British Columbia, Canada. At Kelowna's WWTP, the wastewater undergoes physical treatment processes (i.e., screening, grit removal, and primary sedimentation) followed by a biological nutrient removal (BNR) system. The WAS produced through the BNR process is collected from the bottom of secondary clarifiers and sent to a dissolved air flotation unit for thickening. The generated TWAS is mixed with the fermented primary sludge (PS) at a ratio of 67%-TWAS to 33%-PS by volume. The mixed sludge is transferred to a centrifuge unit and dewatered to produce the DWSC.

Table 2. The characteristics of the municipal sludge used in this research *.

Description	Thickened Waste-Activated Sludge	Dewatered Sludge Cake
pH	6.5 ± 0.1	5.8 ± 0.2
TS (% *w*/*w*)	3.5 ± 0.2	19.2 ± 0.34
VS (% *w*/*w*)	2.7 ± 0.2	16.8 ± 0.35
VS/TS (%)	77.4	87.6 ± 0.24
Total COD (mg/L)	37,420 ± 574	265,702 ± 9422
Soluble COD (mg/L)	1740 ± 350	11,991 ± 591
Total VFAs	309 ± 23	1857 ± 36
Ammonia (mg/L)	201± 17	678 ± 82
Alkalinity (mg/L as $CaCO_3$)	632 ± 128	2145 ± 327

* TS: Total solids; VS: Volatile solids; COD: Chemical oxygen demand; VFAs: Volatile fatty acids as summation of acetic, propionic, and butyric acids.

3.2. Thermal Pretreatment Systems

3.2.1. CH Pretreatment System

Figure 6 shows the configuration and the major components of the three electricity-based thermal pretreatment systems compared in this research. As shown in Figure 1a, the CH system consists a custom-built pressure-sealed vessel. The other components of the CH system include a thermocouple (type K), safety valve, pressure gauge (Winters PEM Series), external fiberglass insulator, DC power supply (Sorensen, Ametek, San Diego, CA, USA), digital multimeter (Agilent, 34401A, Santa Clara, CA, USA), control software, and safety shield. The pressure-sealed vessel was made of a copper cylinder with height, diameter, and thickness of 9.2, 3.8, and 0.32 cm, respectively. The copper vessel was wrapped with 1.5 m of a 0.3 mm-diameter nichrome wire (#80/20) and had a total electrical resistance of 500 Ω. The voltage of the DC power supply was controlled with a computer equipped with a custom-developed LabVIEW program. The heating profile was controlled by changing the DC voltage applied to the nichrome heater.

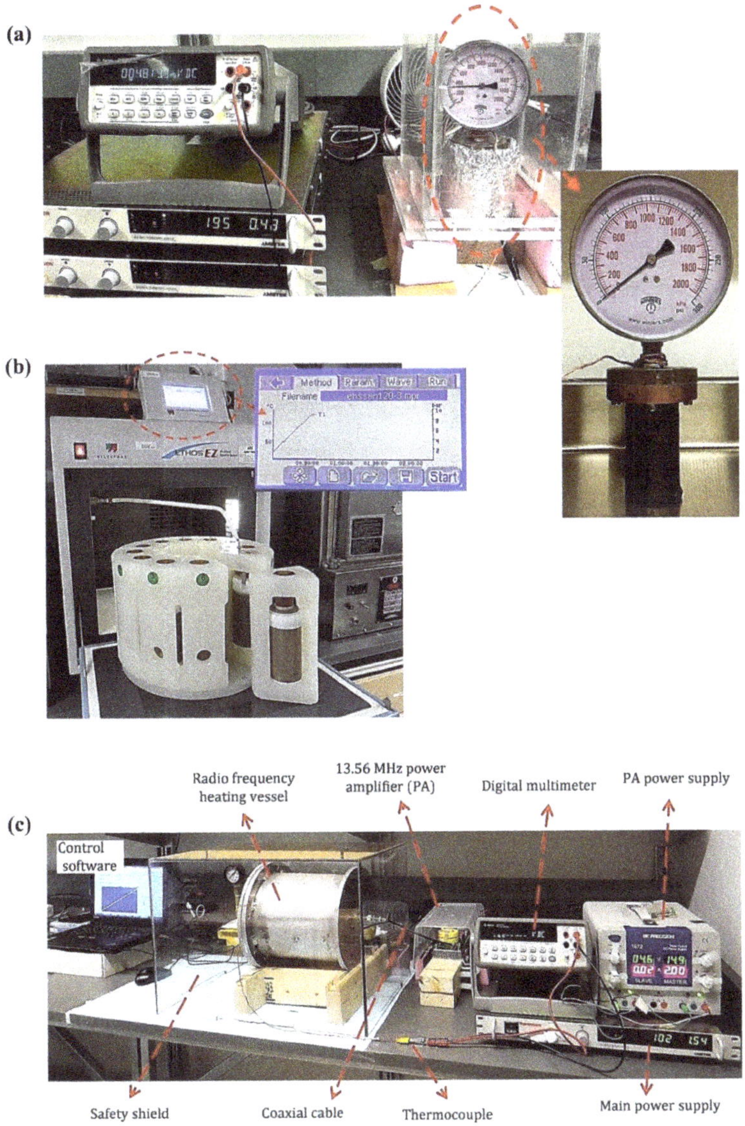

Figure 6. Thermal hydrolysis systems; (**a**) conventional (conductive) heating system (**b**) 2.45 GHz microwave oven; (**c**) 13.56 MHz radio frequency heating system.

3.2.2. MW Pretreatment System

As shown in Figure 6b, a bench-scale 1.2 kW oven operated at a commonly used frequency of 2.45 GHz (ETHOS-EZ, Milestone Inc., Sorisole, Italy) was used for MW pretreatment. The MW system was capable of heating 1.2 L of sludge to a maximum temperature and pressure of 300 °C and 35 bar, respectively. The heating profile in the MW oven was controlled by measuring the temperature of the load using an ATC-400-CE thermocouple.

3.2.3. RF Pretreatment System

The RF heating system is shown in Figure 6c and was custom-designed based on the electrical properties of municipal sludge [36]. The RF heating vessel consisted of a parallel plate structure enclosed in a dielectric cylinder. The cylinder was machined from a solid piece of Teflon which has very low dielectric loss and the parallel plate structure created a uniform electric field throughout the load volume. The Teflon vessel was surrounded by an aluminum cylinder to provide RF shielding from the electric field. A 1 kW RF generator operating at a frequency of 13.56 MHz was connected to the RF heating applicator. The system could heat 400 mL of sludge up to a temperature of 160 °C and under heating rates up to 15 °C/min. A closed loop control system was used to control the thermal profile in the load. A thermocouple was immersed in the load cylinder and the RF power applied to the load was controlled by changing the DC supply voltage to the generator. A software program running in LabVIEW periodically sampled the load temperature and adjusted the DC voltage to maintain a specific software defined thermal profile. The software provided a convenient way to control thermal ramp rates and final load temperatures to match heating profiles used in CH and MW experiments. Further details on the electrical design of the RF heating system are available in other publications [37–39].

3.3. *Experimental Design*

3.3.1. CH vs. MW Comparison

The comparison of the thermal hydrolysis systems was made through a series of solubilization tests followed by AD assessment. Table 3 shows the independent variables and their levels included in the design of the experiments. For the CH vs. MW comparison study (Table 3a), the experimental design included a wide range of final temperatures (80, 120 and 160 °C) and heating rates (3, 6 and 11 °C/min). As listed in Table 3a, fourteen different combinations of the independent variables (pretreatment method, heating rate, and final temperature) were evaluated in addition to one control scenario (without pretreatment). After the solubilization study, a fully randomized half-factorial design was used to define the experimental combinations for the mesophilic and thermophilic batch AD. As a result, 27 mesophilic batch digesters (including triplicates) with pretreated sludge and inoculum were set up. The same number of digesters (27) were also set up under the thermophilic condition. Also, one set of blank digesters (only set up with inoculum) and one set of control digesters (with non-pretreated sludge and inoculum) were included in the experiment. A total of 66 batch digesters (including triplicates) were operated simultaneously.

3.3.2. MW vs. RF Comparison

The comparison of CH and MW pretreatment methods showed that there was no statistically significant difference between the two methods in terms of sludge solubilization and digester performance under identical thermal profiles. Based on this outcome, the next set of experiments compared MW and RF heating methods. MW heating was conducted using the same apparatus and the same experimental methodology in both sets of experiments (CH vs. MW and MW vs. RF), and the MW heating results obtained in both sets of experiments were consistent. Therefore, although the experiments described in the paper were carried out in two phases, the methodology was identical and the outcomes of the experiments were compared. Further, both sets of experiments included control digesters to provide benchmarks for comparison with and without thermal pretreatment.

For the RF vs. MW comparison study (Table 3b), the experimental design included one control and 18 combinations of three independent variables including pretreatment method (RF vs. MW), final temperature (60, 90 and 120 °C), and holding time (0, 60, and 120 min). Although a temperature of 60 °C was not expected to have a significant effect on sludge solubilization and subsequent AD processes, it was included in the experimental design to investigate any possible non-thermal (athermal) effects of the MW and RF pretreatments. Experimental results to compare CH with MW heating, which are summarized in Table 3a, show that a low thermal ramp rate of 3 °C/min resulted in the best sludge disintegration and biogas production. Based on the outcome of the first experiments comparing CH and MW heating, subsequent experiments to compare heating methods, including MW and RF heating, used a heating rate of 3 °C/min. Following the solubilization study, 63 mesophilic batch digesters were set up to compare the effect of the two pretreatment methods (MW vs. RF) on AD performance.

Table 3. The experimental design used for comparison of CH, MW, and RF pretreatment systems.

(a) CH vs. MW Experimental Design				**(b) MW vs. RF Experimental Design**			
Method	Temperature (°C)	Rate (°C/min)	Digester type	Method	Temperature (°C)	Holding time (min)	Digester type
CH	80	11	Batch mesophilic	MW	60	0	Batch mesophilic
	120	6				60	
		11				120	
	160	3			90	0	
MW	80	3				60	
		6				120	
	120	3			120	0	
	160	6				60	
		11				120	
CH	80	6	Batch thermophilic	RF	60	0	
		11				60	
	120	3				120	
	160	3			90	0	
		6				60	
MW	80	3				120	
	120	6			120	0	
		11				60	
	160	11				120	

3.4. Sludge Disintegration Study

The effects of CH, MW, and RF pretreatments on sludge disintegration were evaluated by comparing the soluble concentration of COD and biopolymers (i.e., sugar, protein, and humic acid) before and after thermal pretreatment. The Standard Methods procedure (Section 5250 D) on the application of closed reflux colorimetric method was followed in measuring the COD concentration [40]. The procedure proposed by Dubois et al. (1956) was used for sugar analysis [41]. The COD and sugar measurement was done using an Evolution 60S UV-Vis spectrophotometer (Thermo Fisher Scientific, Inc., Waltham, MA, USA) at the wavelength of 600 and 490 nm, respectively. Protein and humic acid quantification was done at a wavelength of 750 nm with a multi-detection microplate spectrophotometer (BioTek Synergy HT, Winooski, VT, USA). The protein and humic acid sample preparation was done following the modified Lowry's method [42].

3.5. Anaerobic Digestion Study

The batch AD experiments were initiated by placing the substrate and inoculum into 160 mL-bottles. The mesophilic inoculum was taken from a pilot-scale digester which has been continuously fed with a mixture of primary and secondary sludge from the Kelowna's WWTP at a sludge retention time (SRT) of 20 d for more than three years. The thermophilic inoculum was taken from a full-scale digester located at the Annacis Island WWTP in Vancouver (BC, Canada) utilizing a mixture of primary and secondary sludge. The substrate/inoculum mixing ratio was calculated based on the food to microorganism ratio (F/M) of 2.1 ± 0.2 (g VS/g VS). To keep the digester pH above 6.5 throughout the digestion process, additional alkalinity (4000 mg/L of $CaCO_3$) was added into each digester in the form of potassium bicarbonate and sodium bicarbonate. The inoculum was degassed for a period of one week prior to the start of the batch assays. Upon mixing the inoculum and substrate, and before sealing, the digesters were purged with nitrogen gas. The mesophilic and thermophilic batch digesters were placed in two separate incubators (Innova, 44R, Eppendorf Canada, Mississauga, ON, Canada) set at 90 rpm and a temperature of 35 °C and 55 °C, respectively. The value of cumulative bioenergy production of the batch digesters with a duration time of 35 d was used to determine the bioenergy production from the batch digesters.

3.6. Energy Analysis

As shown in Equation (2), the net energy of an advanced AD system (pretreatment +AD) can be determined by subtracting the amount of energy consumed during the sludge pretreatment (input energy) from the amount of energy generated as methane (output energy).

$$E_{net} = E_{out} - E_{in} \quad (2)$$

where, E_{net}, E_{out} and E_{in} are the system net energy, output energy and input energy, respectively. In this research, the E_{out} of the digesters was determined considering the methane energy content of 55.6 $\frac{\text{kJ}}{\text{g CH}_4}$ and the density of 0.715 $\frac{\text{g}}{\text{L}}$ at standard temperature and pressure (0 °C, 1 atm) [5]. To determine the E_{in}, the voltage and the current supplied to the pretreatment systems (CH, MW, and RF) were continuously recorded during the entire pretreatment period. The input power (P_t) was then determined by multiplying the recorded current and voltage. The total energy consumption (E_{in}) of the systems was then calculated by integrating the power over the entire heating time as follows:

$$E_{in} = \int_0^T P_t dt \quad (3)$$

For the CH and RF pretreatment systems, the current and voltage of the DC power supply were automatically recorded by a computer equipped with a custom-developed LabVIEW program. However, for the MW system, it was more convenient to measure the input voltage and current to the MW oven using an oscilloscope which was connected to the AC line input to the oven. It is noteworthy that a complete energy analysis of advanced AD system would need to consider other processes in the AD such as mixing or the energy to thermally regulate the temperature of the sludge inside the digester. However, these additional energy factors are assumed to the same for non-pretreated (control) and thermally-pretreated AD systems and have therefore been excluded from the energy equation.

3.7. Analytical Method

The total and volatile solids (TS and VS) concentration were determined following the procedures of the Standard Methods (Sections 2540 B and 2540 E) [40]. The ammonia (NH_3-N) analysis was conducted using an electrode connected to a dual channel pH/ion meter (Accumet Excel XL25). The total volatile fatty acids (VFAs) were measured in the form of acetic, propionic, and butyric acids by injecting the samples into an Agilent 7890A gas chromatograph (GC) using an autosampler. The GC utilized an Agilent 19091F-112 capillary column with a length 25 m and a diameter of 320 μm. It was also equipped with a flame ionization detector (carrier gas flow rate: 25 mL·He/min; oven, inlet, and outlet temperatures: 200, 220 and 300 °C, respectively). Before injecting samples into the GC, the samples were centrifuged for 20 min at 10,000 rpm and then filtered through 0.45 μm membrane filters. The volume of the biogas was measured using a manometer. The biogas composition was determined in the form of CH_4, CO_2, and N_2 gases using an Agilent 7820A GC equipped with an

Agilent G3591-8003/80002 packed column and a thermal conductivity detector (oven, inlet, and outlet temperatures: 70, 100 and 150 °C, respectively).

3.8. Statistical Analysis

The statistically significant effects of the input parameters (i.e., pretreatment method, temperature, heating rate, etc.) were evaluated by multi-factor ANOVA at a 95% confidence level ($\alpha = 0.05$) using Minitab Software 17 (Minitab Inc., State College, PA, USA). The Fisher's least significant difference test was applied to compare all pairs of means. The Anderson-Darling test was used to judge if the data follow normality distributions. The sample preparation was done randomly following a randomized experimental order determined by Design-Expert 9 software.

4. Conclusions

According to the results and analyses, under identical thermal profiles, the method of thermal pretreatment (CH, MW, and RF) was not a significant factor determining the sludge disintegration and AD performance. The input energy measurements revealed that the CH and MW pretreatment methods consumed 100–440% more electrical energy than the RF heating system to achieve the same pretreatment conditions. The RF heating system used in this study was designed to heat municipal sludge efficiently and therefore it demonstrates the importance of the heating applicator design. Based on the results of the energy analysis, the energy consumption during pretreatment using all thermal hydrolysis methods (CH, MW, and RF) was higher than the increase in the net bioenergy which was achieved during the AD process in form of methane. This resulted in higher net energy production in the control (non-pretreated) digester compared to the sequential pretreatment/AD process. As a way of reducing the input energy per unit dry mass of the sludge, thermal hydrolysis can be applied on thickened sludge at higher solids concentrations (i.e., > 10% TS). In addition, a more representative energy analysis can be carried out on the data collected from larger scale continuously fed digesters which simulate full-scale AD. Given these results, the authors are currently conducting a more comprehensive energy analysis from continuous-flow AD studies incorporating RF pretreatment.

Author Contributions: All the authors conceived and designed the experiments. All the authors analyzed and interpreted the results. E.H.K. performed the experiments and prepared the manuscript draft. T.J. and C.E. reviewed and revised the manuscript. All authors have read and approved the final version of the manuscript.

Funding: This research was funded by Natural Science and Engineering Research Council of Canada (NSERC) through the Strategic Project Grant No. 396519-10.

Acknowledgments: The authors would like to thank the NSERC for its financial support. The support from the City of Kelowna and BC Ministry of Environment is also acknowledged. The authors also thank Saimoom Ferdous for his contributions during his M.Sc. studies and Jake Bobowski for his consultation during the project.

Conflicts of Interest: The authors declare no conflict of interest.

References

1. Spinosa, L.; Vesilind, P.A. *Sludge into Biosolids: Processing, Disposal and Utilization*; IWA Publishing: London, UK, 2001.
2. *Biosolids Generation, Use and Disposal in the United States*; EPA530-R-99-009; U.S. Environmental Protection Agency, Municipal and Industrial Solid Waste Division, Office of Solid Waste: Washington, DC, USA, 1999.
3. Pe´rez-Elvira, S.I.; Nieto Diez, P.; Fdz-Polanco, F. Sludge minimisation technologies. *Rev. Environ. Sci. Bio/Technol.* **2006**, *5*, 375–398. [CrossRef]
4. Appels, L.; Baeyens, J.; Degre've, J.; Dewil, R. Principles and potential of the anaerobic digestion of waste-activated sludge. *Prog. Energy Combust. Sci.* **2008**, *34*, 755–781. [CrossRef]
5. *Metcalf; Eddy, Wastewater Engineering: Treatment and Reuse*, 5th ed.; McGraw-Hill: New York, NY, USA, 2013.
6. Tyagi, V.K.; Lo, S.-L. Microwave irradiation: A sustainable way for sludge treatment and resource recovery. *Renew. Sust. Energy. Rev.* **2013**, *18*, 288–305. [CrossRef]

7. Park, W.J.; Ahn, J.H. Optimization of microwave pretreatment conditions to maximize methane production and methane yield in mesophilic anaerobic sludge digestion. *Environ. Technol.* **2011**, *32*, 1533–1540. [CrossRef] [PubMed]
8. Mehdizadeh, N.S.; Eskicioglu, C.; Bobowski, J.; Johnson, T. Conductive heating and microwave hydrolysis under identical heating profiles for advanced anaerobic digestion of municipal sludge. *Water Res.* **2013**, *47*, 5040–5051. [CrossRef] [PubMed]
9. Sólyom, K.; Mato, R.B.; Pérez-Elvira, S.I.; Cocero, M.J. The influence of the energy absorbed from microwave pretreatment on biogas production from secondary wastewater sludge. *Bioresour. Technol.* **2011**, *102*, 10849–10854. [CrossRef] [PubMed]
10. Appels, L.; Degrève, J.; Van der Bruggen, B.; Van Impe, J.; Dewil, R. Influence of low temperature thermal pre-treatment on sludge solubilisation, heavy metal release and anaerobic digestion. *Bioresour. Technol.* **2010**, *101*, 5743–5748. [CrossRef] [PubMed]
11. Eskicioglu, C.; Kennedy, K.J.; droste, R.L. Characterization of soluble organic matter of waste activated sludge before and after thermal pretreatment. *Water Res.* **2006**, *40*, 3725–3736. [CrossRef] [PubMed]
12. Incropera, F.P.; DeWitt, D.P.; Bergman, T.L.; Lavine, A.S. *Introduction to Heat Transfer*; John Wiley & Sons, Inc.: Hoboken, NJ, USA, 2007.
13. Climent, M.; Ferrer, I.; Mar Baeza, M.; Artola, A.; Vazquez, F.; Font, X. Effects of thermal and mechanical pretreatments of secondary sludge on biogas production under thermophilic conditions. *Chem. Eng. J.* **2007**, *133*, 335–342. [CrossRef]
14. Park, W.; Ahn, J. Effects of microwave pretreatment on mesophilic anaerobic digestion for mixture of primary and secondary sludges compared with thermal pretreatment. *Environ. Eng. Res.* **2011**, *16*, 103–109. [CrossRef]
15. Pino-Jelcic, S.A.; Hong, S.; Park, J.K. Enhanced anaerobic biodegradability and inactivation of fecal coliforms and Salmonella spp. in wastewater sludge by using microwaves. *Water Environ. Res.* **2006**, *78*, 209–216. [CrossRef] [PubMed]
16. Koupaie, E.H.; Johnson, T.; Eskicioglu, C. Advanced anaerobic digestion of municipal sludge using a novel and energy-efficient radio frequency pretreatment system. *Water Res.* **2017**, *118*, 70–81. [CrossRef] [PubMed]
17. Eskicioglu, C.; Terzian, N.; Kennedy, K.J.; Droste, R.L.; Hamoda, M. Athermal microwave effects for enhancing digestibility of waste activated sludge. *Water Res.* **2007**, *41*, 2457–2466. [CrossRef] [PubMed]
18. Beszédes, S.; László, Z.; Szabó, G.; Hodúr, C. Examination of the effect of microwave heating on the biodegradable and soluble fraction of organic matter of sludge. *J. Eng. Ann. Fac. Eng. Hunedoara* **2009**, *7*, 87–90.
19. Beszédes, S.; László, Z.; Horváth, Z.H.; Szabó, G.; Hodúr, C. Comparison of the effects of microwave irradiation with different intensities on the biodegradability of sludge from the dairy- and meat-industry. *Bioresour. Technol.* **2011**, *102*, 814–821. [CrossRef] [PubMed]
20. Hong, S.M.; Park, J.K.; Lee, Y.O. Mechanism of microwave irradiation involved in the destruction of fecal coliforms from biosolids. *Water Res.* **2004**, *38*, 1615–1625. [CrossRef] [PubMed]
21. Kuglarz, M.; Karakashev, D.; Angelidaki, I. Microwave and thermal pretreatment as methods for increasing the biogas potential of secondary sludge from municipal wastewater treatment plants. *Bioresour. Technol.* **2013**, *134*, 290–297. [CrossRef] [PubMed]
22. Hosseini Koupaie, E.; Eskicioglu, C. Below and above boiling point comparison of microwave irradiation and conductive heating for municipal sludge digestion under identical heating/cooling profiles. *Bioresour. Technol.* **2015**, *187*, 235–245. [CrossRef] [PubMed]
23. Hosseini Koupaie, E.; Eskicioglu, C. Conventional heating vs. microwave sludge pretreatment comparison under identical heating/cooling profiles for thermophilic advanced anaerobic digestion. *Waste Manag.* **2016**, *53*, 182–195. [CrossRef] [PubMed]
24. Eskicioglu, C.; Kennedy, K.J.; Droste, R.L. Initial examination of microwave pretreatment on primary, secondary and mixed sludges before and after anaerobic digestion. *Water Sci. Technol.* **2008**, *57*, 311–317. [CrossRef] [PubMed]
25. Ji Park, W.; Ahn, J.-H.; Hwang, S.; Lee, C.-K. Effect of output power, target temperature, and solid concentration on the solubilization of waste activated sludge using microwave irradiation. *Bioresour. Technol.* **2010**, *101*, 13–16. [CrossRef] [PubMed]
26. Park, W.; Ahn, J.; Lee, C. Effect of temperature-increase rate and terminal temperature on the solubilization of sewage sludge using microwave irradiation. *Environ. Eng. Res.* **2009**, *14*, 48–52. [CrossRef]

27. Toreci, I.; Kennedy, K.J.; Droste, R.L. Effect of high-temperature microwave irradiation on municipal thickened waste activated sludge solubilization. *Heat Trans. Eng.* **2011**, *31*, 766–773. [CrossRef]
28. Vergine, P.; Zábranská, J.; Canziani, R. Low temperature microwave and conventional heating pre-treatments to improve sludge anaerobic biodegradability. *Water Sci. Technol.* **2014**, *69*, 518–524. [CrossRef] [PubMed]
29. Wanner, O.; Panagiotidis, V.; Clavadetscher, P.; Siegrist, H. Effect of heat recovery from raw wastewater on nitrification and nitrogen removal in activated sludge plants. *Water Res.* **2005**, *39*, 4725–4734. [CrossRef] [PubMed]
30. Gasafi, E.; Reinecke, M.-Y.; Kruse, A.; Schebek, L. Economic analysis of sewage sludge gasification in super critical water forhydrogen production. *Biomass Bioenergy* **2008**, *32*, 1085–1096. [CrossRef]
31. Akgul, D.; Cella, M.A.; Eskicioglu, C. Influences of low-energy input microwave and ultrasonic pretreatments on single-stage and temperature-phased anaerobic digestion (TPAD) of municipal wastewater sludge. *Energy* **2017**, *123*, 271–282. [CrossRef]
32. Lu, J.; Gavala, H.N.; Skiadas, I.; Mladenovska, Z.; Ahring, B. Improving anaerobic sewage sludge digestion by implementation of a hyperthermophilic prehydrolisis step. *J. Environ. Manag.* **2008**, *88*, 881–889. [CrossRef] [PubMed]
33. Neyens, E.; Baeyens, J. A review of thermal sludge pre-treatment processes to improve dewaterability. *J. Hazard. Mater.* **2003**, *98*, 51–67. [CrossRef]
34. Yin, F.; Li, Z.; Wang, D.; Ohlsen, T.; Dong, H. Performance of thermal pretreatment and mesophilic fermentation system on pathogen inactivation and biogas production of faecal sludge: Initial laboratory results. *Biosyst. Eng.* **2016**, *151*, 171–177. [CrossRef]
35. Cano, R.; Pérez-Elvira, S.I.; Fdz-Polanco, F. Energy feasibility study of sludge pretreatments: A review. *Appl. Energy* **2015**, *149*, 176–185. [CrossRef]
36. Bobowski, J.S.; Johnson, T.; Eskicioglu, C. Permittivity of waste-activated sludge by open-ended coaxial line. *Prog. Electromagn. Res. Lett.* **2012**, *29*, 139–149. [CrossRef]
37. Ferdous, M.S.; Hosseini Koupaie, E.; Eskicioglu, C.; Johnson, T. An experimental 13.56 MHz radio frequency heating system for efficient thermal pretreatment of wastewater sludge. *Prog. Electromagn. Res. B* **2017**, *79*, 83–101. [CrossRef]
38. Hosseini Koupaie, E. Radio Frequency Sludge Hydrolysis as an Energy Efficient Alternative to Microwave and Conductive Heating for Advanced Anaerobic Digestion. Ph.D. Thesis, University of British Columbia Okanagan, Kelowna, BC, Canada, 2017.
39. Ferdous, M.S. Design of a Radio Frequency Heating System for Electrolytic Liquids and Sludges. Master' Thesis, University of British Columbia Okanagan, Kelowna, BC, Canada, 2015.
40. Apha, A. *WPCF, Standard Methods for the Examination of Water and Wastewater*, 20th ed.; American Public Health Association: Washington, DC, USA, 2005.
41. Dubois, M.; Gilles, K.A.; Hamilton, J.K.; Rebers, P.A.; Smith, F. Colorimetric method for the determination of sugar and related substances. *Anal. Chem. Anal. Chem.* **1956**, *28*, 350–356. [CrossRef]
42. Frolund, B.; Griebe, T.; Nielsen, P.H. Enzymatic activity in the activated-sludge floc matrix. *Appl. Microbiol. Biotechnol.* **1995**, *43*, 755–761. [CrossRef] [PubMed]

Sample Availability: Samples of the compounds are not available from the authors.

molecules

Article

Impact of Alkaline H_2O_2 Pretreatment on Methane Generation Potential of Greenhouse Crop Waste under Anaerobic Conditions

N. Altınay Perendeci [1,*], Sezen Gökgöl [1] and Derin Orhon [2,3]

1 Environmental Engineering Department, Akdeniz University, 07058 Antalya, Turkey; sezengokgol@hotmail.com
2 ENVIS Energy and Environmental Systems Research and Development Ltd., ITU ARI Technocity, Maslak, 34469 Istanbul, Turkey; orhon@itu.edu.tr
3 Environmental Engineering Department, Near East University, Near East Boulevard, 99138 Nicosia/TRNC Mersin 10, Turkey
* Correspondence: aperendeci@akdeniz.edu.tr; Tel.: +90-242-3106334

Received: 25 June 2018; Accepted: 16 July 2018; Published: 20 July 2018

Abstract: This paper intended to explore the effect of alkaline H_2O_2 pretreatment on the biodegradability and the methane generation potential of greenhouse crop waste. A multi-variable experimental design was implemented. In this approach, initial solid content (3–7%), reaction time (6–24 h), H_2O_2 concentration (1–3%), and reaction temperature (50–100 °C) were varied in different combinations to determine the impact of alkaline H_2O_2 pretreatment. The results indicated that the alkaline H_2O_2 pretreatment induced a significant increase in the range of 200–800% in chemical oxygen demand (COD) leakage into the soluble phase, and boosted the methane generation potential from 174 $mLCH_4/g$ of volatile solid (VS) to a much higher bracket of 250–350 $mLCH_4/gVS$. Similarly, the lignocellulosic structure of the material was broken down and hydrolyzed by H_2O_2 dosing, which increased the rate of volatile matter utilization from 31% to 50–70% depending on selected conditions. Alkaline H_2O_2 pretreatment was optimized to determine optimal conditions for the enhancement of methane generation assuming a cost-driven approach. Optimal alkaline H_2O_2 pretreatment conditions were found as a reaction temperature of 50 °C, 7% initial solid content, 1% H_2O_2 concentration, and a reaction time of six h. Under these conditions, the biochemical methane potential (BMP) test yielded as 309 $mLCH_4/gVS$. The enhancement of methane production was calculated as 77.6% compared to raw greenhouse crop wastes.

Keywords: alkaline H_2O_2 pretreatment; breakdown of lignocellulosic structure; greenhouse crop waste; methane generation; process optimization

1. Introduction

In the last few decades, there was a drastic change in the conceptual understanding of waste management. Waste is no longer considered as matter to be disposed of at the expense of additional cost, but as a resource. Perhaps the most significant resource component is energy in view of the present and future energy shortages expected, due to demands of rapid population expansion and escalating industrial activities in the world. Therefore, energy recovery from waste is now a hot topic, both in terms of scientific efforts and practical applications.

Recently, renewable energy sources, such as solar energy, wind energy, and geothermal energy, are now being largely explored and exploited. Among these categories, biomass energy should be given specific emphasis mainly due to its accessibility; the energy recovery from biomass is also quite sustainable as the proper disposal of biomass requires costly technical processes. Agricultural waste is an important component of the wide spectrum of waste sources considered within the scope of biomass energy [1]. This study focused on greenhouse agriculture, a significant agricultural practice in areas with a suitable climate, like the Antalya region on the southern coast of Turkey. The Mediterranean region is one of the most important areas in terms of protected cultivation because the mild winter makes production under simple structures possible [2]. Greenhouses provide a protected growing environment that can be controlled during the year. This allows intensive culture with annual yields many times higher than that of field production [3]. Turkey holds an important place in the world for the production of fresh fruit and vegetables, having close to 752,000 decares of greenhouse-covered land, placing it fifth in the world after China, South Korea, Spain, and Japan. About 278,000 decares of greenhouse land is located in the Antalya province, which corresponds to approximately 36.97% of greenhouse land in Turkey. Furthermore, 51% of Turkey's greenhouse vegetable production (3.2 million tons) is provided by Antalya. Greenhouse agriculture is very significant in the districts of Alanya, Aksu, Elmali, Gazipaşa, Kepez, Korkuteli, Kumluca, Manavgat, and Serik.

While total greenhouse production (tomato, pepper, cucumber, eggplant, and zucchini) was 2,256,325 tons, 1,087,247.75 tons of greenhouse crop waste was produced in the production year of 2005–2006 [4]. Unfortunately, greenhouse cultivation waste lignocellulosic residue is improperly disposed into the environment in Turkey. The conventional disposal methods for most of this waste, such as unconfined storage in forests and road edges, landfilling, and uncontrolled burning, cause significant environmental problems [5]. A limited quantity of greenhouse crop waste is also used for mulching. However, growers prefer not to apply mulching, due to the spread of some diseases and the transfer of non-biodegraded pesticides, herbicides, and others for the subsequent cultivation period. Landfilling is the most applied waste management practice, and results in the release of CH_4 which is around 20 times more potent as a greenhouse gas (GHG) than CO_2. Landfilling was shown to be the greatest source of GHG emissions, contributing more than 75% of total emissions associated with waste management [6]. Uncontrolled burning and/or incineration of greenhouse crop waste emits CO_2 and N_2O, a GHG gas 310 times more powerful in atmospheric warming than CO_2. In addition, uncontrolled burning and/or incineration diverts waste from landfill, reducing the amount of methane generated. However, combustion also produces waste in the form of ash. Eventually, waste crops disposed from greenhouses were found to be a renewable and cost-free source of lignocellulosic biomass, whose management is necessary to prevent environmental pollution and to gain an alternative utilization as a fuel biogas. Greenhouse crop waste involves all parts left in the field after the harvest, including roots, stems, leaves, rotten/spoiled vegetables, etc. What makes this category of agricultural waste interesting is its complex lignocellulosic structure, whereby the residue contains cellulose (35–50%), hemicellulose (20–35%), lignin (10–25%), and minor fractions of proteins, oils, and ash [7,8] in such a way that the cellulose is embedded in a lignin–polysaccharide sheet [9]. This structure resists microbial destruction and hydrolysis, and requires pretreatment before an energy recovery process.

Many pretreatment technologies were suggested in the literature, such as physical pretreatment, which generally involves mechanical methods such as shredding and grinding [10,11]. Ultrasonic and microwave methods were also tested [12], but were not recommended due to phenolic by-products and the high energy costs involved [13]. Some physico-chemical methods, based on pretreatment with ammonia [14], hot water, and steam explosion [15,16] were reported, all claiming success; however, they also depend on conditions consuming high energy. Pretreatment conducted under acidic and alkaline conditions [17,18] was also found to be effective in breaking down the lignocellulosic structure.

The delignification process as a means of lignin removal is widely used to bleach high-lignin wood pulps in the pulp and paper industry [19,20]. The application of alkaline H_2O_2 is one of the most effective chemical pretreatment approaches for energy recovery from wastes and residues with a lignocellulosic structure. During the alkaline H_2O_2 pretreatment, while H_2O_2 plays the role of an oxidant, the role of alkaline is to reduce or remove lignin, acetyl, and other uronic substitutions in the hemicellulosic portions of the biomass via swelling, salvation, and saponification, so that the accessibility and digestibility of holocellulose is enhanced [19]. Thus, theH_2O_2 delignification of agricultural wastes is strongly pH-dependent, with an optimal pH of 11.5 for the dissociation reaction of H_2O_2. During the treatment, alkaline H_2O_2 reacts rapidly with lignin to form low-molecular-weight, water-soluble oxidation products. The lignin-oxidizing species is a highly reactive hydroxyl radical ($HO\cdot$), formed during the degradation of H_2O_2 in a reaction with the hydroperoxy anion ($HCOO^-$). $HCOO^-$ is the active species and is responsible for the bleaching action of H_2O_2 under alkaline conditions. On the other hand, hydroperoxyl and hydroxyl radicals generated by the decomposition of H_2O_2 are responsible for solubilizing hemicelluloses [21]. This process also has the advantage of not leaving H_2O_2 residue, and it is considered as an environmentally friendly and low-cost application [22]. While a large number of studies were conducted using alkaline H_2O_2 pretreatment on various types of agricultural waste, such as corn stover, wood waste, soft wood, cashew apple bagasse, energy crops, sugar cane bagasse, agricultural crop stalks, and cotton stalks [19,22–29], this method, although quite promising, remains untested for greenhouse crop wastes.

In this context, the main objective of the study was to carry out an experimental assessment of the effect of alkaline H_2O_2 pretreatment on the biodegradability and the methane generation potential of greenhouse crop wastes. A central composite design (CCD) of response surface methodology (RSM) was applied to determine the optimal process conditions of alkaline H_2O_2 pretreatment for maximum biogas production in the most cost-effective way. H_2O_2 concentration, initial solid content, reaction temperature, and reaction time were selected as independent variables. The effects of these four independent variables on soluble chemical oxygen demand (COD), soluble reducing sugar, total lignin on an extractives free bases, and methane generation potential were investigated in detail. The alkaline H_2O_2 pretreatment process was optimized to enhance methane production assuming a cost-driven approach. The effects of the alkaline H_2O_2 pretreatment process on the molecular-bond characterization and surface properties of greenhouse crop waste were also examined via Fourier-transform infrared spectroscopy (FTIR) and scanning electron microscopy (SEM). To the best of our knowledge, this is the first study on biogas production from greenhouse crop waste with the integration of an alkaline H_2O_2 pretreatment process.

2. Results and Discussion

2.1. Chemical Composition

The greenhouse crop waste used in the experiment contained around 13.6% dry matter, indicating an average moisture content of more than 86%. The organic fraction of the dry solids, i.e., volatile solids (VS), was measured as 68.7%, mostly composed of lignocellulosic material. The characteristics of the greenhouse crop waste, expressed in terms of major parameters, are presented in Table 1. The cellulose, hemicellulose, lignin, and soluble matter contents of the fresh greenhouse crop waste were measured as 19.49%, 3.89%, 0.03%, and 76.58%, respectively. The elemental composition of the fresh greenhouse crop waste was found to be 29.23% C, 4.89% H, and 2.96% N. The general composition profile reflected in the Table 1 is different from a previous assessment of the same waste [5], which had a different composition. While the composition of mixed greenhouse crop waste was 61.71% tomato, 22.44% cucumber, 7.92 % eggplant, 5.72 % pepper, and 2.21% zucchini in the previous work [5], the composition in this work was 72% tomato, 14.31% cucumber, 5.11% eggplant, 6.69% pepper, and 1.88% zucchini. Furthermore, the green house crop waste used in the previous study [5] was obtained from the Kumluca region, located in west Antalya. On the other hand, the green house crop waste in this study was acquired from the Gazipaşa region, located in east Antalya. Conclusively, even though the sampling period was the same, the location and composition of the collected greenhouse crop waste was different. Specifically, the cellulose and hemicellulose contents, together with the carbon content, were found to be lower. The reason is most likely due to sampling done from different cultivation areas, with a different sample composition.

Table 1. Average characteristics of the greenhouse crop waste.

Parameter	Result
Total solid, TS (g/kg)	136.53
Volatile solid, VS (g/kg)	93.9
Total Kjeldahl nitrogen, TKN (mg/gVS)	6.75
Protein (mg/gVS)	60
Chemical oxygen demand, COD (mg/gVS)	1494.1
Soluble chemical oxygen demand, sCOD (mg/gVS)	60.88
Soluble reducing sugar, sRedSugar (mg/gVS)	7.59
Extractable material and lipids * (%)	0.14
Van Soest fractionation	
Soluble matter (%)	76.58
Hemicellulose (%)	3.89
Cellulose (%)	19.49
Lignin (%)	0.03
Total lignin on an extractive free bases (%)	19.39
Acid-insoluble (%)	17.25
Acid-soluble (%)	2.11
Elemental Analysis	
C (%)	29.23
H (%)	4.89
N (%)	2.96
S (%)	1.1

* Determined in extractives soluble in water.

The total COD equivalent of the organic matter in the crop waste was determined as 1.49 gCOD/gVS. This is a significant stoichiometric ratio, quite similar to the f_X value of 1.4 gCOD/gVS, characteristic of biomass in activated sludge systems. This ratio corresponds to the traditional empirical formula of $C_5H_7NO_2$, which is still in use for the basic stoichiometry of activated sludge [30]. While noting that the measured nitrogen content remains somewhat lower, it would be acceptable to adopt this simplified formula for the COD–VS relationship in greenhouse crop waste.

Table 1 also indicates the magnitude of COD leakage into the solution (soluble COD (sCOD), S_T) as 61 mgCOD/gVS, and the soluble reducing sugar (sRedSugar) content in this leakage as 7.6 mgCOD/gVS. It should be noted that the soluble sugar component is basically the same as the readily biodegradable COD fraction (S_S) identified in wastewater [31,32]. It is interesting to note that Sözen et al. [33] reported 5250 mg of COD leakage from 90 g of domestic sludge, quite similar to the 58 mg of S_T per g of dry sludge in "eluate tests" performed for evaluating compliance with the limitation of dissolved organic carbon for the landfilling of municipal treatment sludge.

2.2. *Effect of Alkaline H_2O_2 Pretreatment*

The directly observable effect of alkaline H_2O_2 treatment was the substantial increase in the magnitude of sCOD, as illustrated in Figure 1a. All values in Figure 1a were compared with the sCOD value of 61 mgCOD/gVS in the original raw greenhouse crop waste, in order to visualize the effect of alkaline H_2O_2 treatment. Basically, Figure 1a shows that (i) sCOD (S_T) was increased above 200 mgCOD/gVS in all tests; (ii) the most noticeable increase was observed in experiments conducted at 100 °C; in a few experimental runs, S_T exceeded 500 mgCOD/gSV, corresponding to more than an 800% increase compared with the initial COD leakage capacity of the greenhouse crop waste; (iii) the sCOD increase always remained higher when the reaction time was raised to 24 h while other parameters remained the same. This observation is particularly important, since it shows that the H_2O_2 dosage was adjusted to increase the amount of sCOD, but not to oxidize and chemically remove the sCOD generated.

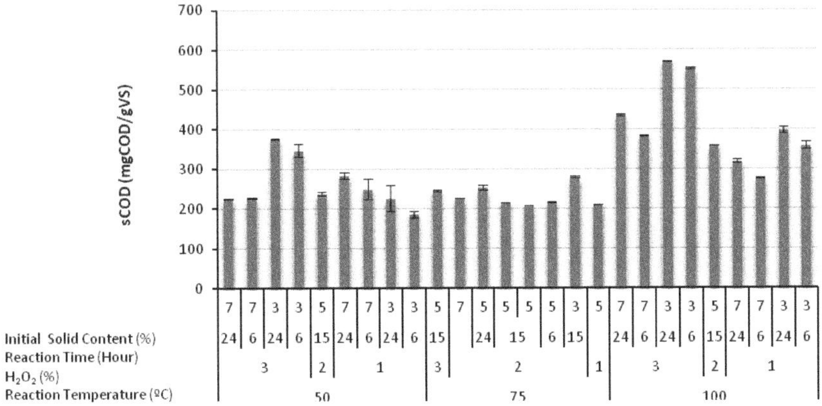

(a)

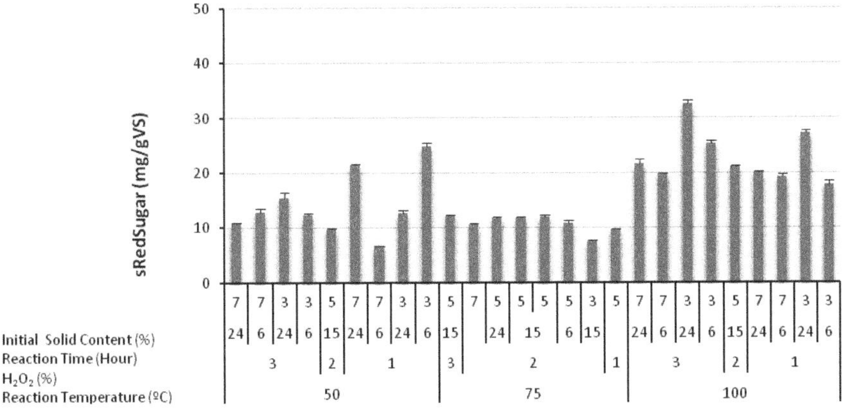

(b)

Figure 1. Increase in soluble chemical oxygen demand (sCOD; (**a**)) and increase in soluble reducing sugar (sRedSugar; (**b**)) due to alkaline H_2O_2 treatment.

Figure 1b shows that alkaline H_2O_2 treatment also increased the soluble sugar (sRedSugar) leakage. The highest sRedSugar concentration was found to be 32.47 mg of glucose/gVS from the greenhouse crop waste pretreated at a reaction temperature of 100 °C, an H_2O_2 concentration of 3%, a reaction time of 24 h, and 3% initial solid content, which are the same pretreatment conditions where the maximum increase in sCOD was observed (Figure 1a). It should be remembered that the sRedSugar/sCOD ratio of the greenhouse crop waste before treatment was 12.4% (Table 1). The values displayed in Figure 1b indicate that, while sRedSugar values also increased with H_2O_2 treatment, the sRedSugar/sCOD ratio decreased from 12.3% to in the range of 3.9–7.8%.

The effect of alkaline H_2O_2 treatment could only be quantified and evaluated in comparison with the methane generation of the raw greenhouse crop waste without pretreatment. The volume of methane produced from the raw greenhouse crop waste was 174 mLCH$_4$/gVS. The experimental outcomes for the biochemical methane potential (BMP) test from the pretreatment experiments are presented in Figure 2. After pretreatment, the highest BMP value was 370.9 mLCH$_4$/gVS, obtained at a reaction temperature of 50 °C, an H_2O_2 concentration of 2%; a reaction time of 15 h, and 5% initial solid content, while the the lowest BMP value (256.6 $mLCH_4/gVS$) was obtained from the greenhouse crop waste pretreated at a reaction temperature of 100 °C, an H_2O_2 concentration of 3%, a reaction time of 24 h, and 3% initial solid content. It can be concluded that the dependent variables of sCOD and sRedSugar, which had the maximum values under these conditions. behaved differently than the variable of BMP.

It should be remembered that an initial sCOD amount of 61 mg/gVS was also measured in the greenhouse crop waste. Based on the ratio of 0.35 $LCH_4/gCOD$, now universally recognized as the relationship between sludge COD utilized and methane generated [34], the utilization of the available sCOD would only correspond to 21 $mLCH_4/gVS$. The generation of the remaining 153 $mLCH_4/gVS$ has to be related to the hydrolysis of the particulate organics, requiring 0.437 g of particulate COD/gVS. This particulate COD consumption may be converted to 0.31 gVS/gVS, using the previously selected ratio of 1.4 gCOD/gVS. In short, biochemical reactions for raw greenhouse crop waste depleted all available sCOD, and broke down/hydrolyzed 31% of the existing volatile solids, converting them into methane.

The increase in magnitude of methane generation was obviously a direct observation of the effect of alkaline H_2O_2 treatment. The first important observation is the escalation in the volume of collected methane to a narrow bracket of 250–350 $mLCH_4/gVS$ as a result of alkaline H_2O_2 treatment. The second is the relatively lower methane volumes of around 250 $mLCH_4/gVS$ associated with the experimental runs conducted at 100 °C, despite much higher sCOD levels achieved in the same experiments.

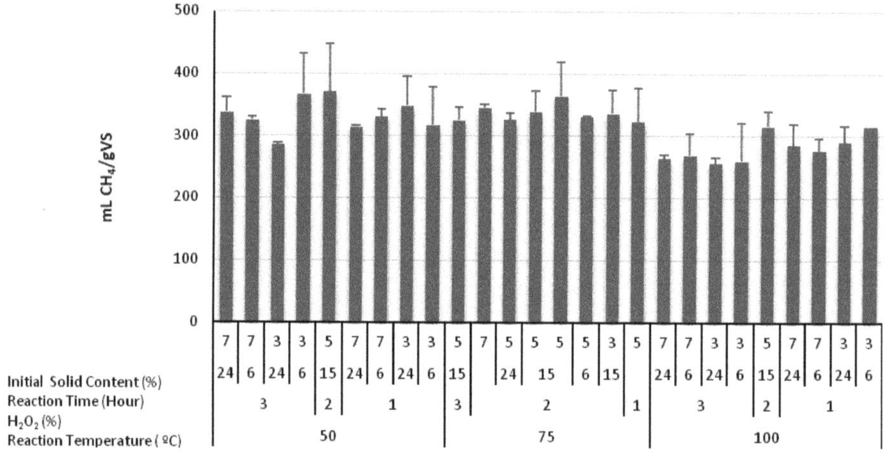

Figure 2. Methane generation due to the impact of alkaline H_2O_2 pretreatment.

This effect may be further evaluated in terms of (i) the increase in the sCOD levels, and (ii) changes in the levels of particulate organic matter hydrolysis for this purpose. The related evaluations are plotted in Figure 3a,b, which show both the relative contributions of sCOD and the particulate matter hydrolysis. From a different perspective, in the experimental conditions describing a reaction temperature of 50 °C, an H_2O_2 concentration of 3%, 7% initial solid content, and a reaction time of 24 h, only 78.8 $mLCH_4/gVS$ was related to the available sCOD, while 258.8 $mLCH_4/gVS$ was produced from the hydrolysis of 52.8% VS. Whereas at a reaction temperature of 100 °C, an H_2O_2 concentration of 3%, 3% initial solid content, and a reaction time of 24 h, the increased amount of sCOD produced 199.7 mL of the 256.6 $mLCH_4/gVS$ generated, while particulate organic matter hydrolysis remained limited to 11.6%. On this basis, the role of the particulate COD breakdown and hydrolysis seemed reversed at high temperatures. The limitation of methane generation under these conditions may be attributed to the formation of inhibitory by-products likely to be formed during H_2O_2 oxidation.

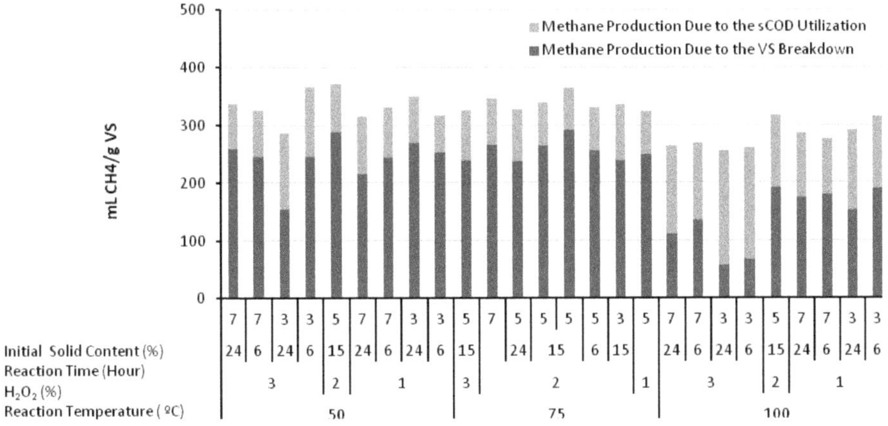

(**a**)

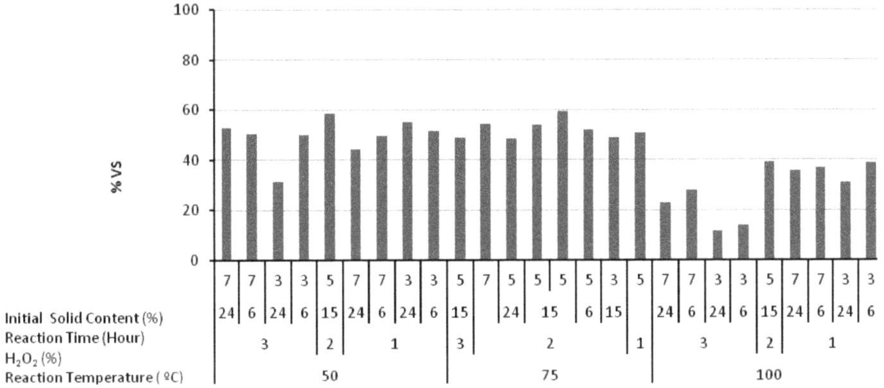

(**b**)

Figure 3. Methane production from sCOD and volatile solid (VS) destruction (**a**) and VS breakdown (**b**) due to the impact of alkaline H_2O_2 treatment.

The utilization rate of particulate organic matter under anaerobic conditions is an important parameter that reflects the biodegradability characteristics of the waste. The chemical structure of the greenhouse crop waste, dominated by lignocellulosic material, is too complex for biodegradation under natural conditions. In fact, the experiments indicated that only 31% of the waste could be utilized to generate methane without any pretreatment. Alkaline H_2O_2 treatment breaks down this complex chemical structure and hydrolyzes it into simple/soluble compounds, detectable by the increase in the magnitude of sCOD. This process significantly affects and increases the biodegradation of the waste. The destruction of the volatile solids takes place in two steps: (i) initial conversion into sCOD, and (ii) partial utilization of volatile solids under anaerobic conditions. For example, at a reaction temperature of 50 °C, an H_2O_2 concentration of 3%, 7% initial solid content, and a reaction time of 24 h, the incremental sCOD increase between the pretreated and raw samples (ΔsCOD) was 164.4 mgsCOD/gVS, corresponding to a VS hydrolysis (ΔVS) of 0.117 gVS/gVS. The generation of

338 mLCH_4/gVS additionally consumed 0.528 gVS/gVS, with an overall VS destruction calculated as 64.6%. Furthermore, at a reaction temperature of 100 °C, an H_2O_2 concentration of 3%, 3% initial solid content, and a reaction time of 24 h, ΔsCOD was measured as 509.8 mg/gVS, representing an initial VS hydrolysis of 0.364 gVS/gVS. An additional amount of volatile solids (ΔVS) of 0.116 gVS/gVS was also converted into methane, resulting in a lower VS destruction of 48%. These values should be compared with the 40–50% volatile matter utilization in the anaerobic digestion of sewage sludge [35]. The VS utilization profile achieved with alkaline H_2O_2 treatment is plotted in Figure 4a. The decrease in utilization rate at high sCOD levels also confirmed the presence and effect of inhibitory oxidation by-products. Furthermore, the experimental outcomes for the total lignin on an extractives free bases are presented in Figure 4b. As plotted in Figure 4b, the lowest total lignin on an extractives free bases was measured as 13.1% from the greenhouse crop waste pretreated at a reaction temperature of 100 °C, an H_2O_2 concentration of 3%, a reaction time of 24 h, and 7% initial solid content. It should be remembered that the second lowest BMP value of 264.2 mLCH_4/gVS was also observed under these conditions.

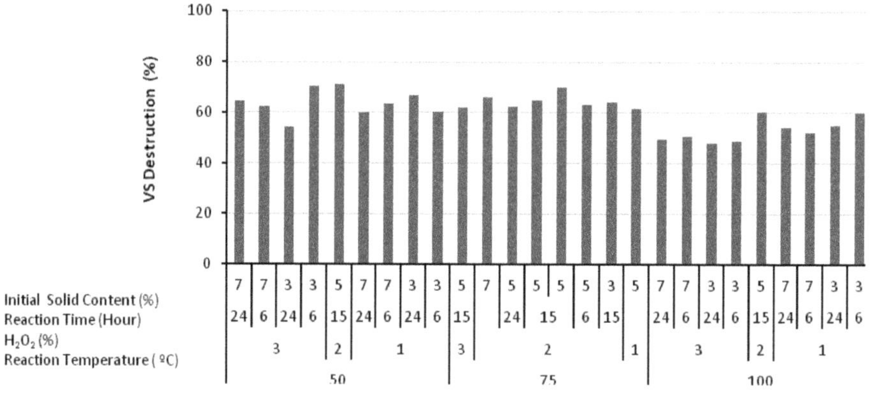

(**a**)

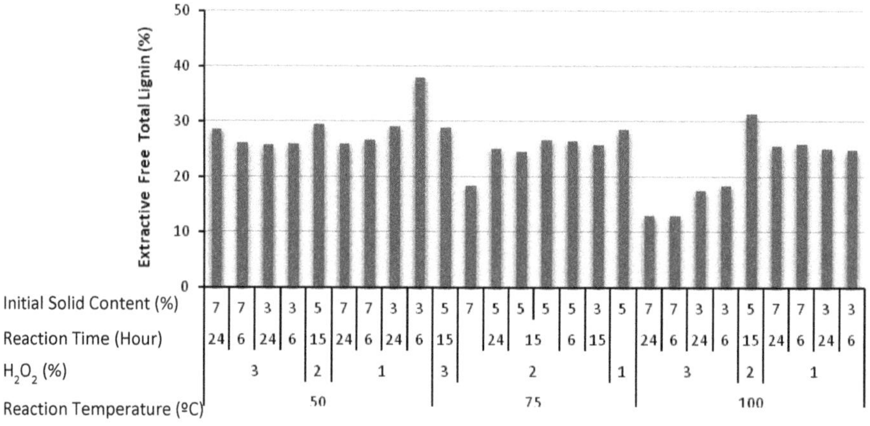

(**b**)

Figure 4. Destruction profile for volatile solids due to the impact of alkaline H_2O_2 treatment (**a**). Experimental outcomes for the total lignin on an extractives free bases (**b**).

2.3. Alkaline H_2O_2 Pretreatment Process Optimization

The accuracy of the models was explained by the determination coefficient (R^2) and coefficient of adjusted determination (Adj-R^2). The R^2 values were found to be 0.9682, 0.7740, 0.8376, and 0.5728 for the sCOD, sRedSugar, total lignin on an extractives free bases, and BMP, respectively, whereas the Adj-R^2 values were calculated as 0.9562, 0.6966, 0.7762, and 0.4112. The R^2 and Adj-R^2 values for the models of sCOD, sRedSugar, and total lignin on an extractives free bases in Table 2 indicated that acceptable fits were obtained between the response and the independent variables. However, only moderate R^2 and Adj-R^2 values were calculated for the BMP model. Quadratic regression models were strongly considerable, as it was apparent from Fisher's F-test with very low probability outcomes (p-value > F = 0.0001 for sCOD, sRedSugar, total lignin on an extractives free bases, and BMP).

Since the objective of alkaline H_2O_2 pretreatment was the enhancement of methane production with a reasonable process cost, process optimization of alkaline H_2O_2 pretreatment was executed based on minimizing the cost of the process (cost-driven approach) using the models developed for sCOD, sRedSugar, total lignin on an extractives free bases, and BMP. In the cost-driven optimization approach, the dependent variables of sCOD and total lignin on an extractives free bases were set in range, whereas sRedSugar (+) and BMP (+) were maximized. On the other hand, the independent variables of reaction temperature (+++++), reaction time (+++++), and H_2O_2 concentration (+++++) were minimized, while VS content (+++++) was maximized.

Optimal alkaline H_2O_2 pretreatment conditions were determined with the highest desirability of 0.917 at a reaction temperature of 50 °C, 7% initial solid content, an H_2O_2 concentration of 1%, and a reaction time of six h under these restraints. The optimal values for sCOD, sRedSugar, total lignin on an extractives free bases, and BMP were predicted to be 296.4 mgsCOD/gVS, 102.1 mg sRedSugar/gVS, 28.7%, and 318.6 mLCH_4/gVS, respectively, using the models. An alkaline H_2O_2 pretreatment experiment using a cost-driven approach conditions was performed for validation of the process optimization. The values of sCOD, sRedSugar, total lignin on an extractives free bases, and BMP were measured as 290.3 mgsCOD/gVS, 106.9 mg sRedSugar/gVS, 28.1%, and 309 mLCH_4/gVS, respectively, supporting the predictive power of the developed models. The BMP enhancement was calculated as 77.6% compared to the raw greenhouse crop waste under the conditions optimized for the process cost.

Table 2. ANOVA results for sCOD, sRedSugar, total lignin on an extractives free bases, and biochemical methane potential (BMP) models.

sCOD Model			
Quadratic model			
Prob > F	<0.0001 Significant	Adj-R^2	0.9562
R^2	0.9682	Pred-R^2	0.9338
Adeq Precision	35.6593	C.V%	8.85
sCOD = +1045.11218 − 24.76191 × Reaction temp. − 44.99164 × Solid content + 88.00049 × H_2O_2 concent. − 3.98184 × Reaction time − 0.64327 × Reaction temp. × Solid content + 1.48441 × Reaction temp. × H_2O_2 concent. + 0.022507 × Reaction temp. × Reaction time − 27.45672 × Solid content × H_2O_2 concent. − 0.000607639 × Solid content × Reaction time − 0.62899 × H_2O_2 concent. × Reaction time + 0.19592 × Reaction Temp.^{2} + 11.76235 × Solid content^2 + 6.39440 × H_2O_2 concent.^{2} + 0.21604 × Reaction time^2.			
sRedSugar Model			
Quadratic model			
Prob > F	<0.0001 Significant	Adj-R^2	0.6966
R^2	0.7740	Pred-R^2	0.5519
Adeq Precision	11.705	C.V%	41.85
sRedSugar = +844.41473 − 18.34946 × Reaction temp. + 16.89274 × Solid content − 136.48065 × H_2O_2 concent. − 13.05242 × Reaction time − 0.17577 × Reaction temp. × Solid content + 1.01831 × Reaction temp. × H_2O_2 concent. + 0.063115 × Reaction temp. × Reaction time − 2.93797 × Solid content × H_2O_2 concent. + 0.27415 × Solid content × Reaction time + 0.12308 × Reaction temp.^{2} − 1.32058 × Solid content^2 + 18.98017 × H_2O_2 concent.^{2} + 0,29090 × Reaction time^2			

Table 2. *Cont.*

Total Lignin on an Extractives Free Bases Model			
Quadratic model			
Prob > F	<0.0001 Significant	Adj-R^2	0.7762
R^2	0.8376	Pred-R^2	0.6727
Adeq Precision	14.903	C.V%	14.18
1/(Lignin) = +0.0736566 + 5.8380149 × 10^{-5} × Reaction temp. − 0.0284772 × Solid content − 7.8491088 × 10^{-3} × H_2O_2 concent. − 4.5014496 × 10^{-4} × Reaction time + 3.4923132 × 10^{-5} × Reaction temp. × Solid content + 2.3900179 × 10^{-4} × Reaction temp. × H_2O_2 concent. − 2.2097257 × 10^{-7} × Reaction temp. × Reaction time + 8.6433922 × 10^{-4} × Solid content × H_2O_2 concent. − 4.3403413 × 10^{-5} × Solid content × Reaction time − 3.95154005 × 10^{-5} × H_2O_2 concent. × Reaction time − 6.01390484 × 10^{-6} × Reaction temp.^{2} + 2.664218431 × 10^{-3} × Solid content^2 − 1.7142995 × 10^{-3} × H_2O_2 concent.^{2} + 2.74301920 × 10^{-5} × Reaction time^2			
BMP Model			
Quadratic model			
Prob > F	<0.0001 Significant	Adj-R^2	0.4112
R^2	0.5728	Pred-R^2	0.1190
Adeq Precision	7.23	C.V%	10.35
1/(BMP) = +4.20476 × 10^{-3} − 1.31145 × 10^{-5} × Reaction temp. − 4.36888 × 10^{-5} × Solid content − 9.28724 × 10^{-4} × H_2O_2 concent. − 4.17111 × 10^{-5} × Reaction time + 4.08924 × 10^{-7} × Reaction temp. × Solid content + 3.96470× 10^{-6} × Reaction temp. × H_2O_2 concent. − 8.50445 × 10^{-8} × Reaction temp. × Reaction time − 3.64937 × 10^{-5} × Solid content × H_2O_2 concent. − 2.27112 × 10^{-6} × Solid content × Reaction time + 4.32086 × 10^{-6} × H_2O_2 concent. × Reaction time + 1.02675 × 10^{-7} × Reaction temp.^{2} + 1.19229 × 10^{-5} × Solid content^2 + 2.09008 × 10^{-4} × H_2O_2 concent.^{2} + 1.83088 × 10^{-6} × Reaction time^2			
sCOD Model			
Quadratic model			
Prob > F	<0.0001 Significant	Adj-R^2	0.9562
R^2	0.9682	Pred-R^2	0.9338
Adeq Precision	35.6593	C.V%	8.85
sCOD = +1045.11218 − 24.76191 × Reaction temp. − 44.99164 × Solid content + 88.00049 × H_2O_2 concent. − 3.98184 × Reaction time − 0.64327 × Reaction temp. × Solid content + 1.48441 × Reaction temp. × H_2O_2 concent. + 0.022507 × Reaction temp. × Reaction time − 27.45672 × Solid content × H_2O_2 concent. − 0.000607639 × Solid content × Reaction time − 0.62899 × H_2O_2 concent. × Reaction time + 0.19592 × Reaction Temp.^{2} + 11.76235 × Solid content^2 + 6.39440 × H_2O_2 concent.^{2} + 0.21604 × Reaction time^2.			
sRedSugar Model			
Quadratic model			
Prob > F	<0.0001 Significant	Adj-R^2	0.6966
R^2	0.7740	Pred-R^2	0.5519
Adeq Precision	11.705	C.V%	41.85
sRedSugar = +844.41473 − 18.34946 × Reaction temp. + 16.89274 × Solid content − 136.48065 × H_2O_2 concent. − 13.05242 × Reaction time − 0.17577 × Reaction temp. × Solid content + 1.01831 × Reaction temp. × H_2O_2 concent. + 0.063115 × Reaction temp. × Reaction time − 2.93797 × Solid content × H_2O_2 concent. + 0.27415 × Solid content × Reaction time + 0.12308 × Reaction temp.^{2} − 1.32058 × Solid content^2 + 18.98017 × H_2O_2 concent.^{2} + 0.29090 × Reaction time^2			
Total Lignin on an Extractives Free Bases Model			
Quadratic model			
Prob > F	<0.0001 Significant	Adj-R^2	0.7762
R^2	0.8376	Pred-R^2	0.6727
Adeq Precision	14.903	C.V%	14.18
1/(Lignin) = +0.0736566 + 5.8380149 × 10^{-4} × Reaction temp. − 0.0284772 × Solid content − 7.8491088 × 10^{-3} × H_2O_2 concent. − 4.5014496 × 10^{-4} × Reaction time + 3.4923132 × 10^{-5} × Reaction temp. × Solid content + 2.3900179 × 10^{-4} × Reaction temp. × H_2O_2 concent. − 2.2097257 × 10^{-7} × Reaction temp. × Reaction time + 8.6433922 × 10^{-4} × Solid content × H_2O_2 concent. − 4.3403413 × 10^{-5} × Solid content × Reaction time − 3.95154005 × 10^{-5} × H_2O_2 concent. × Reaction time − 6.01390484 × 10^{-6} × Reaction temp.^{2} + 2.664218431 × 10^{-3} × Solid content^2 − 1.7142995 × 10^{-3} × H_2O_2 concent.^{2} + 2.74301920 × 10^{-5} × Reaction time^2			
BMP Model			
Quadratic model			
Prob > F	<0.0001 Significant	Adj-R^2	0.4112
R^2	0.5728	Pred-R^2	0.1190
Adeq Precision	7.23	C.V%	10.35

Table 2. *Cont.*

$1/(\mathrm{BMP}) = +4.20476 \times 10^{-3} - 1.31145 \times 10^{-5} \times$ Reaction temp. $- 4.36888 \times 10^{-5} \times$ Solid content $- 9.28724 \times 10^{-4} \times H_2O_2$ concent. $- 4.17111 \times 10^{-5} \times$ Reaction time $+ 4.08924 \times 10^{-7} \times$ Reaction temp. $\times$ Solid content $+ 3.96470 \times 10^{-6} \times$ Reaction temp. $\times H_2O_2$ concent. $- 8.50445 \times 10^{-8} \times$ Reaction temp. $\times$ Reaction time $- 3.64937 \times 10^{-5} \times$ Solid content $\times H_2O_2$ concent. $- 2.27112 \times 10^{-6} \times$ Solid content $\times$ Reaction time $+ 4.32086 \times 10^{-6} \times H_2O_2$ concent. $\times$ Reaction time $+ 1.02675 \times 10^{-7} \times$ Reaction temp.$^2 + 1.19229 \times 10^{-5} \times$ Solid content$^2 + 2.09008 \times 10^{-4} \times H_2O_2$ concent.$^2 + 1.83088 \times 10^{-6} \times$ Reaction time^2

Three-dimensional (3D) graphs were employed to emphasize the impacts of independent variables under optimal conditions. The effects of independent variables on BMP are demonstrated in Figure 5a–f. In Figure 5a, BMP decreased due to increasing H_2O_2 concentration at a reaction temperature of 100 °C, whereas BMP increased due to decreasing reaction temperature (from 100 °C to 50 °C) within the range of 1–3% H_2O_2 concentration. A maximum predicted BMP enhancement of 106.9% compared to the raw greenhouse crop waste was observed at a reaction temperature of 68 °C and an H_2O_2 concentration of 2%. In Figure 5b, c, BMP decreased when the reaction temperature was increased to 100 °C at a reaction time of 24 h and 7% initial solid content. When the reaction time was maintained at 24 h, a decrease in BMP was observed when the temperature was increased to 100 °C. Similarly, when the initial solid content was kept constant at 7%, the decrease in BMP was temperature has a negative impact on BMP. Furthermore, as seen in Figure 5d–f, BMP was not affected by the interactive effects of H_2O_2 concentration with initial solid content, reaction time with initial solid content, and reaction time with H_2O_2 concentration. A maximum BMP was obtained at 4–6% initial solid content, H_2O_2 concentrations of 1.5–2.5%, and reaction times of 10–18 h.

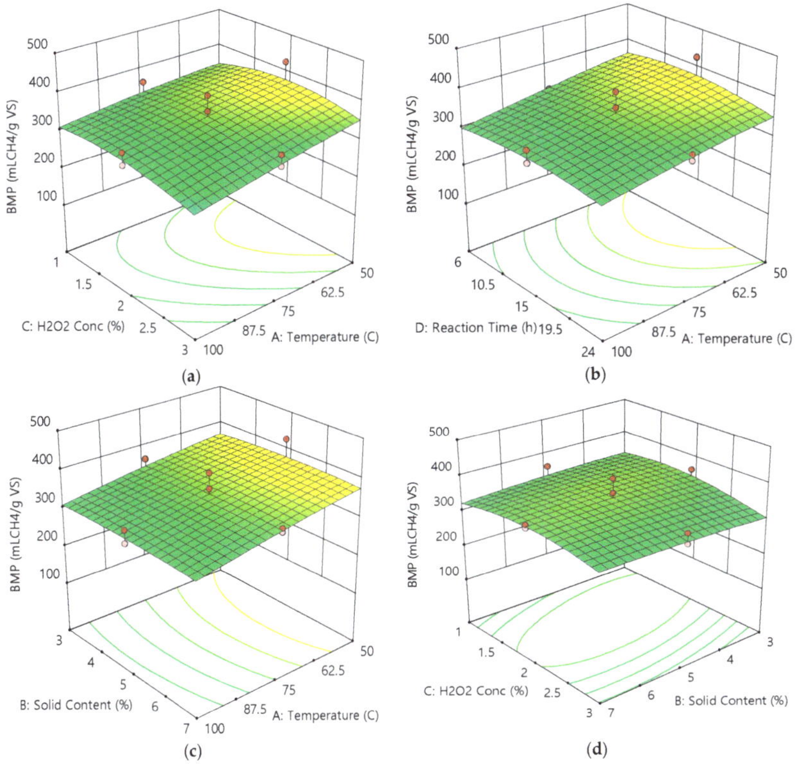

Figure 5. *Cont.*

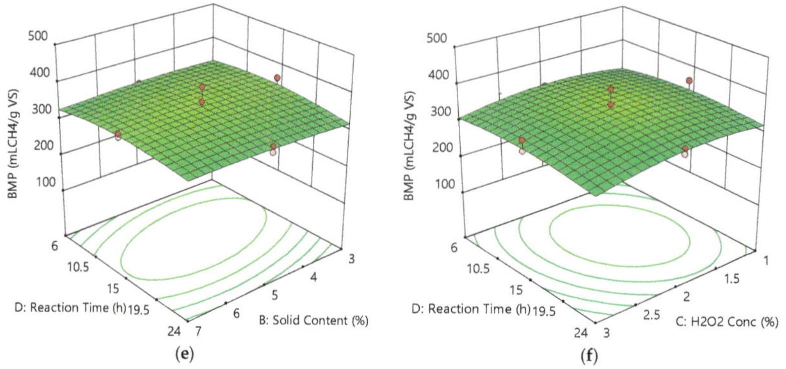

Figure 5. Effects of independent variables on biochemical methane potential (BMP). (**a**) H_2O_2 concentration and temperature; (**b**) reaction time and temperature; (**c**) solid content and temperature; (**d**) H_2O_2 concentration and solid content; (**e**) reaction time and solid content; (**f**) reaction time and H_2O_2 concentration.

2.4. Chemical Structure and Morphological Changes of Biomass

The FTIR spectra and SEM images of greenhouse crop waste pretreated with alkaline H_2O_2 under different conditions (50 °C, 5% VS, 15 h, 2% H_2O_2 for maximum CH_4 production; 100 °C, 3% VS, 24 h, 3% H_2O_2 for maximum sCOD and sRedSugar production, along with minimum CH_4 production; and 50 °C, 7% VS, 6 h, 1% H_2O_2 for cost optimization) compared to those of the raw greenhouse crop waste are presented in Table 3 and Figure 6.

As seen in Figure 6, the spectral profiles and relative intensities of the bands belonging to the raw greenhouse crop waste and that pretreated with alkaline H_2O_2 were found to be very similar under conditions of 50 °C, 7% VS, 6 h, and 1%H_2O_2 for cost optimization. On the other hand, the spectral profiles were different from the raw greenhouse crop waste for that pretreated with alkaline H_2O_2 under conditions of 50 °C, 5% VS, 15 h, and 2% H_2O_2 for maximum CH_4 production, and that pretreated with alkaline H_2O_2 under conditions of 100 °C, 3% VS, 24 h, and 3% H_2O_2 for maximum sCOD and sRedSugar production, along with minimum CH_4 production. New peaks were observed after alkaline H_2O_2 pretreatment, indicating that the chemical composition of greenhouse crop waste changed. In particular, the prominent absorbances at 895–900, 1050, 1270, 1430–1460, 1510–1600, 2920–2925, 3420, and 3446 cm^{-1} in the spectra were relatively different from the spectrum of raw greenhouse crop waste. As clearly seen in Table 3, the lignin-related absorbance values observed at 1270, 1430–1460, and 1510–1600 cm^{-1} revealed that the alkaline H_2O_2 pretreatment was effective on lignin disintegration. Sun et al. [19] also stated that the delignification of agricultural crop stalks could occur during the alkaline H_2O_2 pretreatment process, while the macromolecular structure of cellulose did not show any noticeable change. Results from this study confirm the findings of Sun et al. [19].

As seen in Figure 6, the raw greenhouse crop waste exhibited a smooth, non-porous, compact, and rigid surface structure. There was no separation of fibers, or ruptures and scars. On the other hand, the pretreated greenhouse crop waste demonstrated a rough and porous structure. In particular, the fibrils of greenhouse crop waste pretreated with alkaline H_2O_2 under conditions of 100 °C, 3% VS, 24 h, and 3% H_2O_2 were completely deformed, and their structural integrity was disrupted. The SEM examination revealed that the morphological changes, along with the tissue damage, resulted from the alkaline H_2O_2 pretreatment. Similar to our findings, Rezende et al. [36] also stated that alkaline and NaCl pretreatment dissolved the inter-fibrillar or bulk lignin, while disrupting the initial fiber structure, leading to the disaggregation of micro-fibrils from their neighboring fibers.

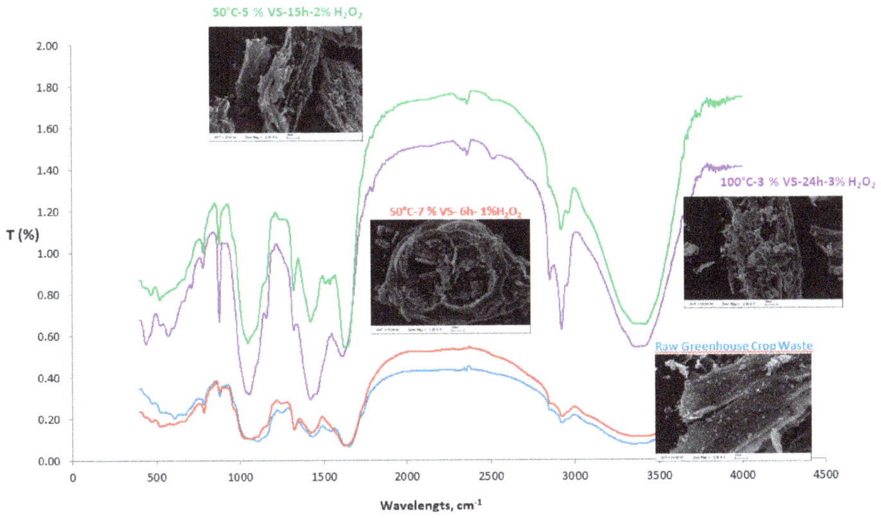

Figure 6. Fourier-transform infrared (FTIR) spectra and SEM images of raw and pretreated greenhouse crop waste.

Table 3. Comparison of Fourier-transform infrared (FTIR) spectra of waste pretreated with alkaline H_2O_2 under different conditions with with that of raw greenhouse crop waste.

Wavelength (cm^{-1})	Region	50 °C, 5% VS, 15 h, 2% H_2O_2	50 °C, 7% VS, 6 h, 1% H_2O_2	100 °C, 3% VS, 24 h, 3% H_2O_2
895–900	Characteristic absorption peak of cellulose associated with the ß-glycosidic bond [19,20]	+++++	+	++++
1050	C–O stretch of the C–O–C in cellulose, hemicellulose, and lignin [19,21]	+++++	+	++
1270	C–O stretch in the guaiacyl aromatic ring associated with lignin [19,37]	+++	++	+++++
1430–1460	Aromatic skeletal vibration combined with C–H in plane deformation associated with lignin [37,38]	++++	+++	+++++
1510–1600	Aromatic skeletal vibration of lignin constituting conjugated C=C, aryl-substituted C=C, and alkenyl C=C stretch [37–39]	+++++	+	+++
2920–2925	C–H vibration of CH_2 and CH_3 groups [19,37]	+++	++	+++++
3420	Inter- and intramolecular hydrogen bonding [19]	++++	+++	+++++
3446	O–H stretch vibration in cellulose [37]	+++	+	++++

+++++ to +: Max to Min.

3. Materials and Methods

3.1. Experimental Rationale

Antalya is the largest area for greenhouse cultivation in Turkey, providing tomato, pepper, cucumber, eggplant, and zucchini. Greenhouse crop waste, consisting of roots, stalks, leaves, and fruits from cultivation, is generated in the region, creating environmental problems. The greenhouse crop waste was supplied by the growers, and fresh waste was sliced into approximately 1cm pieces, and was stored in sealed plastic bags at −20 °C until used for composition analyses, alkaline H_2O_2 pretreatment experiments, and methane generation potential tests.

The first phase of the experiments involved the characterization of the greenhouse crop waste in terms of the parameters that would be used as major indicators for the extent of energy recovery achieved by means of alkaline H_2O_2 treatment.

Analyses of the total solids (TS) and volatile solids (VS) were performed based on standard methods 2540C [40]. Analyses of the total chemical oxygen demand (COD) were done according to standard methods 5220B [40]. The Kjeldahl nitrogen was determined using a Kjeldahl nitrogen analyzer (Büchi Digest Automat K-438, Büchi Auto Kjeldahl Unit K-370 and Radiometer TitraLab 840, Büchi, Flawil, Switzerland). The contents of lignin, cellulose, hemicellulose, and soluble matter were determined according to the Van Soest procedure [41] using a Gerhard FBS6 (Gerhard, Königswinter, Germany). Analyses of the total free lignin of extractives (acid-insoluble and acid-soluble) were performed according to the "Determination of Structural Carbohydrates and and Lignin in Biomass, NREL/TP-510-42618" [42]. The protein concentration was determined using the Lowry method [43]. The extractive matter and lipid contents of samples were determined using Soxhlet extraction [44]. The soluble chemical oxygen demand (sCOD) was determined using a Hach-Lange DR5000 spectrophotometer (Hach Lange GmbH, Duesseldorf, Germany) and a Lange LT200 (Grasscht, Germany) with COD kits. The concentrations of soluble reducing sugar (sRedSugar) were determined via the Dinitrosalicylic acid (DNS) method [45]. The elemental composition of the greenhouse crop waste was identified using a CHNS elemental analyzer (LECO, CHNS-932, St. Joseph, MI, USA). All composition analyses were executed in triplicate, and the quotable outcomes are demonstrated as means.

3.2. *Alkaline H_2O_2 Pretreatment Experiments*

The greenhouse crop waste was pretreated in a Parr reactor (Parr Instrument Company) with a 200 mL working volume. The independent variables with a potential impact on alkaline H_2O_2 pretreatment were selected as reaction temperature (50–100 °C), H_2O_2 concentration (1–3%), reaction time (6–24 h), and initial solid content of greenhouse crop waste (3–7%). The pretreatment experiments were done in duplicate under each condition. The calculated amount of fresh greenhouse crop waste and H_2O_2 solution (w/w) was loaded into the pretreatment reactor, and initial pH values were set to 11.5 using 6M NaOH solution, with the reactors heated to the appropriate reaction temperature. When the predetermined temperature was attained, the experiment time was started. After reaching the determined reaction time, the reactor was put into ice and a water bath to cool down and stop the reaction. The pretreatment process was evaluated according to sCOD, sRedSugar, total free lignin of extractives, and BMP as objective functions related to pretreatment yield. The samples were centrifuged at 15,000 rpm for 10 min for the sCOD and sRedSugar analyses. The amount of sCOD was determined using a Hach-Lange DR5000 spectrophotometer and a Lange LT200 (Grasshut, Germany) with COD kits. The sRedSugar concentrations were determined via the DNS method [45]. Analyses of the total free lignin of extractives (acid-insoluble and acid-soluble) were performed according to the "Determination of Structural Carbohydrates and and Lignin in Biomass, NREL/TP-510-42618" [42] using the solid phase of the pretreated samples. The remaining pretreated samples containing solid and liquid fractions were stored at −20 °C for the subsequent methane generation potential experiment.

3.3. *Methane Generation Potential Experiment*

The efficiency of alkaline H_2O_2 pretreatment was determined using a biochemical methane potential test (BMP) based on methane production. The samples, including macro and micro nutrients, were incubated in a closed glass reactor with a specific quantity of seed sludge (inoculum). Mesophilic conditions (35 °C) were preferred for the BMP tests. The BMP protocol according to Carrère et al. and Us & Perendeci [5,46] was implemented. For the BMP tests, 500 mL glass reactors with a working volume of 400 mL were filled with sample, seed sludge, nutrients, and a tampon solution. All BMP reactors were loaded with seed sludge from the anaerobic reactor of an Antalya city wastewater treatment plant. Fifty-six glass reactors were used in the study, and two of them were fed with only seed sludge and nutrients to specify the methane potential of seed sludge on its own. The 52 glass reactors were used with different pretreated samples, and two reactors containing raw greenhouse crop waste were used as controls. After the optimization of conditions for alkaline H_2O_2 pretreatment, the BMP test was also conducted under optimal conditions with two duplicates for validation of the model. The food-to-microorganism ratio (F/M) was fixed at 0.5 (gVS waste/gVS inoculum) for the glass reactors. The initial pH was set to neutral for all reactors. To keep anaerobic conditions in the reactors, a gas mixture of N_2/CO_2 (70/30%) was flushed. The BMP test lasted for 62 days. The produced biogas was measured based on a gas-water displacement method. The biogas composition was ascertained using gas chromatography (GC; Varian 4900). A standard gas consisting of 60% (*v*/*v*) CH_4 and 40% CO_2 was used for the calibration of gas chromatography. The gas production of seed sludge was counted in the computation of biogas production of the samples. The methane production was estimated as mL of methane per g of VS ($mLCH_4/gVS$) added to the reactor.

3.4. Optimization of the Alkaline H_2O_2 Pretreatment Process

The pretreatment process was optimized using a CCD of RSM. Three levels of four independent variables were applied for the CCD, using the Design-Expert® software (Minneapolis, MN, USA). The ranges of each independent variable were established based on information in the literature and on our previous experimental experience. The levels of the independent variables were coded as −1 and +1. The four independent variables were changed within the following ranges: 50–100 °C (reaction temperature), 6–24 h (reaction time), 1–3% (H_2O_2 concentration), and 3–7% (initial solid content). A total of 52 runs, including four runs at the design center and duplicates of each run, were determined using a CCD.

The performance of the alkaline H_2O_2 pretreatment process was evaluated based on sCOD, sRedSugar, total free lignin of extractives, and the BMP test as dependent variables. The outcomes from the pretreatment experiments were modeled using the Design-Expert® software (Minneapolis, MN, USA). Analyses of the regression coefficients, variance (ANOVA), and the *p*- and *F*-values were preferred for the model assessment. The adequacy of the model fit was presented by the coefficient of determination (R^2) and the adjusted determination coefficient (Adj-R^2).

The alkaline H_2O_2 pretreatment process was also optimized using the optimization module of the Design-Expert® software (Minneapolis, MN, USA). The optimization of the alkaline H_2O_2 pretreatment process was executed using the models developed for sCOD, sRedSugar, total free lignin of extractives, and BMP. The goal settings were carried out using the plus (+) symbols in the Design-Expert® program (Minneapolis, MN, USA).

3.5. Fourier-Transform Infrared (FTIR) Spectroscopy and Scanning Electron Microscopy (SEM)

Changes in the molecular-bond characterization of greenhouse crop waste were evaluated using an ATR-FTIR-Varian 1000 model FTIR spectrometer. The measurements were analyzed by averaging the signal of 16 scans across the range of 500 cm^{-1} to 4000 cm^{-1} with a spectral resolution of 4 cm^{-1}. The evaluation of deformations on the surface of the greenhouse crop waste was also investigated, using a Zeiss Leo 1430 scanning electron microscope at a voltage of 15 kV.

4. Conclusions

In the light of the experimental results and evaluations reported in the preceding sections, a number of concluding remarks could be drawn for this study.

The alkaline H_2O_2 pretreatment partially destroyed the complex lignocellulosic structure of the greenhouse crop waste. The organic matter was initially broken down and then hydrolyzed into simple, soluble compounds. On this basis, the alkaline H_2O_2 pretreatment induced a significant increase in the range of 200–800% in COD leakage into the soluble phase, and boosted the methane generation potential from 174 $mLCH_4/gVS$ to a much higher bracket of 250–350 $mLCH_4/gVS$. Similarly, the volatile matter utilization increased from 31% in the waste material before treatment to 50–70% after treatment, depending on the selected experimental conditions.

The alkaline H_2O_2 pretreatment was optimized to determine the optimal conditions for the enhancement of methane generation assuming a cost-driven approach. The optimal alkaline H_2O_2 pretreatment conditions were found to be a reaction temperature of 50 °C, 7% initial solid content, an H_2O_2 concentration of 1%, and a reaction time of six h. Under these conditions, the BMP test yielded a production of 309 $mLCH_4/gVS$. The enhancement of methane production was calculated as 77.6% compared to raw greenhouse crop waste.

The results obtained provide an optimistic perspective for the possibility of energy recovery from complex waste such as greenhouse crop waste. It is recommended that future studies be directed toward testing new pretreatment processes, as well as toward novel energy recovery technologies such as pyrolysis, instead of traditional anaerobic digestion.

Author Contributions: Conceptualization, N.A.P.; Methodology, S.G. and N.A.P.; Software, N.A.P. and S.G.; Formal Analysis, S.G. and N.A.P.; Investigation, N.A.P., S.G. and D.O. Writing-Original Draft Preparation, N.A.P. and D.O.; Writing-Review & Editing, N.A.P. and D.O.

Funding: This research was funded by Scientific Research Projects Unit of Akdeniz University (Grant Number FLY 2015-623).

Conflicts of Interest: The authors declare no conflict of interest.

References

1. Herzog, A.V.; Lipman, T.E.; Kammen, D.M. Renewable Energy Sources. Encyclopedia of Life Support Systems (EOLSS) Forerunner: Paris, France. Available online: https://pdfs.semanticscholar.org/57ad/8c86294ae7af352177d11c28e6dad12db7de.pdf (accessed on 24 May 2018).
2. Tüzel, Y.; Öztekin, G.B. Protected cultivation in Turkey. *Chron. Horticult.* **2015**, *55*, 21–26.
3. Kacıra, M.; Sase, S.; Kacıra, O.; Okushima, L.; Ishii, M.; Kowata, H.; Moriyama, H. Status of greeenhouse production in Turkey: Focusing on vegetable and floriculture production. *J. Agric. Meteorol.* **2004**, *60*, 115–122. [CrossRef]
4. Perendeci, N.A. *Valorization of Agricultural Residues from Greenhouses and Sewage Sludge into Biogas by Anaerobic Digestion Combined with Pre-Treatments*; TÜBİTAK Project Report; TÜBİTAK: Ankara, Turkey, 2010.
5. Us, E.; Perendeci, A. Improvement of methane production from greenhouse residues: Optimization of thermal and H_2SO_4 pretreatment process by experimental design. *Chem. Eng. J.* **2012**, *181*, 120–131. [CrossRef]
6. Mohareb, E.A.; MacLean, H.L.; Kennedy, C.A. Greenhouse gas emissions from waste management-Assesment of quantification methods. *J. Air Waste Manag. Asooc.* **2011**, *61*, 480–493. [CrossRef]
7. Conde-Mejiaa, C.; Jiménez-Gutiérreza, A.; Halwagi, M.E. A comparison of pretreatment methods for bioethanol production from lignocellulosic materials. *Process Saf. Environ. Prot.* **2012**, *90*, 189–202. [CrossRef]
8. Limayem, A.; Ricke, S.C. Lignocellulosic biomass for bioethanol production: Current perspectives, potential issues and future prospects. *Prog. Energy Combust. Sci.* **2012**, *38*, 449–467. [CrossRef]
9. Lee, H.V.; Hamid, S.B.A.; Zain, S.K. Conversion of lignocellulosic biomass to nanocellulose: Structure and chemical process. *Sci. World J.* **2014**. [CrossRef] [PubMed]
10. Galbe, M.; Zacchi, G. Pretreatment of lignocellulosic materials for efficient bioethanol production. *Biofuels* **2007**, *108*, 41–65.
11. Taherzadeh, M.J.; Karimi, K. Pretreatment of lignocellulosic wastes to improve ethanol and biogas production: A review. *Int. J. Mol. Sci.* **2008**, *9*, 1621–1651. [CrossRef] [PubMed]
12. Chandra, R.; Takeuchi, H.; Hasegawa, T. Methane production from lignocellulosic agricultural crop wastes: A review in context to second generation of biofuel production. *Renew. Sustain. Eng. Rev.* **2012**, *16*, 1462–1476. [CrossRef]

13. Sambusiti, C. Physical, Chemical and Biological Pretreatments to Enhance Biogas Production from Lignocellulosic Substrates. Ph.D. Thesis, Politecnico di Milano, Milano, Italy, 2013. Available online: https://www.politesi.polimi.it/bitstream/10589/74843/3/2013_03_PhD_Sambusiti.pdf (accessed on 24 May 2018).
14. Alvira, P.; Tomás-Pejó, E.; Ballesteros, M.; Negro, M.J. Pretreatment technologies for an efficient bioethanol production process based on enzymatic hydrolysis: A review. *Bioresour. Technol.* **2010**, *101*, 4851–4861. [CrossRef] [PubMed]
15. Mosier, N.; Wyman, C.; Dale, B.; Elander, R.; Lee, Y.Y.; Holtzapple, M.; Ladisch, M. Features of promising technologies for pretreatment of lignocellulosic biomass. *Bioresour. Technol.* **2005**, *96*, 673–686. [CrossRef] [PubMed]
16. Kurabi, A.; Berlin, A.; Gilkes, N.; Kilburn, D.; Bura, R.; Robinson, J.; Markov, A.; Skomarovsky, A.; Gusakov, A.; Okunev, O.; et al. Enzymatic hydrolysis of steam-exploded and ethanol organasolv pretreated Douglas-Firby novel and commercial fungal cellulases. *Appl. Biochem. Biotechnol.* **2005**, *121*, 219–230. [CrossRef]
17. Sun, Y.; Cheng, J.J. Dilute acid pretreatment of rye straw and Bermuda grass for ethanol production. *Bioresour. Technol.* **2005**, *96*, 1599–1606. [CrossRef] [PubMed]
18. Silverstein, R.A.; Chen, Y.; Sharma-Shivappa, R.R.; Boyette, M.D.; Osborne, J. A comparison of chemical pretreatment methods for improving saccharification of cotton stalks. *Bioresour. Technol.* **2007**, *98*, 3000–3011. [CrossRef] [PubMed]
19. Sun, C.; Liu, R.; Cao, W.; Yin, R.; Mei, Y.; Zhang, L. Impacts of alkaline hydrogen peroxide pretreatment on chemical composition and biochemical methane potential of agricultural crop stalks. *Energy Fuels* **2015**, *29*, 4966–4975. [CrossRef]
20. Qing, Q.; Zhou, L.; Huang, M.; Guo, Q.; He, Y.; Wang, L.; Zhang, Y. Improving enzymatic saccharification of bamboo shoot shell by alkaline salt pretreatment with H_2O_2. *Bioresour. Technol.* **2016**, *201*, 230–236. [CrossRef] [PubMed]
21. Fang, J.M.; Sun, R.C.; Salisbury, D.; Fowler, P.; Tomkinson, J. Comparative study of hemicelluloses from wheat straw by alkali and hydrogen peroxide extractions. *Polym. Degrad. Stab.* **1999**, *66*, 423–432. [CrossRef]
22. Michalska, K.; Ledakowicz, S. Alkaline hydrogen peroxide pretreatment of energy crops for biogas production. *Chem. Pap.* **2013**, *68*, 913–922. [CrossRef]
23. Banerjee, G.; Car, J.; Scott-Craig, J.S.; Hodge, D.B.; Walton, J.B. Alkaline peroxide pretreatment of corn stover: Effects of biomass, peroxide, and enzyme loading and composition on yields of glucose and xylose. *Biotechnol. Biofuels* **2011**, *4*, 16. [CrossRef] [PubMed]
24. Ayeni, A.O.; Hymore, F.K.; Mudliar, S.N.; Deshmukh, S.C.; Satpute, D.B.; Omoleye, J.A.; Pandey, R.A. Hydrogen peroxide and lime based oxidative pretreatment of wood waste to enhance enzymatic hydrolysis for a biorefinery: Process parameters optimization using response surface methodology. *Fuel* **2011**, *106*, 187–194. [CrossRef]
25. Vasco, C.A.; Zhang, X. Alkaline hydrogen peroxide pretreatment of softwood: Hemicellulose degradation pathways. *Bioresour. Technol.* **2013**, *150*, 321–327. [CrossRef] [PubMed]
26. Correia, J.A.C.; Junior, J.E.M.; Gonçalves, L.R.B.; Rocha, M.V.P. Alkaline hydrogen peroxide pretreatment of cashew apple bagasse for ethanol production: Study of parameters. *Bioresour. Technol.* **2013**, *139*, 249–256. [CrossRef] [PubMed]
27. Rabelo, S.C.; Andrade, R.R.; Filho, R.M.; Costa, A.C. Alkaline hydrogen peroxide pretreatment, enzymatic hydrolysis and fermentation of sugarcane bagasse to ethanol. *Fuel* **2014**, *136*, 349–357. [CrossRef]
28. Su, Y.; Du, R.; Gu, H.; Ca, M.; Wu, Q.; Su, R.; Qi, W.; He, Z. Fractional pretreatment of lignocellulose by alkaline hydrogen peroxide: Characterization of its major components. *Food Bioprod. Process* **2015**, *94*, 322–330. [CrossRef]
29. Zhang, H.; Ning, Z.; Khalid, H.; Zhang, R.; Liu, G.; Chen, C. Enhancement of methane production from cotton stalk using different pretreatment techniques. *Sci. Rep.* **2018**, *8*, 3463. [CrossRef] [PubMed]
30. Orhon, D. Evolution of the activated sludge process: The first fifty years. *J. Chem. Technol. Biotecnol.* **2015**, *90*, 608–640. [CrossRef]
31. Henze, M. Characterization of wastewater for modelling of activated sludge process. *Water Sci. Technol.* **1992**, *25*, 1–15. [CrossRef]

32. Başaran, S.T.; Aysel, M.; Kurt, H.; Ergal, I.; Kumru, M.; Akarsubaşı, A.; Sözen, S.; Orhon, D. Removal of readily biodegradable substrate in super gas membrane reactor. *J. Membr. Sci.* **2012**, *423*, 477–486. [CrossRef]
33. Sözen, S.; Çokgör, E.U.; Insel, G.; Taş, D.O.; Dülkadiroğlu, H.; Karaca, C.; Filibeli, A.; Meriç, S.; Orhon, D. Scientific basis of dissolved organic carbon limitation for landfilling of municipal treatment sludge—Is it attainable and justifiable? *Waste Manag.* **2014**, *34*, 1657–1666. [CrossRef] [PubMed]
34. Karaca, C.; Sözen, S.; Orhon, D.; Okutan, H. High temperature pyrolysis of sewage sludge as a sustainable process for energy recovery. *Waste Manag.* **2018**, *78*, 217–228. [CrossRef]
35. Svardal, K.; Kroiss, H. Energy requirements for waste water treatment. *Water Sci Technol.* **2011**, *64*, 1355–1361. [CrossRef] [PubMed]
36. Rezende, C.A.; de Lima, M.A.; Maziero, P.; deAzevedo, E.R.; Garcia, W.; Polikarpov, I. Chemical and morphological characterization of sugarcane bagasse submitted to a delignification process for enhanced enzymatic digestibility. *Biotechnol. Biofuels* **2011**, *4*, 1–18. [CrossRef] [PubMed]
37. Koutsianitis, D.; Mitani, C.; Giagli, K.; Tsalagkas, D.; Halász, K.; Kolonics, O.; Gallis, C.; Csóka, L. Properties of ultrasound extracted bicomponent lignocellulose thin films. *Ultrason. Sonochem.* **2015**, *23*, 148–155. [CrossRef] [PubMed]
38. Gabhane, J.; William, S.P.M.P.; Vaidya, A.N.; Das, S.; Wate, S.R. Solar assisted alkali pretreatment of garden biomass: Effects on lignocellulose degradation, enzymatic hydrolysis, crystallinity and ultra-structural changes in lignocellulose. *Waste Manag.* **2015**, *40*, 92–99. [CrossRef] [PubMed]
39. Lin, R.; Cheng, J.; Song, W.; Ding, L.; Xie, B.; Zhou, J.; Cen, K. Characterization of water hyacinth with microwave-heated alkali pretreatment for enhanced enzymatic digestibility and hydrogen/methane fermentation. *Bioresour. Technol.* **2015**, *182*, 1–7. [CrossRef] [PubMed]
40. Eaton, A.D.; Clesceri, L.S.; Rice, E.W.; Greenberg, A.E. *Standard Methods for the Examination of Water and Wastewater*; American Public Health Association; American Water Works Association; Water Environment Federation: Washington, DC, USA, 2005.
41. Van Soest, P.J. Use of detergent in the analysis of fibrous feeds. A rapid method for the determination of fibre and lignin. *J. Assoc. Off. Anal. Chem.* **1963**, *46*, 829–835.
42. Sluiter, A.; Hames, B.; Ruiz, R.; Scarlata, C.; Sluiter, J.; Templeton, D.; Crocker, D. Determination of Structural Carbohydrates and and Lignin in Biomass, 2012. Available online: https://www.nrel.gov/docs/gen/fy13/42618.pdf (accessed on 24 May 2015).
43. Lowry, O.H.; Rosebrough, N.J.; Fau, A.L.; Randall, R.J. Protein measurement with the Folin reagent. *J. Biol. Chem.* **1951**, *193*, 265–275. [PubMed]
44. Bridoux, G.; Dhulster, P.; Manem, J. Grease analysis on municipal waste water treatment. *Plants Tech. Sci. Methods* **1994**, *5*, 257–262.
45. Miller, G.L. Use of dinitrosalicylic acid reagent for determination of reducing sugar. *Anal. Chem.* **1959**, *31*, 426–428. [CrossRef]
46. Carrere, H.; Sialve, B.; Bernet, N. Improving pig manure conversioninto biogas by thermal and thermo-chemical pretreatments. *Bioresour. Technol.* **2009**, *100*, 3690–3694. [CrossRef] [PubMed]

Sample Availability: Samples of the compounds are not available from the authors.

molecules

MDPI

Article

Acid Assisted Organosolv Delignification of Beechwood and Pulp Conversion towards High Concentrated Cellulosic Ethanol via High Gravity Enzymatic Hydrolysis and Fermentation

Konstantinos G. Kalogiannis [1,*], Leonidas Matsakas [2,*], James Aspden [2], Angelos A. Lappas [1], Ulrika Rova [2] and Paul Christakopoulos [2]

1 Chemical Process and Energy Resources Institute (CPERI), Centre for Research and Technology Hellas (CERTH), 6th km Harilaou-Thermi Rd, 57001 Thessaloniki, Greece; angel@cperi.certh.gr

2 Biochemical Process Engineering, Division of Chemical Engineering, Department of Civil, Environmental and Natural Resources Engineering, Luleå University of Technology, 971-87 Luleå, Sweden; jamesaspden95@gmail.com (J.A.); ulrika.rova@ltu.se (U.R.); paul.christakopoulos@ltu.se (P.C.)

* Correspondence: kkalogia@cperi.certh.gr (K.G.K.); leonidas.matsakas@ltu.se (L.M.); Tel.: +30-2310498357 (K.G.K.); +46-(0)920-493043 (L.M.)

Received: 19 June 2018; Accepted: 4 July 2018; Published: 5 July 2018

check for updates

Abstract: Background: Future biorefineries will focus on converting low value waste streams to chemical products that are derived from petroleum or refined sugars. Feedstock pretreatment in a simple, cost effective, agnostic manner is a major challenge. Methods: In this work, beechwood sawdust was delignified via an organosolv process, assisted by homogeneous inorganic acid catalysis. Mixtures of water and several organic solvents were evaluated for their performance. Specifically, ethanol (EtOH), acetone (AC), and methyl- isobutyl- ketone (MIBK) were tested with or without the use of homogeneous acid catalysis employing sulfuric, phosphoric, and oxalic acids under relatively mild temperature of 175 °C for one hour. Results: Delignification degrees (DD) higher than 90% were achieved, where both AC and EtOH proved to be suitable solvents for this process. Both oxalic and especially phosphoric acid proved to be good alternative catalysts for replacing sulfuric acid. High gravity simultaneous saccharification and fermentation with an enzyme loading of 8.4 mg/g_{solids} at 20 wt.% initial solids content reached an ethanol yield of 8.0 w/v%. Conclusions: Efficient delignification combining common volatile solvents and mild acid catalysis allowed for the production of ethanol at high concentration in an efficient manner.

Keywords: beechwood; organosolv delignification; ethanol fermentation; enzymatic hydrolysis; high gravity

1. Introduction

Lignocellulosic feedstocks have attracted a lot of interest for the production of biofuels and other high added-value bio-based chemicals and materials. Production of biofuels from lignocellulosic biomass waste streams, such as agricultural or forestry residues, comprises the following steps: pretreatment, enzymatic saccharification, and microbial conversion of sugars to biofuels. Pretreatment is the first step towards overcoming the complexity and recalcitrance of lignocellulosic biomass, aiming to make cellulose susceptible to enzymatic hydrolysis [1]. The pretreatment process, aiming at removing lignin, is considered to be the costliest and most challenging part of the lignocellulose conversion scheme. Lignin, which is a polyphenolic polymer surrounds the cellulose and hemicellulose, and it is essentially responsible for making biomass highly recalcitrant to pathogens, microorganisms, and enzymes [2]. Hence, a pretreatment step is required in order to disrupt the carbohydrate–lignin complex and to allow for the hydrolytic enzymes to gain access to the carbohydrates [3–5]. Hydrothermal pretreatment, without the use of chemicals, efficiently degrades hemicelluloses and increases the biomass porosity, which, in turn, enhances enzymatic hydrolysis of the pretreated solids [6]. However, the lignin that cannot be removed via hydrothermal pretreatment is partly rearranged on the surface of the lignocellulosic biomass exhibiting an inhibitory effect on downstream enzymatic hydrolysis [7].

Organosolv pretreatment has attracted an increased research interest, as it offers an effective method to remove lignin with the use of organic solvents. Organosolv employs aqueous-organic solvent mixtures, with high solvent concentration (30–70%) at temperatures of 100–220 °C, with or without the addition of catalysts [8]. One of the main benefits of organosolv pretreatment is the isolation of high-quality lignin and high-purity lignin-free cellulose [9,10]. The lignin recovered is sulfur free, while the organic solvents used (ethanol, acetone, formic, and acetic acid, etc.) can easily be recovered which is a significant advantage for small scale biorefinery plants [11]. The addition of an organic solvent allows for better mass transfer and the dissolution of lignin [12], reducing its recondensation on the external surface area of the pulp [13]. In addition, organosolv pulps have bleachability and viscosity retention when compared to cellulose soda and kraft pulps [14].

For these reasons, there is significant research interest in investigating the best pretreatment method for lignocellulosic materials. Sequential hot water pretreatment for hemicelluloses depolymerization and organosolv delignification for the removal of lignin and the production of high purity pulps have been published [10,15,16]. These studies investigated the effect of the different pretreatment techniques on the physical and chemical properties of the pulps, together with the saccharification effect of the residual solid. The existence of a two-stage sequential pretreatment method has a negative impact in the economic feasibility of the process when compared with the one-stage pretreatment methods.

Typically, both hydrothermal and organosolv pretreatments are catalytically assisted with mineral acids and bases, such as NaOH, H_2SO_4, etc. Despite their wide use, there are some limitations; they are not environmentally friendly, they generate large quantities of acid wastes and require high energy inputs, thus increasing overall process cost [17–19]. For the above reasons, an effort is being made to replace or exclude highly corrosive mineral acids such as H_2SO_4. Use of milder acids, such as H_3PO_4 or even O_2 combining organosolv and oxidation processes, are considered as interesting alternatives [20].

To make the production of ethanol economically viable and at the same time reduce the environmental impact of the process, the use of high solid concentration (high gravity—HG) during saccharification and fermentation can serve as a solution. The use of high solids concentration during saccharification can result in high glucose concentration in the broth and in turn in high ethanol production. It has been already argued that an ethanol content of at least 4% w/w is required for an economically feasible ethanol distillation [21]. Moreover, HG processes are advantageous from a water economy point of view [22]. Despite the obvious advantages of HG processes, they also present several challenges during their implementation. The high solids content create a very viscous material, practically without any free water, which is hard to mix and pump, leading to insufficient mass and heat transfer [23]. Various alternatives have been proposed to overcome these issues and achieve efficient saccharification of lignocellulosic biomass under HG conditions, such as fed-batch hydrolysis [24]. Towards this direction, Luleå University of Technology (LTU) group has previously developed and implemented a free-fall mixing reactor that was successfully used for the saccharification of various lignocellulosic materials, such as sweet sorghum bagasse [25], food waste [26], corn stover [27], wheat straw [28], and beech wood [20] at high solids content prior to ethanol fermentation. Other groups have also developed high gravity processes, successfully fermenting steam pretreated spruce to ethanol [29] or beechwood to biobutanol and dicarboxylic acids in a Terrafors reactor [30].

In this work, different organic solvents were tested for the pretreatment of beechwood sawdust in an effort to efficiently delignify the biomass. The pretreatment conditions were optimized by studying the effect of the organic solvent, concentration, and type of acidic catalyst. The aim was to maximize lignin removal, while achieving high cellulose purity and recovery in the resulting pulps. The pulps were tested for their potential in enzymatic release of glucose. The materials demonstrating the highest saccharification yields were used in HG saccharification and fermentation at a solid content of 20 wt.%. Saccharification was done in a HG custom made reactor and it resulted in the production of an aqueous solution containing up to 8.0 wt.% ethanol in the subsequent fermentation. In addition, the removed lignin was easily recovered via solvent distillation and precipitation, and found to be potentially of high quality, being suitable for further conversion towards added value products.

2. Results and Discussion

2.1. *Effect of the Type of Organic Solvent*

Table 1 presents the experimental conditions of all runs conducted, while Table 2 presents the lignin, cellulose, and hemicellulose content of the pretreated pulp, together with the recoveries of each individual component into the pretreated pulps. It should be noted that, in some cases, the recoveries of the constituents are calculated at above 100%, due to the experimental errors of the analytical methods.

Table 1. Experimental conditions for the organosolv pretreatment.

Run No. *	Solvent	Solvent, vol.%	Catalyst, wt.% on Dry Basis
1	Ethanol	60	-
2	Ethanol	60	H_2SO_4, 1.0%
3	MIBK	60	-
4	MIBK	60	H_2SO_4, 1.0%
5	Acetone	25	-
6	Acetone	60	-
7	Acetone	60	H_2SO_4, 1.0%
8	Ethanol	60	H_3PO_4, 1.0%
9	Ethanol	60	H_3PO_4, 5.6%
10	Ethanol	60	$C_2H_2O_4$, 1.0%
11	Ethanol	60	$C_2H_2O_4$, 2.6%

* Reaction temperature: 175 °C, reaction time: 60 min, liquid to solid (LSR) ratio: 10.

Table 2. Biomass constituents pulp content and % retrieved in the solid pulp.

Run No.	Cellulose (%)	Hemicellulose (%)	Lignin (%)	Cellulose Retrieved (%)	Hemicellulose Retrieved (%)	Lignin Retrieved (%)
Initial Biomass *	43.1	20.2	24.2	-	-	-
1	60.0	23.8	17.8	101.5	85.8	53.5
2	86.0	9.0	5.7	92.5	20.5	11.0
3	65.7	6.4	25.2	92.1	19.0	62.8
4	72.4	1.4	24.5	83.8	3.4	50.6
5	69.8	9.1	19.9	94.4	26.2	47.9
6	59.5	19.1	17.2	102.3	70.0	52.7
7	89.7	6.3	4.2	91.7	13.6	7.6
8	80.2	16.6	10.9	85.0	37.5	20.6
9	85.1	10.0	4.6	93.6	23.3	9.1
10	79.2	15.5	10.6	102.0	42.4	24.2
11	80.0	15.2	10.7	103.2	41.7	24.5

* Untreated Lignocel extractives are 9.1%.

Varying degrees of delignification were achieved depending on the solvent used and the presence or absence of acids that act as catalysts. Figure 1 presents graphically the pulp compositions and biomass constituents' recoveries when employing different organic solvents with and without homogeneous acidic catalysis.

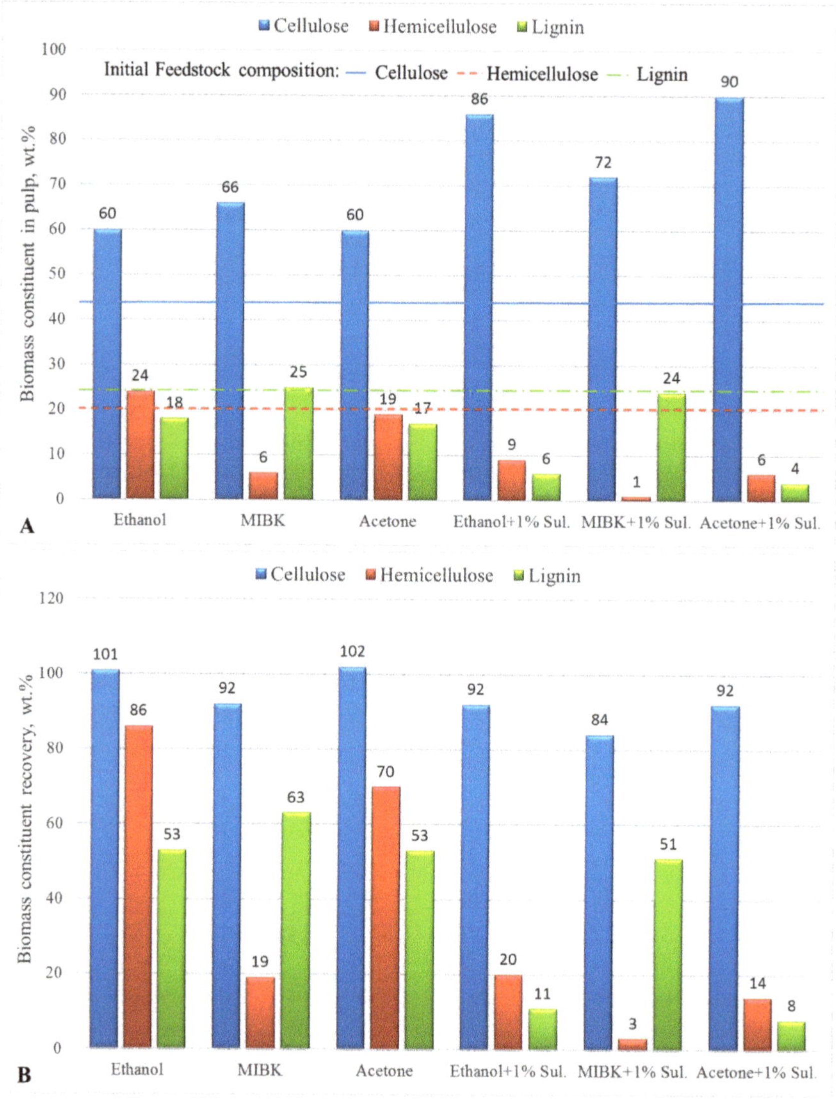

Figure 1. Pulp compositions (**A**) and biomass constituents' recoveries (**B**) in solid pulps in batch autoclave runs at 175 °C, 1 h reaction time, LSR = 10, effect of organic solvents without and with use of 1 wt.% H_2SO_4, data labels have been rounded for clarity of presentation.

Organosolv pretreatment was found to be very efficient in the pretreatment of beech wood biomass, as high cellulose low lignin content was achieved in all the treatment conditions. In some cases, the cellulose content exceeded 80 wt.% (runs 2, 7, 8, 9, and 11). Accordingly, the lignin content was very low, ranging from 4.2 to 10.7 wt.%. The pairs of runs 1–2, 3–4, and 6–7 employed different organic solvents, specifically, ethanol (EtOH), methyl- isobutyl- ketone (MIBK), and acetone (AC), without and with the use of 1 wt.% H_2SO_4 as catalyst. EtOH and AC are water miscible solvents, typically used in organosolv processes. MIBK is water immiscible, forming a biphasic system with water, which on one hand, can impact the fractionation efficiency of the system, but on the other hand, can significantly simplify the separation process of the organic lignin rich fraction from the aqueous carbohydrates rich fraction. MIBK has been used in biphasic systems for production of chemicals from biomass [31], as a co-solvent during fractionation of organosolv lignin in single phase systems [32,33] and as an extracting agent for the isolation of lignin from liquors rich in lignin and hemicellulose [34].

Regardless of the solvent used, the use of H_2SO_4 resulted in lower hemicellulose and lignin content and higher cellulose content in all of the pulps. Clearly, the hydrolyzing effect of the catalyst allowed for easier and more effective hemicellulose hydrolysis and removal. This, in turn, made the removal of lignin easier since it is closely connected to hemicellulose through a variety of bonds, such as ether and hydrogen bonds [35,36]. Among the three solvents used, both AC and EtOH proved to be effective in delignifying the biomass. AC was slightly more effective probably due to its higher solvent strength. On the other hand, MIBK was not as effective in delignifying the biomass. Compared to AC and EtOH, MIBK has a Hildebrand solubility parameter of 8.4, which is lower than typical lignin solvents, which is in the 10.5–12.5 range [32]. In addition, MIBK is not soluble in water, making this a two liquid phase reaction system. MIBK's insolubility in water is responsible for its poor performance, however it is this property that makes it very interesting for use as delignifying agent. Since the organic phase, which contains the dissolved lignin, and the aqueous phase, which contains the hemicellulose hydrolysate, can be very easily separated by spontaneous phase separation, this simplifies the separation process, and in turn, reduces the energy demands for lignin recovery. Hence, the 50% delignification degree (DD) achieved, although low, is satisfactory enough to justify further investigation in future work. Teng et al. [37] used the H_2O/MIBK biphasic system successfully to delignify different biomasses such as corn cob and rice straw. They found that the use of acidic ionic liquids (IL) was significantly more efficient when compared to the use of mineral acids. Use of H_2SO_4 achieved a DD of 61.5%, while the use of the IL [$C_4H_8SO_3$Hmim]HSO_4 resulted in a DD of 76.3%. Pretreatment without the use of any catalyst resulted in poor delignification with a DD of 24% for corncob. They attributed the lower efficiency of the mineral acids to their miscibility in MIBK, which resulted in a reduction of their actual concentration in the aqueous solution, lowering their catalytic efficiency. In our work, use of mineral acids in the case of MIBK increased the DD from 37.2% to 49.5%, which is in accordance to the findings by Teng et al. [37]. Another interesting note is that in runs 3 and 4 where MIBK is used, there is a significant reduction of the hemicellulose that is retrieved in the solid pulp without H_2SO_4 (19 wt.%) and with H_2SO_4 (3.4 wt.%) when compared to EtOH (85.8 and 20.5 wt.%, respectively) and AC (70 and 13.6 wt.%, respectively). The immiscibility of the MIBK with water resulted in the stronger solvent power and hydrolysis effect of the water towards the biomass hemicellulose. In contrast, EtOH and AC that are water miscible act as antisolvents, in part reducing the hydrolysis achieved by H_2O. This is also validated by the cellulose recovery in the pulps, which in the case of runs 3 and 4 with MIBK drops to 92.1 and 83.8 wt.% without and with H_2SO_4 respectively. Cellulose, which is much more recalcitrant compared to hemicellulose [31], is not that affected, but part of it is solubilised in the aqueous fraction, especially when H_2SO_4 is employed. Apparently, the use of one phase systems with EtOH and AC results in even lower solubilisation of cellulose, hence most of it is recovered in the solid pulp.

2.2. *Effect of Catalyst Type and Concentration*

Runs 8–11 along with runs 1 and 2 aimed at understanding the effect that homogeneous catalysis can have on the removal of lignin and the depolymerization and hydrolysis of hemicellulose in the bid to produce a high cellulose pulp. For this purpose, three different types of acids were investigated and their effect on the composition of the pulps is graphically presented in Figure 2. H_2SO_4 was tested as a base case scenario, since it is the most used acid for biomass pretreatment [38]. H_3PO_4 was tested as an inorganic acid alternative. Its main advantages are the fact that it is much less corrosive, easier to recycle, and can yield more amorphous cellulose pulp [39]. Oxalic acid was tested as an organic acid alternative. Dicarboxylic acids exhibit some advantageous characteristics, such as controlled stepwise acidity, biodegradability, and diminished corrosivity. In addition, they can be produced from bio-based and renewable resources, making them particularly attractive catalysts for biomass conversion [40].

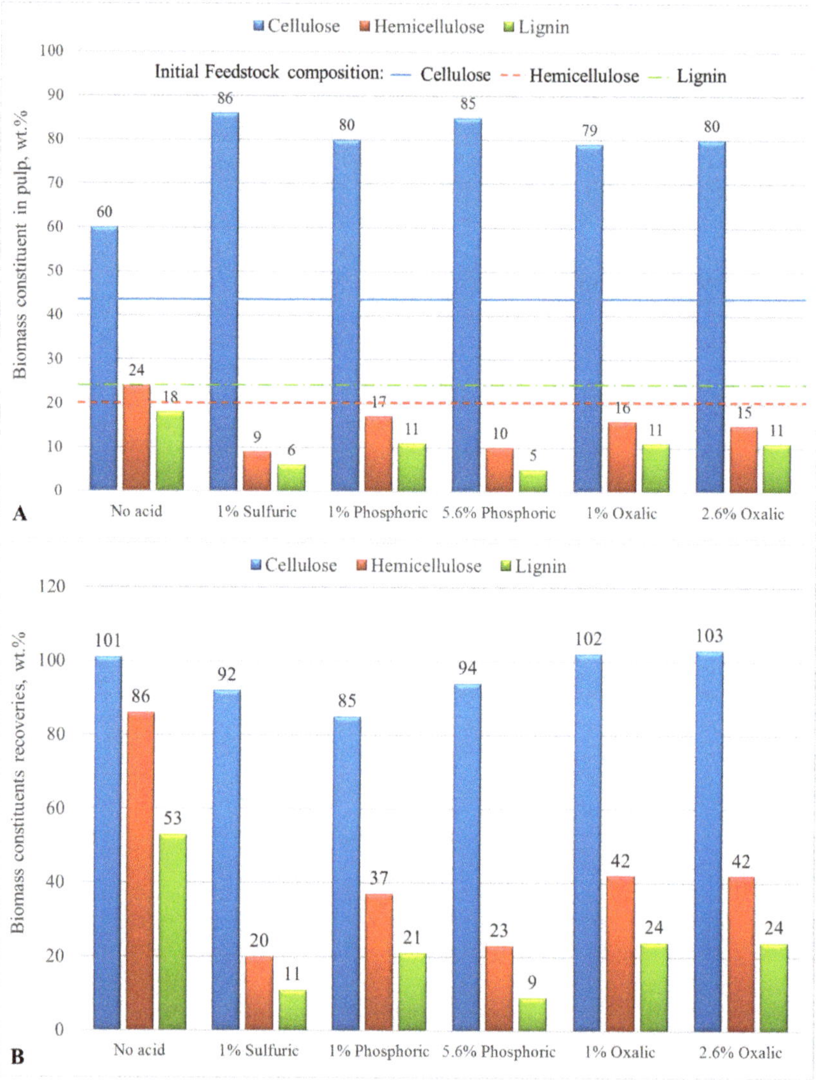

Figure 2. Pulp compositions (**A**) and biomass constituents recoveries (**B**) in solid pulps in batch autoclave runs at 175 °C, 1 h reaction time, LSR = 10, effect of homogeneous catalysis, data labels have been rounded for clarity of presentation.

Comparing run 1 and 2, the use of H_2SO_4 at 1 wt.% on dry biomass basis as catalyst has a pronounced effect, increasing the removal of hemicellulose and lignin from 14 and 46 to 80 and 89%, respectively. As expected, it enhanced hemicellulose hydrolysis, which also facilitated the removal of lignin, since these two components are connected via ether bonds, removing one can significantly boost the removal efficiency of the other. Both phosphoric and oxalic acids were also tested as catalysts. Run 8 and 10 employed 1 wt.% of each acid on a biomass basis, while run 9 and 11 used 5.6 and

2.6 wt.% of phosphoric and oxalic, respectively. This was done in order to reach the same pH as in the case of 1 wt.% H_2SO_4, so as to test the three different catalysts at the same severity. Phosphoric acid proved to be quite efficient in enhancing hemicellulose hydrolysis and lignin removal, at the 5.6 wt.% addition it was marginally better when compared to H_2SO_4 for delignification. The addition of oxalic acid also increased the efficiency of delignification when compared to the treatment without acid catalysis. However increasing its concentration had no further effect. Stein et al. [41] achieved delignification using oxalic acid as catalyst in a water/2-methyltetrahydrofuran (2-MTHF) biphasic system. Oxalic acid has been previously used to depolymerize the hemicellulosic part of biomass [41], leaving the cellulosic crystalline part intact even at temperatures as high as 180 °C [42]. The above is in accordance with our work. Cellulose recovery in the solid pulp was 100% when oxalic acid was added, however hemicellulose recovery in solid form dropped from ~86% of initial hemicellulose when no oxalic acid was used to ~42% with oxalic acid catalysis. Lignin was also successfully removed, its recovery in the solid pulp dropped from 53.5% to ~24% (run 1, 10, 11, in Table 1).

2.3. Pulp and Lignin Quality

Apart from the composition of the resulting pulps, their crystallinity index (CrI) was determined as an attempt to evaluate the effect of the pretreatment on the pretreated solids and their potentials for enzymatic saccharification. Table 3 presents the CrI of all the produced pulps.

Table 3. Crystallinity index (CrI) of pretreated pulps.

Run No.	Crystallinity Index CrI (%) *
1	68.8
2	77.5
3	74.1
4	78.2
5	74.4
6	69.0
7	78.1
8	75.1
9	77.3
10	73.2
11	72.9

* Standard deviation for CrI was ± 1.3%.

As expected, there is an overall trend that resulted in the increase of the CrI as the cellulose content in the pulp increased due to the inherent crystallinity of the cellulosic part of the biomass. Run No. 1, for example, had cellulose content of 60% corresponding to a CrI of 68.8%, while runs 7 and 9 with increased cellulose contents of 89 and 85% had CrI at around 78%. In addition, it is noted that it is the presence of hemicellulose rather than lignin in the pulp that lowers the CrI. Pulps with high hemicellulose content had lower CrI due to the hemicellulose amorphous regions. Figure 3 presents SEM images of the initial biomass and pulps retrieved from run 7 and 9, which employed H_2SO_4 and H_3PO_4, respectively.

It appears that the removal of lignin and hemicellulose results in the partial change in the fiber morphology. Untreated beechwood (Figure 3A) has a relatively smooth surface, while AC-1%H_2SO_4 and EtOH-5.6%H_3PO_4 pulps have rougher surface. Especially in the case of EtOH-5.6%H_3PO_4, the pulp appeared to be partially defibrilated and individual cellulose fibers were exposed (Figure 3C). The surface area of the pulps was slightly increased when compared to the untreated beechwood. More specifically, untreated beechwood had surface area of 0.27 m^2/g, while for pulps that are produced from run 7 and 9, this increased to 1.18 m^2/g and 1.08 m^2/g, respectively. This is a small increase in surface area but has been found to positively affect the enzymes' efficiency. Arantes et al. concluded that the topology/porosity of the pulp can limit protein penetration into the microfibril pores of the pulp, and hence affect the enzyme efficiency [43]. This is in agreement with the findings of Thygesen et al. who showed that the enzymes first penetrated into the porous regions of the pulp, and subsequently hydrolysed the cellulosic parts towards mono and oligomeric sugars [44].

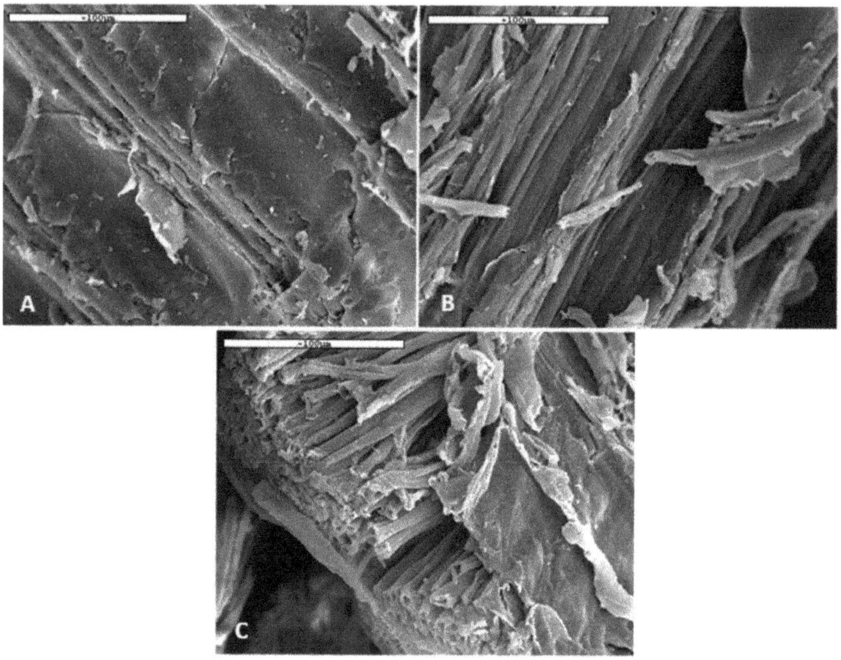

Figure 3. Scanning electron microscopy (SEM) images, bar scale of 100 μm (**A**) untreated beechwood, (**B**) AC-1%H_2SO_4, and (**C**) EtOH-5.6%H_3PO_4.

Lignins were retrieved from all runs and some selected samples were analysed via NREL to evaluate their purity. The lignins from run 7 and 9, which were found to be the most suitable for biomass delignification, were found to have very high lignin content at >94.5 wt.% and 92 wt.% purity, respectively. Lignin from run 7 had 0 wt.% cellulose content and only 0.8 wt.% hemicellulose content. For comparison, lignin from run 6 had lignin content of around 89 wt.% and hemicellulose content around 4.2 wt.%. The lack of an acid catalyst in the case of run 6 led to the sedimentation of some hemicellulose oligo- and poly- saccharides. The use of the severe H_2SO_4 in the case of run 7 hydrolyzed hemicellulose to such an extent that none was retrieved in the solid fraction of lignin. Run 9, on the other hand, had 2 wt.% and 1.8 wt.% cellulose and hemicellulose content, respectively. The milder acidity of H_3PO_4 was enough to solubilize a small part of cellulose and leave some hemicellulose intact, so as to receive it in the solid lignin. Overall, all of the lignins retrieved were very pure and well fractionated. Finally, the lignins from run 7 and 9 were also analysed via FTIR (Figure 4).

From the spectra, it appears that the delignification treatment did not degrade the recovered lignin. The FTIR graphs have peaks at characteristic wavelengths below 1500 cm^{-1}, corresponding to guaiacyl, syringyl, and some methyl- and methylene- side chains that are typically found at 1385, 1420, and 1463 cm^{-1} [45]. Wavelengths at 1216, 1271, and 1328 cm^{-1}, corresponding to stretching of C–C and C–O bonds in guaiacyl oligomers and condensed syringyl and guaiacyl rings typical of hardwood lignin are also detected [46], suggesting that the structures of the lignins remain intact. This is a very important finding, since this pure lignin product, which is easily recovered from the solvent mixture, could be upgraded to high value chemicals towards the establishment of a holistic biorefinery approach.

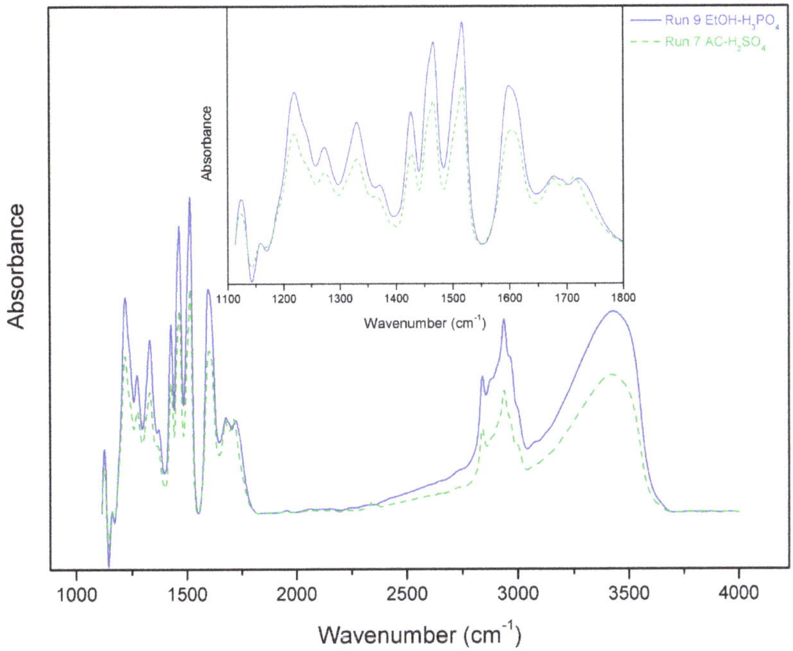

Figure 4. Fourier Transform Infrared Spectroscopy (FTIR) spectra of lignins retrieved from beechwood delignification from run 7 (AC-1%H_2SO_4) and 9 (EtOH-1%H_3PO_4).

2.4. Enzymatic Saccharification of Pretreated Pulps

To evaluate the potential of the pretreated pulps as raw materials for ethanol production, their susceptibility to enzymatic saccharification was assessed under low solids content. Table 4 presents the cellulose conversion after 24 and 48 h of enzymatic saccharification. The numbers in parentheses in the 24 h column indicate how much of the total glucose production occurred in the first 24 h, which is an important parameter and is indicative of the conversion speed.

Table 4. Enzymatic hydrolysis to glucose at 24 and 48 h.

Run No.	24 h * (wt.% on Feed Cellulose)	48 h (wt.% on Feed Cellulose)
1	11.6 (70.0)	16.5
2	50.9 (94.6)	53.8
3	14.6 (70.5)	20.7
4	14.9 (50.8)	29.3
5	15.5 (48.9)	31.7
6	14.4 (82.3)	17.5
7	57.0 (100.9)	56.5
8	36.8 (80.6)	45.6
9	60.8 (89.3)	68.1
10	38.7 (86.8)	44.6
11	39.3 (74.2)	53.0

* numbers in parentheses depict the percentage of the amount of cellulose hydrolyzed to glucose in 24 h to the amount hydrolyzed in 48 h.

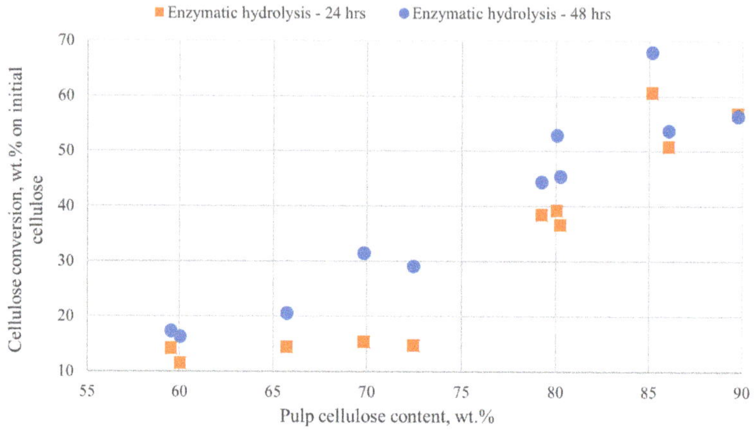

Figure 5. Cellulose conversion to glucose via enzymatic hydrolysis at 24 and 48 h vs pulp cellulose content.

An overall trend is noted where the higher the cellulose content of the pulp, the higher the cellulose conversion percentage was achieved (Figure 5). This is attributed to the lower lignin content of the high cellulose content pulps. Lignin has been known to have significant impact on the enzymes used for cellulose hydrolysis, inhibiting the depolymerisation of cellulose and the production of monomeric sugars [2]. In addition, some interesting observations can be deduced from the combination of Table 4 and Figure 5. More specifically, run 2 and 7–11, where homogeneous acidic catalysis was employed, produced pulps that were enzymatically hydrolysed to glucose easier (higher conversion after 48 h), but also more rapidly (higher % of conversion in first 24 h). Run 2 and 7 have the highest conversion rates; ~95 and 100% of overall cellulose to glucose conversion occurs in the first 24 h, respectively. This can be attributed not only to the high DD achieved, but also to a partial depolymerization of the cellulose to lower molar mass cellulose that can be enzymatically hydrolysed more rapidly. Run 9, where 5.6 wt.% H_3PO_4 was used, had the highest conversion of cellulose at 24 and 48 h, higher than that of run 7 at roughly the same lignin content. Work in the literature suggests that treating biomass with concentrated H_3PO_4 results in the swelling of the fibres and the reduction of the cellulose crystallinity [47,48]. In our work, the CrI increased as a consequence of the increased cellulose content of highly delignified pulp. Pulps produced with the aid of H_2SO_4 or H_3PO_4 catalysis had no significant differences in the CrI at similar cellulose and lignin contents (runs 2, 7, 9). Sathitsuksanoh et al. treated biomasses with concentrated H_3PO_4 and found that the CrI values varied greatly, depending on several parameters, such as measurement techniques, calculation approaches, and sample drying conditions. They concluded that the effects of CrI data obtained from dried samples on enzymatic hydrolysis should be interpreted with caution. On the other hand, they suggested that increase of the fibres surface area through lignin and hemicellulose removal and disruption of the hydrogen bonds found in crystalline cellulose could significantly increase the hydrolysis rates and efficiencies [49]. Hence, a possible explanation for the hydrolysability of pulps produced with H_3PO_4 assisted catalysis is the disruption in part of hydrogen bonding, which is not necessarily depicted as a reduction in the CrI. Enzymatic hydrolysis proved to be dependent mostly on cellulose and lignin content and was irrelevant of the CrI. Lignin, which has been found to be a major inhibitor in cellulose saccharification should therefore be removed to achieve high glucose production [50,51].

2.5. High Solids Hydrolysis and Fermentation

Based on the results from the saccharification at low solid content, two different delignified pulps, specifically from run 7 and 9, which employed AC with 1 wt.% H_2SO_4 and EtOH with 5.6 wt.%

H_3PO_4, respectively, were selected for evaluation under high solids hydrolysis and fermentation towards ethanol. The pulps from run 7 and 9 were found to have the highest DD, lowest lignin content, and over 90 wt.% cellulose recovery in the solid pulp. They were thus deemed suitable for high solids simultaneous saccharification and fermentation (SSF). As noted in the Methods section, the liquefaction/saccharification duration was 8 h at an enzyme loading of 8.4 mg/g_{solids}. After 8 h of pre-liquefaction/saccharification, the concentration of glucose was 63.8 g/L and 74.7 g/L, corresponding to 32.1% and 39.5% cellulose saccharification for the H_2SO_4 and H_3PO_4 assisted runs, respectively. Efficient glucose production in the first 8 h meant that ethanol concentrations higher than 40 g/L could be reached; a required minimum for downstream low-cost distillation [17]. Figure 6 presents the evolution of ethanol concentration for a six-day period for both delignified pulps.

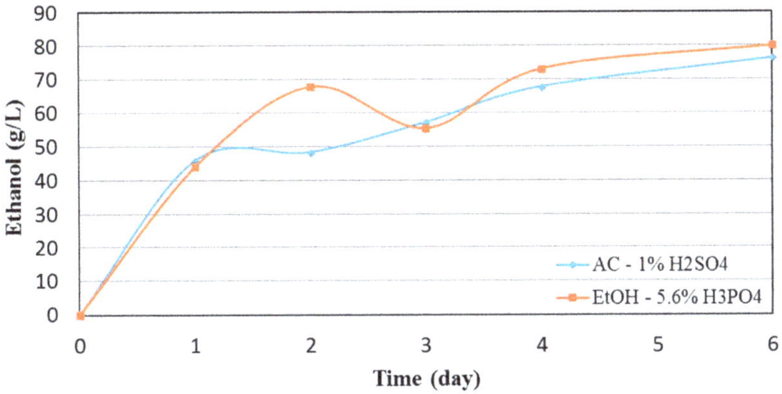

Figure 6. Ethanol concentration in high solids simultaneous saccharification and fermentation (SSF), an 8 h hydrolysis step preceded the SSF.

Both pulps reached the 40 g/L ethanol concentration threshold in less than 24 h of SSF. The AC–H_2SO_4 delignified pulp produced slightly more EtOH the first 24 h reaching a concentration of ~46 g/L. Afterwards, the ethanol production gradually leveled off to a final concentration of 76.3 g/L after six days of SSF, which is equal to approximately 75% of the maximum theoretical ethanol yield that could be attained for the cellulose content of the pulp. The EtOH-H_3PO_4 (pulp No. 9), on the other hand, had a slightly lower production rate in the first 24 h, however it retained its high production rate for up to 48 h, reaching an ethanol concentration of 68.7 g/L after the first 48 h of SSF. After six days of SSF, the ethanol concentration reached a maximum of 80 g/L, which is equal to approximately 83% of the maximum theoretical ethanol yield. Pulp No. 7 demonstrated a slightly higher productivity during the first 24 h of fermentation. This difference in the initial ethanol productivity, can be attributed to a minor inhibition of the fermentation process by the higher initial glucose concentration; behavior that has also been observed elsewhere [27]. Pulp No. 9, which used H_3PO_4, had a steadier fermentation rate for up to 48 h. Even though its cellulose content was slightly lower when compared to pulp No. 7, it achieved higher final ethanol concentration. As explained above, the H_3PO_4 may have disrupted in part the hydrogen bonds allowing for more efficient cellulose hydrolysis and consequently fermentation towards ethanol. The HG results are in good agreement with the initial enzymatic hydrolysis evaluation runs, where pulp No. 7 was found to quickly reach its maximum conversion in the first 24 h, while pulp No. 9 gave higher overall conversion in the 48 h period. Table 5 summarizes some of the work that has been done in HG SSF of different types of lignocellulosic feeds for the production of ethanol. The ethanol concentration of 80 g/L, as reported in our work, is one of the highest achieved in the literature.

Table 5. Work found in the literature on high gravity (HG) SSF for ethanol production.

WIS (%)	Material	Pre-Treatment	Enzyme Loading	Ethanol (g/L)	Time (h)	Reference
20	Beechwood	Organosolv with acetone and sulphuric acid	8.4 mg/g	76.3	144	Current work
20	Beechwood	Organosolv with ethanol and phosphoric acid	8.4 mg/g	80	144	Current work
20	Beechwood	Acetone/water oxidation	8.4 FPU/g	75.9	120	[20]
36	Bermudagrass	Phosphoric acid-acetone	25 FPU/g cellulose	56.1	96	[52]
20	Birch	Hybrid organosolv–steam explosion	18.5 FPU/g	80	192	[53]
20	Birch	Steam pre-treated	20 FPU/g	14.4	144	[54]
20	Corn stover	Steam explosion	17.7 FPU/g	59.8	192	[55]
20	Eastern redcedar	Acid bisulfite	46 FPU/g glucan	52	42	[56]
15	Eucalyptus	Organosolv	20 FPU/g	42	72	[57]
20	Rapeseed straw	Dilute acid	15 FPU/g	39.9	24	[58]
25	Pine	Sulfite	15 FPU/g	82	24	[59]
20	Spruce	Steam pre-treated	22.5 FPU/g	40	96	[60]
10	Spruce	Steam pre-treated	30 FPU/g glucan	45	100	[61]
10	Spruce	Steam pre-treated	20 FPU/g	45.8	96	[62]
25	Wheat straw	Steam explosion	15 FPU/g	58.6	80	[63]

3. Materials and Methods

3.1. *Raw Materials*

Commercially available beechwood sawdust with particle size 150–500 μm (Lignocel® HBS 150–500) and moisture content 8 wt.% was used as biomass feedstock. It was handled, as described by Kalogiannis et al. [45].

3.2. *Strains and Enzymes*

The *Saccharomyces cerevisiae* strain Ethanol Red® was used as fermenting microorganism during the current work. This specific strain was developed by Fermentis (Marcq-en-Barœl, France) for industrial fuel ethanol production, and therefore it exhibits high ethanol tolerance, making it suitable for use in HG fermentation processes. The commercial enzyme solution Cellic® CTec2 from Novozymes A/S (Bagsværd, Denmark) was used for the saccharification trials under low solids content and at HG conditions. The protein content of the enzyme solution was 100 mg/mL, as determined by using the Bradford assay [64]. All the other chemicals and reagents were of analytical grade.

3.3. *Organosolv Pretreatment*

Organosolv pretreatment of Lignocel was performed in metallic cylinders of 2.5 L size, which were placed in an air-heated multidigester apparatus [27] at 175 °C for 60 min. During the pretreatment, 110 g of biomass were mixed with 1.1 L of solvent-aqueous mixture. The following solvents were tested: ethanol, acetone, and methyl-isobutyl-ketone at a content of 60% v/v (with the acetone to be also tested to an acetone content of 25% v/v) with or without the addition of sulfuric acid (1 wt.% on dry biomass) as acidic catalyst. Replacement of sulfuric acid with phosphoric acid and oxalic acid was also tested with ethanol as the solvent. In that case, the concentration of the acid catalysts was either similar to sulfuric acid (1 wt.% on dry biomass) or was fixed to achieve the same pH as the sulfuric acid during the pretreatment (phosphoric acid, 5.6 wt.% on dry biomass; oxalic acid, 2.6 wt.% on dry biomass). At the end of the pretreatment, the pretreated solids were separated from the pretreatment liquor by vacuum filtration, washed with the same solvent used during the pretreatment, air dried, and stored until further use. The weight of the pretreated solids was measured to determine biomass solubilization and the composition of the solids was determined, as described in the *Analytical Methods* section.

The pretreatment liquor was collected and the solvent was evaporated (when ethanol and acetone were used) under vacuum in order to reduce lignin solubility. Lignin was then separated from the liquid by centrifugation (14,000 rpm, 29,416× g, at 4 °C for 15 min), and finally air-dried [53]. When MIBK was used as solvent, a different lignin isolation process was followed. MIBK is water immiscible at room temperature, resulting in phase separation with the lignin being recovered in the solvent phase. The solvent was then evaporated under vacuum, leading to the recovery of the solid lignin.

All of the experimental conditions are presented in Table 1. The resulting pulps were dried and weighed, while the original biomass and the resulting pulps were analysed by the NREL method to determine (see analytical methods paragraph) cellulose, hemicellulose, and lignin content. The delignification degree (DD) can be calculated as 100%-lignin recovery (%).

3.4. Enzymatic Saccharification Trials

The pretreated solids were assessed for their enzymatic saccharification yields under a solid content of 2 wt.% in 50 mM citrate buffer (pH 5). The enzyme load was 8.4 mg/g_{solids} of the commercial enzyme solution Cellic® CTec 2. Sodium azide at a concentration of 0.02 wt.% was added in the solution to prevent microbial contaminations. Incubation took place in 2 mL Epperdorf tubes containing 1.5 mL of the solution in ThermoMixer C (Eppendorf, Hamburg, Germany) at 50 °C and 1200 rpm for 48 h. Samples were withdrawn at 0 h, 24 h, and 48 h, and analyzed for glucose concentration. All of the enzymatic hydrolysis trials were performed in duplicates. The enzymatic saccharification yield was expressed as the percentage of cellulose converted to glucose and was calculated according to the following equation:

$$Saccharification\ yield = \left(\frac{C_{glucose} \times V_{liquid} \times 0.90}{m_{solids} \times x_{cellulose}} \right) \times 100$$

where $C_{glucose}$ is the concentration of glucose, V_{liquid} is the volume of the liquid during the trials, 0.90 is the correction factor for the conversion of glucose to cellulose, m_{solids} is the mass of the dry solids, and $x_{cellulose}$ is the cellulose content of solids.

3.5. High Gravity Saccharification and Fermentation

The two most promising materials were further subjected to high gravity saccharification and fermentation trials. Saccharification took place at a freefall mixing saccharification reactor, as previously described [25]. More specifically, the dry material content used was 20 wt.% in 50 mM citrate buffer with an enzyme load of 8.4 mg/g_{solids}. Saccharification took place at 50 °C for 8 h. At the end of the saccharification the slurry was collected and supplemented with nutrients for the yeast growth at a final concentration of 1 g/L yeast extract, 0.5 g/L $(NH_4)_2HPO_4$, and 0.025 g/L $MgSO_4{\cdot}7H_2O$. The fermentation was initiated by inoculation with *S. cerevisiae* suspension (that was grown overnight at YPD media at 35 °C and 180 rpm) to achieve an initial dry cell weight concentration of 1 g/L. Incubation was carried out at 35 °C and 120 rpm, and the samples were withdrawn daily, diluted, filtered through a 0.2 μm syringe filter, and analyzed for ethanol. The fermentations were performed in duplicates.

3.6. Analytical Methods

The cellulose, hemicellulose, lignin and ash contents of lignocellulosic biomass were determined, according to the procedures provided by National Renewable Energy Laboratory (NREL; Golden, CO, USA) [65]. The sugars were analysed at a high pressure liquid chromatography (HPLC) apparatus, coupled with a refractive index detector equipped an Aminex HPX-87P column (Bio-Rad, Hercules, CA, USA). Analysis performed at 85 °C, with ultrapure water as mobile phase at a flow rate of 0.6 mL/min. Ethanol produced during the SSF was analysed by the same HPLC apparatus equipped with an Aminex HPX-87H (Bio-Rad, Hercules, CA, USA) chromatography column. The column was kept at 40 °C and the mobile phase was 5 mM sulphuric acid in degassed HPLC grade water at a flow rate of 0.6 mL/min.

Fourier Transform Infrared Spectroscopy (FTIR) (Nicolet 5700, Thermo Electron Corporation, Waltham, MA, USA) analysis was employed for further characterization of the lignin samples' structure.

Details may be found elsewhere [46]. X-ray Diffraction analysis was done on a Siemens D500, copper ray with a Nickel filter (λ = 15,406 Å, voltage 40 KV, intensity 30 mA) (Bruker, Wien, Austria). The angle 2θ was between 5° and 50° with a step 0.04 and step time 2 s. Surface area of the pulps was measured on a Micromeritics Tristar 3000 (Micromeritics, Norcross, GA, USA) via the BET method after outgassing the biomass samples at 25 °C for 72 h. Scanning electron microscopy (SEM) images were obtained on a Jeol JSM-6300 microscope (Jeol, Peabody, MA, USA).

For the determination of the surface area (BET method), pore volume, and pore size distribution (BJH method) of the catalyst samples, N_2 adsorption/desorption measurements were carried out at −196 °C, using an Autosorb-1MP Automatic Volumetric Sorption Analyzer (Quantachrome, Boynton Beach, FL, USA).

4. Conclusions

In the present work, the efficiency of organosolv pretreatment on lignin and hemicellulose removal and its effect on the downstream biochemical conversion of the solid pulp to ethanol were evaluated. A hardwood feedstock, more specifically beechwood, was treated with mixtures of water and different organic solvents, namely AC, EtOH, and MIBK. In addition, the effect of homogeneous catalysis was investigated. Sulfuric, phosphoric, and oxalic acids were tested at different concentrations and their effect on the DD and the hydrolysability and fermentability of the resulting pulps was evaluated. Both AC and EtOH, which are water miscible, were found to be very efficient in removing lignin and hemicellulose from the initial feedstock. Both were able to remove almost 50% of the lignin found in the feedstock. MIBK, on the other hand, behaved poorly due to its non-miscibility in water. Use of sulfuric acid as catalyst significantly improved the DD; more than 90% of initial lignin was removed and pulps with high cellulose content (>85%) were produced. Phosphoric and oxalic acid were used as alternative catalysts and were both found to enhance lignin removal. In the case of phosphoric acid, partial defibrillation and exposure of the cellulose fibrils was also noted. Moreover, the lignin retrieved from the solvent system was found to be intact and of high purity and quality making it a valuable potential feedstock for production of bio-based chemicals and materials. High gravity SSF at 20 wt.% solids yielded highly concentrated ethanol solutions (8 wt.%), which is one of the highest reported in the literature for beechwood feedstock and stresses the potential of combining organosolv pretreatment with high solids fermentation on the basis of a biorefinery approach.

Author Contributions: Conceptualization, K.G.K. and L.M.; Methodology, K.K. and L.M.; Formal Analysis, X.X.; Investigation, K.G.K., L.M. and J.A.; Resources, P.C. and U.R.; Data Curation, K.G.K., L.M., P.C. and A.A.L.; Writing-Original Draft Preparation, K.G.K. and L.M.; Writing-Review & Editing, P.C., U.R. and A.A.L.; Supervision, P.C., U.R. and A.A.L.; Project Administration, K.G.K., P.C. and U.R.; Funding Acquisition, K.G.K., L.M., P.C., U.R. and A.A.L.

Funding: The authors wish also to acknowledge the support from COST Action FP1306 via the Short Term Scientific Mission of K. Kalogiannis in LTU.

Acknowledgments: The authors would like to thank Novozymes A/S, Denmark, for providing the Cellic® CTec2 enzyme solution and Lesaffre Advanced Fermentations, France, for providing the Ethanol Red® that were used during this work. L.M., U.R. and P.C. would like to thank Bio4Energy, a strategic research environment appointed by the Swedish government, for supporting this work.

Conflicts of Interest: The authors declare no conflict of interest.

References

1. Zhao, X.; Cheng, K.; Liu, D. Organosolv pretreatment of lignocellulosic biomass for enzymatic hydrolysis. *Appl. Microbiol. Biotechnol.* **2009**, *82*, 815–827. [CrossRef] [PubMed]
2. Studer, M.H.; DeMartini, J.D.; Davis, M.F.; Sykes, R.W.; Davison, B.; Keller, M.; Tuskan, G.A.; Wyman, C.E. Lignin content in natural Populus variants affects sugar release. *Proc. Natl. Acad. Sci. USA* **2011**, *108*, 6300–6305. [CrossRef] [PubMed]
3. Zhang, B.; Shahbazi, A. Recent developments in pretreatment technologies for production of lignocellulosic biofuels. *J. Pet. Environ. Biotechnol.* **2011**, *2*. [CrossRef]

4. Yang, B.; Wyman, C.E. Pretreatment: The key to unlocking low-cost cellulosic ethanol. *Biofuels Bioprod. Biorefin.* **2008**, *2*, 26–40. [CrossRef]
5. Wyman, C.E.; Dale, B.E.; Elander, R.T.; Holtzapple, M.; Ladisch, M.R.; Lee, Y.Y. Coordinated development of leading biomass pretreatment technologies. *Bioresour. Technol.* **2005**, *96*, 1959–1966. [CrossRef] [PubMed]
6. Zhang, J.; Tang, M.; Viikari, L. Xylans inhibit enzymatic hydrolysis of lignocellulosic materials by cellulases. *Bioresour. Technol.* **2012**, *121*, 8–12. [CrossRef] [PubMed]
7. Kristensen, J.B.; Thygesen, L.G.; Felby, C.; Jorgensen, H.; Elder, T. Cell wall structural changes in wheat straw pretreated for bioethanol production. *Biotechnol. Biofuels* **2008**, *1*, 5. [CrossRef] [PubMed]
8. Taherzadeh, M.; Karimi, K. Pretreatment of lignocellulosic wastes to improve ethanol and biogas production: A review. *Int. J. Mol. Sci.* **2008**, *9*, 1621–1651. [CrossRef] [PubMed]
9. Zhang, K.; Pei, Z.; Wang, D. Organic solvent pretreatment of lignocellulosic biomass for biofuels and biochemicals: A review. *Bioresour. Technol.* **2016**, *199*, 21–33. [CrossRef] [PubMed]
10. Huijgen, W.J.J.; Smit, A.T.; de Wild, P.J.; den Uil, H. Fractionation of wheat straw by prehydrolysis, organosolv delignification and enzymatic hydrolysis for production of sugars and lignin. *Bioresour. Technol.* **2012**, *114*, 389–398. [CrossRef] [PubMed]
11. Vallejos, M.E.; Zambon, M.D.; Area, M.C.; da Silva Curvelo, A.A. Low liquid-solid ratio fractionation of sugarcane bagasse by hot water autohydrolysis and organosolv delignification. *Ind. Crops Prod.* **2015**, *65*, 349–353. [CrossRef]
12. Oliet, M.; García, J.; Rodríguez, F.; Gilarrranz, M. Solvent effects in autocatalyzed alcohol–water pulping. *Chem. Eng. J.* **2002**, *87*, 157–162. [CrossRef]
13. Oliet, M.; Rodríguez, F.; García, J.; Gilarranz, M.A. The effect of autocatalyzed ethanol pulping on lignin characteristics. *J. Wood Chem. Technol.* **2001**, *21*, 81–95. [CrossRef]
14. Shatalov, A.A.; Pereira, H. Polysaccharide degradation during ozone-based TCF bleaching of non-wood organosolv pulps. *Carbohydr. Polym.* **2007**, *67*, 275–281. [CrossRef]
15. Garrote, G.; Falqué, E.; Domínguez, H.; Parajó, J.C. Autohydrolysis of agricultural residues: Study of reaction byproducts. *Bioresour. Technol.* **2007**, *98*, 1951–1957. [CrossRef] [PubMed]
16. Amendola, D.; De Faveri, D.M.; Egües, I.; Serrano, L.; Labidi, J.; Spigno, G. Autohydrolysis and organosolv process for recovery of hemicelluloses, phenolic compounds and lignin from grape stalks. *Bioresour. Technol.* **2012**, *107*, 267–274. [CrossRef] [PubMed]
17. García, A.; González, M.; Llano-Ponte, R.; Labidi, J. Energy and economic assessment of soda and organosolv biorefinery processes. *Biomass Bioenergy* **2011**, *35*, 516–525. [CrossRef]
18. Amiri, H.; Karimi, K.; Zilouei, H. Organosolv pretreatment of rice straw for efficient acetone, butanol, and ethanol production. *Bioresour. Technol.* **2014**, *152*, 450–456. [CrossRef] [PubMed]
19. Ruiz Cuilty, K.; Ballinas-Casarrubias, L.; Rodríguez de San Miguel, E.; de Gyves, J.; Robles-Venzor, J.C.; González-Sánchez, G. Cellulose recovery from Quercus sp. sawdust using Ethanosolv pretreatment. *Biomass Bioenergy* **2018**, *111*, 114–124. [CrossRef]
20. Katsimpouras, C.; Kalogiannis, K.G.; Kalogianni, A.; Lappas, A.A.; Topakas, E. Production of high concentrated cellulosic ethanol by acetone/water oxidized pretreated beech wood. *Biotechnol. Biofuels* **2017**, *10*. [CrossRef] [PubMed]
21. Zacchi, G.; Axelsson, A. Economic evaluation of preconcentration in production of ethanol from dilute sugar solutions. *Biotechnol. Bioeng.* **1989**, *34*, 223–233. [CrossRef] [PubMed]
22. Koppram, R.; Tomás-Pejó, E.; Xiros, C.; Olsson, L. Lignocellulosic ethanol production at high-gravity: Challenges and perspectives. *Trends Biotechnol.* **2014**, *32*, 46–53. [CrossRef] [PubMed]
23. Cardona, M.J.; Tozzi, E.J.; Karuna, N.; Jeoh, T.; Powell, R.L.; McCarthy, M.J. A process for energy-efficient high-solids fed-batch enzymatic liquefaction of cellulosic biomass. *Bioresour. Technol.* **2015**, *198*, 488–496. [CrossRef] [PubMed]
24. Gao, Y.; Xu, J.; Yuan, Z.; Zhang, Y.; Liu, Y.; Liang, C. Optimization of fed-batch enzymatic hydrolysis from alkali-pretreated sugarcane bagasse for high-concentration sugar production. *Bioresour. Technol.* **2014**, *167*, 41–45. [CrossRef] [PubMed]
25. Matsakas, L.; Christakopoulos, P. Fermentation of liquefacted hydrothermally pretreated sweet sorghum bagasse to ethanol at high-solids content. *Bioresour. Technol.* **2013**, *127*, 202–208. [CrossRef] [PubMed]

26. Matsakas, L.; Kekos, D.; Loizidou, M.; Christakopoulos, P. Utilization of household food waste for the production of ethanol at high dry material content. *Biotechnol. Biofuels* **2014**, *7*, 4. [CrossRef] [PubMed]
27. Katsimpouras, C.; Zacharopoulou, M.; Matsakas, L.; Rova, U.; Christakopoulos, P.; Topakas, E. Sequential high gravity ethanol fermentation and anaerobic digestion of steam explosion and organosolv pretreated corn stover. *Bioresour. Technol.* **2017**, *244*, 1129–1136. [CrossRef] [PubMed]
28. Paschos, T.; Xiros, C.; Christakopoulos, P. Simultaneous saccharification and fermentation by co-cultures of Fusarium oxysporum and Saccharomyces cerevisiae enhances ethanol production from liquefied wheat straw at high solid content. *Ind. Crops Prod.* **2015**, *76*, 793–802. [CrossRef]
29. Hoyer, K.; Galbe, M.; Zacchi, G. The effect of prehydrolysis and improved mixing on high-solids batch simultaneous saccharification and fermentation of spruce to ethanol. *Process. Biochem.* **2013**, *48*, 289–293. [CrossRef]
30. Tippkötter, N.; Duwe, A.-M.; Wiesen, S.; Sieker, T.; Ulber, R. Enzymatic hydrolysis of beech wood lignocellulose at high solid contents and its utilization as substrate for the production of biobutanol and dicarboxylic acids. *Bioresour. Technol.* **2014**, *167*, 447–455. [CrossRef] [PubMed]
31. Saha, B.; Abu-Omar, M.M. Advances in 5-hydroxymethylfurfural production from biomass in biphasic solvents. *Green Chem.* **2014**, *16*, 24–38. [CrossRef]
32. Bozell, J.J.; Black, S.K.; Myers, M.; Cahill, D.; Miller, W.P.; Park, S. Solvent fractionation of renewable woody feedstocks: Organosolv generation of biorefinery process streams for the production of biobased chemicals. *Biomass Bioenergy* **2011**, *35*, 4197–4208. [CrossRef]
33. Katahira, R.; Mittal, A.; McKinney, K.; Ciesielski, P.N.; Donohoe, B.S.; Black, S.K.; Johnson, D.K.; Biddy, M.J.; Beckham, G.T. Evaluation of clean fractionation pretreatment for the production of renewable fuels and chemicals from corn stover. *ACS Sustain. Chem. Eng.* **2014**, *2*, 1364–1376. [CrossRef]
34. Brudecki, G.; Cybulska, I.; Rosentrater, K. Optimization of clean fractionation process applied to switchgrass to produce pulp for enzymatic hydrolysis. *Bioresour. Technol.* **2013**, *131*, 101–112. [CrossRef] [PubMed]
35. Kobayashi, T.; Kohn, B.; Holmes, L.; Faulkner, R.; Davis, M.; Maciel, G.E. Molecular-level consequences of biomass pretreatment by dilute sulfuric acid at various temperatures. *Energy Fuels* **2011**, *25*, 1790–1797. [CrossRef]
36. Sturgeon, M.R.; Kim, S.; Lawrence, K.; Paton, R.S.; Chmely, S.C.; Nimlos, M.; Foust, T.D.; Beckham, G.T. A mechanistic investigation of acid-catalyzed cleavage of aryl-ether linkages: Implications for lignin depolymerization in acidic environments. *ACS Sustain. Chem. Eng.* **2014**, *2*, 472–485. [CrossRef]
37. Teng, J.; Ma, H.; Wang, F.; Wang, L.; Li, X. Catalytic fractionation of raw biomass to biochemicals and organosolv lignin in a methyl isobutyl ketone/H_2O biphasic system. *ACS Sustain. Chem. Eng.* **2016**, *4*, 2020–2026. [CrossRef]
38. Kumar, A.K.; Sharma, S. Recent updates on different methods of pretreatment of lignocellulosic feedstocks: A review. *Bioresour. Bioprocess.* **2017**, *4*, 7. [CrossRef] [PubMed]
39. Klemm, D.; Philipp, B.; Heinze, T.; Heinze, U.; Wagenknecht, W. *Comprehensive Cellulose Chemistry. Volume 1. Fundamentals and Analytical Methods*; Wiley: Weinheim, Germany, 1998; 260p.
40. Demain, A.L.; Adrio, J.L. Contributions of microorganisms to industrial biology. *Mol. Biotechnol.* **2008**, *38*, 41–55. [CrossRef] [PubMed]
41. Vom Stein, T.; Grande, P.M.; Kayser, H.; Sibilla, F.; Leitner, W.; Domínguez de María, P. From biomass to feedstock: One-step fractionation of lignocellulose components by the selective organic acid-catalyzed depolymerization of hemicellulose in a biphasic system. *Green Chem.* **2011**, *13*, 1772. [CrossRef]
42. Lee, J.-W.; Rodrigues, R.C.L.B.; Kim, H.J.; Choi, I.-G.; Jeffries, T.W. The roles of xylan and lignin in oxalic acid pretreated corncob during separate enzymatic hydrolysis and ethanol fermentation. *Bioresour. Technol.* **2010**, *101*, 4379–4385. [CrossRef] [PubMed]
43. Arantes, V.; Saddler, J.N. Cellulose accessibility limits the effectiveness of minimum cellulase loading on the efficient hydrolysis of pretreated lignocellulosic substrates. *Biotechnol. Biofuels* **2011**, *4*, 3. [CrossRef] [PubMed]
44. Thygesen, L.G.; Hidayat, B.J.; Johansen, K.S.; Felby, C. Role of supramolecular cellulose structures in enzymatic hydrolysis of plant cell walls. *J. Ind. Microbiol. Biotechnol.* **2011**, *38*, 975–983. [CrossRef] [PubMed]

45. Kalogiannis, K.G.; Stefanidis, S.; Marianou, A.; Michailof, C.; Kalogianni, A.; Lappas, A. Lignocellulosic biomass fractionation as a pretreatment step for production of fuels and green chemicals. *Waste Biomass Valorization* **2015**, *6*. [CrossRef]
46. Kalogiannis, K.G.; Stefanidis, S.D.; Michailof, C.M.; Lappas, A.A.; Sjöholm, E. Pyrolysis of lignin with 2DGC quantification of lignin oil: Effect of lignin type, process temperature and ZSM-5 in situ upgrading. *J. Anal. Appl. Pyrolysis* **2015**, *115*, 410–418. [CrossRef]
47. Takata, E.; Tsutsumi, K.; Tsutsumi, Y.; Tabata, K. Production of monosaccharides from napier grass by hydrothermal process with phosphoric acid. *Bioresour. Technol.* **2013**, *143*, 53–58. [CrossRef] [PubMed]
48. Kang, P.; Zheng, Z.; Qin, W.; Dong, C.; Yang, Y. Efficient fractionation of Chinese white poplar biomass with enhanced enzymatic digestibility and modified acetone-soluble lignin. *BioResources* **2011**, *6*, 4705–4720. [CrossRef]
49. Sathitsuksanoh, N.; Zhu, Z.; Wi, S.; Percival Zhang, Y.-H. Cellulose solvent-based biomass pretreatment breaks highly ordered hydrogen bonds in cellulose fibers of switchgrass. *Biotechnol. Bioeng.* **2011**, *108*, 521–529. [CrossRef] [PubMed]
50. Min, D.; Yang, C.; Shi, R.; Jameel, H.; Chiang, V.; Chang, H. The elucidation of the lignin structure effect on the cellulase-mediated saccharification by genetic engineering poplars (*Populus nigra* L. × *Populus maximowiczii* A.). *Biomass Bioenergy* **2013**, *58*, 52–57. [CrossRef]
51. De Souza, A.P.; Kamei, C.L.A.; Torres, A.F.; Pattathil, S.; Hahn, M.G.; Trindade, L.M.; Buckeridge, M.S. How cell wall complexity influences saccharification efficiency in *Miscanthus sinensis*. *J. Exp. Bot.* **2015**, *66*, 4351–4365. [CrossRef] [PubMed]
52. Li, H.; Kim, N.J.; Jiang, M.; Kang, J.W.; Chang, H.N. Simultaneous saccharification and fermentation of lignocellulosic residues pretreated with phosphoric acid-acetone for bioethanol production. *Bioresour. Technol.* **2009**, *100*, 3245–3251. [CrossRef] [PubMed]
53. Matsakas, L.; Nitsos, C.; Raghavendran, V.; Yakimenko, O.; Persson, G.; Olsson, E.; Rova, U.; Olsson, L.; Christakopoulos, P. A novel hybrid organosolv: Steam explosion method for the efficient fractionation and pretreatment of birch biomass. *Biotechnol. Biofuels* **2018**, *11*, 160. [CrossRef] [PubMed]
54. Wang, R.; Koppram, R.; Olsson, L.; Franzén, C.J. Kinetic modeling of multi-feed simultaneous saccharification and co-fermentation of pretreated birch to ethanol. *Bioresour. Technol.* **2014**, *172*, 303–311. [CrossRef] [PubMed]
55. Liu, Z.H.; Qin, L.; Zhu, J.Q.; Li, B.Z.; Yuan, Y.J. Simultaneous saccharification and fermentation of steam-exploded corn stover at high glucan loading and high temperature. *Biotechnol. Biofuels* **2014**, *7*, 167. [CrossRef] [PubMed]
56. Ramachandriya, K.D.; Wilkins, M.; Atiyeh, H.K.; Dunford, N.T.; Hiziroglu, S. Effect of high dry solids loading on enzymatic hydrolysis of acid bisulfite pretreated Eastern redcedar. *Bioresour. Technol.* **2013**, *147*, 168–176. [CrossRef] [PubMed]
57. Yáñez-S, M.; Rojas, J.; Castro, J.; Ragauskas, A.; Baeza, J.; Freer, J. Fuel ethanol production from Eucalyptus globulus wood by autocatalized organosolv pretreatment ethanol-water and SSF. *J. Chem. Technol. Biotechnol.* **2013**, *88*, 39–48. [CrossRef]
58. López-Linares, J.C.; Romero, I.; Cara, C.; Ruiz, E.; Moya, M.; Castro, E. Bioethanol production from rapeseed straw at high solids loading with different process configurations. *Fuel* **2014**, *122*, 112–118. [CrossRef]
59. Dong, C.; Wang, Y.; Zhang, H.; Leu, S.Y. Feasibility of high-concentration cellulosic bioethanol production from undetoxified whole Monterey pine slurry. *Bioresour. Technol.* **2018**, *250*, 102–109. [CrossRef] [PubMed]
60. Koppram, R.; Olsson, L. Combined substrate, enzyme and yeast feed in simultaneous saccharification and fermentation allow bioethanol production from pretreated spruce biomass at high solids loadings. *Biotechnol. Biofuels* **2014**, *7*, 54. [CrossRef] [PubMed]
61. Bertilsson, M.; Olofsson, K.; Lidén, G. Prefermentation improves xylose utilization in simultaneous saccharification and co-fermentation of pretreated spruce. *Biotechnol. Biofuels* **2009**, *2*, 8. [CrossRef] [PubMed]
62. Frankó, B.; Galbe, M.; Wallberg, O. Influence of bark on fuel ethanol production from steam-pretreated spruce. *Biotechnol. Biofuels* **2015**, *8*, 15. [CrossRef] [PubMed]
63. Alvira, P.; Moreno, A.D.; Ibarra, D.; Sáez, F.; Ballesteros, M. Improving the fermentation performance of *saccharomyces cerevisiae* by laccase during ethanol production from steam-exploded wheat straw at high-substrate loadings. *Biotechnol. Prog.* **2013**, *29*, 74–82. [CrossRef] [PubMed]

64. Bradford, M.M. A rapid and sensitive method for the quantitation of microgram quantities of protein utilizing the principle of protein-dye binding. *Anal. Biochem.* **1976**, *72*, 248–254. [CrossRef]
65. Sluiter, A.; Hames, B.; Ruiz, R.; Scarlata, C.; Sluiter, J.; Templeton, D.; Nrel, D.C. *Determination of Structural Carbohydrates and Lignin in Biomass*; National Renewable Energy Laboratory: Golden, CO, USA, 2012.

Sample Availability: Samples of the materials produced in the current work are available from the authors at a reasonable request.

molecules

MDPI

Article

Lime Pretreatment of Miscanthus: Impact on BMP and Batch Dry Co-Digestion with Cattle Manure

Hélène Laurence Thomas, Jordan Seira, Renaud Escudié and Hélène Carrère *

LBE, University of Montpellier, INRA, 102, Avenue des Etangs, 11100 Narbonne, France; helene.thomas@inra.fr (H.L.T.); jordanseira@hotmail.fr (J.S.); renaud.escudie@inra.fr (R.E.)
* Correspondence: helene.carrere@inra.fr; Tel.: +33-468-425-168

Received: 18 May 2018; Accepted: 28 June 2018; Published: 2 July 2018

check for updates

Abstract: In Europe, the agricultural biogas sector is currently undergoing fast developments, and cattle manure constitutes an important feedstock. Batch dry digester processes with leachate recirculation prove to be particularly interesting for small-scale plants. However, their startup being relatively slow, the process could be facilitated by co-digestion with energy crops. In this study, *Miscanthus x giganteus* was chosen for its high biomass yields and low input requirements. The carbohydrate accessibility of this lignocellulosic biomass is limited but may be improved with alkali pretreatment. The efficiency of lime (CaO) pretreatment with low water addition on the biochemical methane potential (BMP) of miscanthus was investigated through two experimental designs (CaO concentrations ranged between 2.5 and 17.5% and pretreatment lasted 1, 3, or 5 days). The pretreated miscanthus was then co-digested with cattle manure in dry leach bed reactors. CaO pretreatments led to a 14–37% improvement of miscanthus BMP, and a 67–227% increase in the first-order kinetics constant; a high contact time was shown to favor methane production. According to these results and to industrial requirements, miscanthus was pretreated with 5 and 10% CaO for 5 days, then co-digested with manure in dry leach bed reactors. Nevertheless, the promising results of the BMP tests were not validated. This could be related to the high water absorption capacity of miscanthus.

Keywords: anaerobic digestion; biogas; lignocellulosic biomass; alkali pretreatment

1. Introduction

Within the context of having to mitigate global warming and reduce greenhouse effect gas emissions, anaerobic digestion (AD), which allows the production of renewable energy from various organic wastes, is undergoing rapid developments. In particular, the French government has set the target of 1500 biogas plants by 2020, including 1000 plants based on agricultural feedstocks [1]. In many agricultural anaerobic digestion plants, manure represents the main part of the feed. Furthermore, cattle manure is available in high quantities all over the country. Its production has been estimated at about 69 MT per year in 2010 [2]. Cattle manure is rich in straw and is characterized by a total solids (TS) content of about 20–30%. It is thus suited for dry AD [3], also called solid-state anaerobic digestion. A process occupying an important part in the development of the agricultural AD sector is the leach bed reactor (LBR) operated in batch mode [4]. In this high-solids process, the solid substrate is loaded into the reactor while a liquid phase, usually stored in a separate container, is regularly sprinkled over the solid bulk and percolates through it. However, batch mono-digestion of cattle manure usually takes time to start up and produces low amounts of biogas [5].

In this view, co-digestion would be a good option for improving biogas production and productivity. For example, Botji et al. (2017) [6] demonstrated how the co-digestion of poultry manure with maize silage improved methane production by 24% relative to mono-digestion. This is presumably due to the improved C/N ratio. Nevertheless, the use of food or feed product-dedicated crops (e.g., cereals) or energy crops as AD plant feedstocks is limited to less than 15% of the total feed ration by the French national legislation [7]. Some exceptions are, however, still possible for catch crops and biomass cultivated on marginal lands that are not in conflict with food and feed production. Among these, miscanthus presents many advantages, including high biomass yields, low input requirements (i.e., water, fertilizers), prolonged soil cover, reduced soil disturbance, and increased soil carbon content [8,9]. This crop can also grow on polluted soils [10,11]. Few studies have used *Miscanthus x giganteus* as a co-substrate for manure AD. Moiceanu et al. (2016) [12] used miscanthus as a reference co-substrate to investigate the influence of different types of manure on biogas production.

Nonetheless, for most lignocellulosic biomasses, carbohydrate accessibility is limited and AD performance can be improved by pretreatment [13]. For example, a 170 °C hydrothermal pretreatment of miscanthus led to a 21% increase in biogas production [14]. In another study, Nges et al. (2016) [15] applied grinding, steam explosion, and acid and alkali pretreatments to *Miscanthus lutarioriparius*. The best result—i.e., 57% increase in methane production—was obtained with a mild alkaline pretreatment. Indeed, a high lignin content and lignin/polysaccharides links have been identified as main bottlenecks for lignocellulosic biomass AD [16]. Among the different kinds of pretreatment techniques (mechanical, biological, chemical) [17], alkali pretreatments have been recognized as the most efficient for degrading lignin [18,19]. Alkali pretreatments generally employ soda. Because digestates from agricultural AD plants are systematically used as organic fertilizers and returned to agricultural soils, sodium spreading into soils should be avoided. The aim of this study is to therefore investigate miscanthus alkali pretreatment with lime.

The first objective was to assess and optimize miscanthus pretreatment conditions compatible with a further application in dry AD (i.e., with low water content). In fact, high solid content pretreatments reduce waste generation, do not require a separation step before further processing, and reduce the environmental impact of the entire process [20]. In order to keep pretreatment costs as low as possible, the conditions were set to ambient temperature. The impact of lime concentration and pretreatment duration on the biochemical methane potential (BMP) of miscanthus was investigated using a response surface methodology. The second objective of this study is to evaluate the impact of a selected lime pretreatment of miscanthus on its batch co-digestion using cattle manure in an LBR. Startup performances, as well as methane production, are reported.

2. Results

2.1. Impact of Lime Pretreatment on BMP

The different pretreatment conditions were carried out at a TS content of 13%, CaO concentrations between 2.5 and 17.5%, and pretreatment durations of 1, 3, or 5 days. Two experimental designs were created consecutively: Design 1 (CaO concentrations between 7.5 and 17.5%, and durations of 1, 3, or 5 days) and Design 2 (CaO concentrations between 2.5 and 12.5%, durations of 1, 3, or 5 days). BMP tests were performed in duplicate using these pretreated substrates. Table 1 reports the BMP and first-order kinetics constant k values. The duplicates revealed a very good repeatability. In comparison with the BMP obtained for the non-pretreated biomass (158 ± 2 $NmL_{CH4}{\cdot}g_{VS}^{-1}$), the effect of the pretreatment was significant (p-value = 9.8×10^{-4}) and positive. An improvement in BMP was observed, ranging from +14% (for 15% CaO; 1 day) to +37% (for 5% CaO; 5 days) for the best-performing condition.

The adjustment of Equation (2) for estimating k was excellent over all experimental conditions ($R^2 > 0.97$; data not shown). With a focus on the first kinetics constant k, a strong and positive effect was also noticed as an improvement, ranging from +63% (for 10% CaO; 1 day) to +221% (for 17.5% CaO; 3 days). This was calculated by comparing k with that obtained for raw biomass (0.024 ± 0.002 $NmL_{CH4}{\cdot}g_{VS}^{-1}{\cdot}d^{-1}$). Even though a clear correlation did not emerge, the evolution of BMP and k presented similar trends, as the highest BMPs were mostly characterized by the highest k values. These results suggest that lime pretreatment of the miscanthus does, indeed, induce an increase for both BMP and k. Despite these positive observations, it was not possible to assess which parameter was most relevant. The effect of each parameter therefore needs to be unraveled using an experimental design.

Table 1. Pretreatment conditions, biochemical methane potential (BMP), and first-order kinetics constant values and their improvement, compared to raw, for Design 1 and Design 2.

Design 1	Variable A: [CaO]		Variable B: Duration		BMP		k	
Run	Coded	%	Coded	d	$NmL_{CH4}{\cdot}g_{VS}^{-1}$	Improvement *	d^{-1}	Improvement *
Raw	-	-	-	-	158 ± 2	-	0.024 ± 0.002	-
1	0	12.5	0	3	196	+24%	0.047 ± 0.001	+96%
1′	0	12.5	0	3	201	+27%	0.048 ± 0.001	+100%
2	1	17.5	0	3	191 ± 2	+21%	0.077 ± 0.005	+221%
3	0.5	15	0.866	5	207 ± 1	+31%	0.075 ± 0.004	+213%
4	−0.5	10	0.866	5	208 ± 4	+32%	0.051 ± 0.004	+113%
5	−1	7.5	0	3	202 ± 4	+28%	0.042 ± 0.001	+75%
6	−0.5	10	−0.866	1	199 ± 2	+26%	0.039 ± 0.002	+63%
7	0.5	15	−0.866	1	179 ± 13	+14%	0.050 ± 0.002	+108%
Design 2								
Raw	-	-	-	-	158 ± 2	-	0.024 ± 0.002	-
1	0	7.5	0	3	204	+29%	0.041 ± 0.001	+71%
1′	0	7.5	0	3	199	+26%	0.043 ± 0.001	+79%
2	1	12.5	0	3	199 ± 4	+26%	0.047 ± 0.001	+96%
3	0.5	10	0.866	5	208 ± 4	+32%	0.051 ± 0.004	+113%
4	−0.5	5	0.866	5	216 ± 1	+37%	0.050 ± 0.001	+108%
5	−1	2.5	0	3	193 ± 3	+22%	0.040 ± 0.003	+67%
6	−0.5	5	−0.866	1	193 ± 1	+22%	0.048 ± 0.003	+100%
7	0.5	10	−0.866	1	199 ± 2	+26%	0.042 ± 0.004	+75%

* by comparison with raw (i.e., non-pretreated) sample.

2.2. *Mathematical Models to Describe Impact of Concentration and Pretreatment Duration on BMP Values*

2.2.1. Experimental Design 1 (CaO Concentration from 7.5 to 17.5% and Duration from 1 to 5 Days)

The effect of variables A (CaO concentration) and B (pretreatment duration) on BMP and first-order kinetics constant k was investigated by statistical analysis based on response surfaces. According to Oliveira et al. (2015) [21], response surface methodology (RSM) is a collection of both mathematical and statistical techniques that involves (i) designing and carrying out experiments with a reduced investment; (ii) building models; (iii) evaluating the relative significance of the studied variables; and (iv) assessing the optimal conditions for a favorable response. Using data displayed in Table 1 and multiple regression analysis, a polynomial equation was determined to predict BMP and k depending on the variables, as well as their interactions (Equation (1)). The different coefficients with their standard deviation, the Fisher value (F-value), and the coefficient of determination R^2 of the models are provided in Table 2 for each design.

Table 2. Coefficients, Fisher value, and R^2 of the two designs.

Coefficient	Design 1		Design 2
	Y_{BMP}	Y_k	Y_{BMP}
a_0	198.5 (±2.7)	0.0474 (±0.0006)	201.5 (±2.3)
a_1	−7.2 (±2.2)	0.0173 (±0.0005)	1.5 (±1.9)
a_2	10.7 (±2.2)	0.0107 (±0.0005)	9.5 (±1.9)
a_{12}	10.4 (±4.4)	0.0076 (±0.0010)	−8.1 (±3.8)
a_{11}	−2.5 (±3.8)	0.0122 (±0.0009)	−5.8 (±3.3)
a_{22}	0.2 (±3.8)	0.0043 (±0.0009)	4.9 (±3.3)
F-value	1.33	0.22	0.75
R^2	0.952	0.999	0.950

The F-value for each case was far less than the Fisher parameter calculated at the 95% confidence level (161.45), thus indicating that the models were significant and fitted the data nicely. Joglekar and May [22] suggested that, for a good fit of a model, the R^2 should be higher than 0.8. The high R^2 ($\geq$0.95) obtained for each case was a strong hint of suitability, as it indicated that 95% of the data were explained by the regressions, even reaching 99.9% for the prediction of k in Design 1. Consequently, all the models were validated.

Standardized Pareto charts displaying the effects of the different terms of the models are provided for BMP in Figure 1a and for the first-order kinetics constant k in Figure 1b. The duration of the pretreatment (variable B) was the only parameter that significantly affected the BMP (Figure 1a). Moreover, the effect being positive, long contact times between CaO and miscanthus should favor methane production. Interestingly, the pretreatment duration also produced a very significant and positive effect on the first-order kinetics constant k, although this effect was less significant than the effect of CaO concentration (Figure 1b). In addition, both quadratic terms proved to be significant for the first-order kinetics constant k, thus implying that the influence of the variables was not necessarily linear. Finally, the interaction term A $\times$ B was also significant, with a positive effect on the first-order kinetics constant. This result could not have been anticipated by using a univariate approach. Although both BMP and k were related, it is noteworthy that both selected variables affected these responses at different levels (positive or negative effect) and with various extents of contribution.

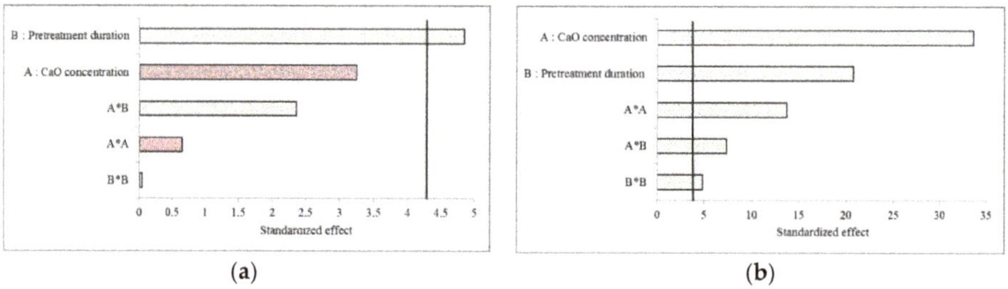

Figure 1. Pareto diagram showing the effect of different coefficient terms on BMP (**a**) and kinetics constant k (**b**) for Design 1. Red bars indicate a negative impact, and green bars show a positive impact. Bars exceeding the vertical line point to the significance of the coefficient terms ($p < 0.05$, corresponding to 4.3 according to Student t-test in our conditions).

Response surfaces were plotted in 3-D for each parameter (i.e., BMP and k) as a function of CaO concentration and pretreatment duration (Figure 2). The response surface plot for BMP (Figure 2a) led to the following observations: (i) the longer the duration of the pretreatment, the better the BMP, which is the same conclusion as that stated in a previous section; (ii) the more the CaO concentration increases within the experimental domain, the more the BMP decreases. Even if it was not significant, the negative effect of this variable could be anticipated from Figure 1. Consequently, in order to favor methane production, a combination of "low" CaO concentration (lower part of the experimental domain) with "high" contact time for pretreatment (upper part of the experimental domain) could be a viable option.

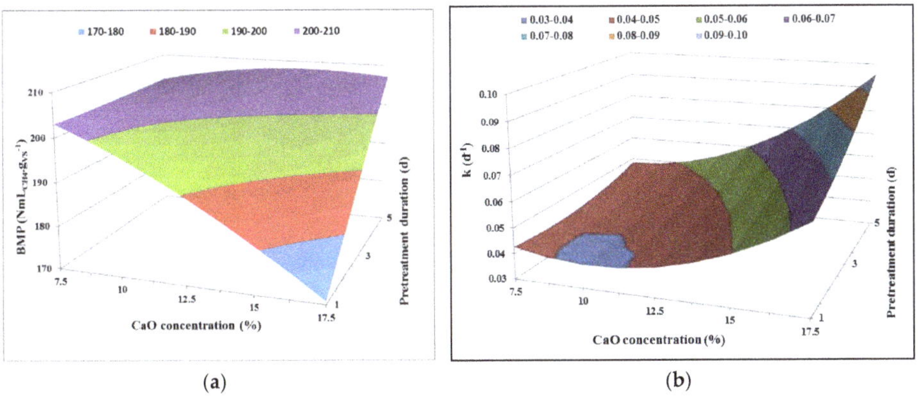

Figure 2. Response surface plots showing the impact of lime concentration and pretreatment duration on BMP (**a**) and kinetics values (**b**) for Design 1.

Focusing on the response surface plot for k (Figure 2b), the coupling of a "high" CaO concentration with a "long" pretreatment duration was linked to a high methane production rate. Nevertheless, the effect was more pronounced in the upper part of the domain (CaO concentration > 16% and pretreatment duration > 3 days) where few experimental data were generated (Table 1). The prediction thus rather relies on extrapolation rather than on interpolation. Therefore, the selection of the experimental domain to be exploited was made where the trend described by the model is well established. In this case, it is the middle-upper part of the domain that was chosen. Application of a "medium" CaO concentration with a "long" contact time pretreatment was hence considered most relevant to achieve a high first-order kinetics constant k.

As the responses for BMP and k reflect different impacts, it is difficult to find a consensus regarding the values of the variables to select (lime concentration and pretreatment duration) in order to optimize both parameters simultaneously. Moreover, no optima could be determined within the investigated domain. Only one extreme stationary point was identified in the lower domain for the kinetics constant k, which is irrelevant when both BMP and k need to be maximized. Owing to its energetic relevance, BMP is the parameter to prioritize in this study. The first-order kinetics constant k will therefore not be discussed anymore in the further section.

2.2.2. Experimental Design 2 (CaO Concentration from 2.5 to 12.5% and Duration from 1 to 5 Days)

According to the trend displayed in Figure 2a, the application of a lower CaO concentration could contribute to enhancement of the BMP and even lead to an optimal result. The experimental domain was thus extended to CaO concentrations ranging from 2.5 to 12.5% for a second design of experiments (DOE). Finally, and even though the duration of pretreatment had a positive effect on BMP, its experimental domain was not extended for the following reasons: (i) longer contact times would not be realistic for applications in full-scale plants; (ii) the extension of a Doehlert design with the reuse of certain points is only possible for one factor; (iii) the possible formation of refractory compounds, which could further impede methane production [23]; and (iv) the possible pre-degradation of accessible substrates by microorganisms already present in the bulk matrix, which could decrease the bioavailable fraction for methane production [24].

The results obtained for the second DOE are summarized in Table 2, and as a Pareto chart and response surface plot in Figure 3a,b. As depicted in Figure 3a, variable B (pretreatment duration) was the only one that significantly and positively impacted methane production, as was previously observed. The response surface (Figure 3b) confirmed this observation, as BMP increased with the application of longer contact times. Regarding variable A (CaO concentration), the effect was minor within the investigated domain (from 2.5 to 12.5%), even though the highest concentrations appear to lead to an increase in BMP. Unfortunately, it was not possible to determine an optimum in the second DOE domain. However, high responses for BMP can be observed in a region of interest. According to these results, the application of a long contact time seems necessary for enhancing methane production. A duration of 5 days for the "pretreatment duration" variable was thus selected. For the CaO concentrations, response surfaces revealed that the additional increase in BMP was negligible when a concentration above 10% was applied, while a detrimental effect was even possible for concentrations below 5% (Figure 3).

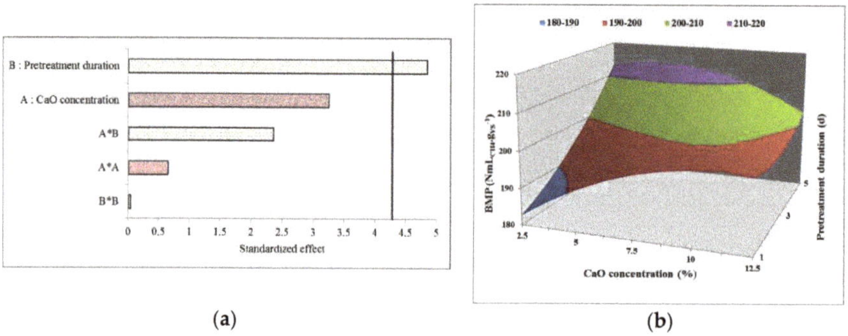

Figure 3. Pareto diagram showing the effect of different coefficient terms on BMP (**a**) and response surface plot showing the impact of lime concentration and pretreatment duration on BMP in Design 2 (**b**).

2.3. Co-Digestion

According to RSM conclusions and due to economic incentives, a concentration of 5% CaO was first selected. In order to confirm the trends displayed by RSM (e.g., increase in k with increasing CaO concentration), it was also worthwhile to consider a higher CaO concentration (10%). As a consequence, the following combinations were retained for application in an LBR at lab scale: 5% CaO during 5 days of pretreatment and 10% CaO during 5 days of pretreatment (performed in duplicate).

The concentration in volatile fatty acids (VFAs) in the leachate and the evolution of pH were measured each day at the beginning and on a regular basis thereafter (Figure 4). For duplicates, the VFA concentrations were similar (Figure 4a). VFAs were rapidly produced, and their maximum value was reached after 1 day for miscanthus pretreated with 5 and 10% of lime for 5 days (4.5 ± 0.5 g·L^{-1} and 4 g·L^{-1}, respectively). For raw miscanthus, the maximum was reached after 2 days (4.2 ± 0.1 g·L^{-1}). For all reactors, pH variations were similar (Figure 4b). The evolution of VFA and pH can be separated into two steps. During the first step, the accumulation of VFA during the first 3 days induces, with a brief delay, a slight decrease in pH down to 6.8. Two hypotheses could explain this observation: either (1) the positive impact of higher alkalinity due to CaO or (2) the buffer capacity of manure. The second hypothesis, manure buffer capacity, seems to be more plausible, given that in the case of co-digestion with the raw substrate, the pH is similar to that of the pretreated substrate. The pH did not present a sharp drop and the VFA concentration was not very high either. Moreover, since the VFA/alkalinity ratio remains below 1 (Table 3), the risk of acidification is avoided [25] and the drop in pH is swiftly reduced. During the second step, the pH stabilized close to 7.3 due to the decrease in VFA concentrations. After 15 days, there were no more VFAs accumulated; they were simultaneously produced and consumed at the same rate.

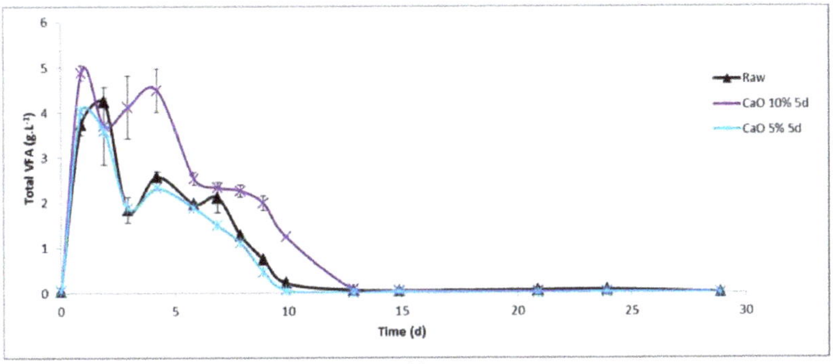

(a)

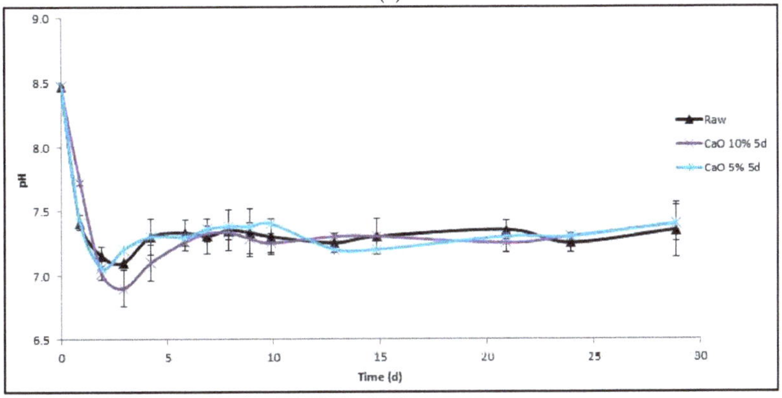

(b)

Figure 4. Volatile fatty acid (VFA) concentration (**a**) and pH variation (**b**) in leachate during the first 30 days.

Table 3. VFA/alkalinity ratio at 3 days, methane production at 6, 10, 15, 24, 29, and 59 days, expected methane production calculated from BMP values, and first-order kinetics constants.

	VFA/Alkalinity ($gHAceq \cdot g_{CaCO3}{}^{-1}$) [1]	Methane Production ($NmL_{CH4} \cdot g_{VS}{}^{-1}$)							k (d^{-1})
day	3	6	10	15	24	38	59	Expected [2]	-
raw	0.35 ± 0.07	36 ± 1	59 ± 4	78 ± 7	109 ± 3	135 ± 4	158 ± 4	181	0.040 ± 0.004
10 % 5 days	0.8 ± 0.14	43 ± 2	69 ± 3	92 ± 3	119 ± 2	145 ± 2	167 ± 2	208	0.049 ± 0.002
5 % 5 days	0.4	43	67	84	109	131	150	207	0.054

[1] HAceq means acetic acid equivalent, [2] from BMP values.

Methane production from co-digestion of raw miscanthus and miscanthus pretreated with lime at 5 and 10% for 5 days with cattle manure was 158 ± 4, 150, and 167 ± 2 $mL \cdot g_{VS}{}^{-1}$ after 59 days, respectively (Table 3). The 59-day period was selected because it corresponds to the duration of a batch in industrial plants. For this time span, there is no significant difference in the methane production between the control and the two different conditions of pretreatment (p-value = 0.7, 0.82) nor between both pretreatments (p-value = 0.82). However, after a shorter time, differences can be observed (Table 3). After 15 days of anaerobic co-digestion in an LBR, the methane production was higher for the miscanthus pretreated at 10% than for the raw miscanthus (+18%). It is related to a higher kinetics constant (+23%).

3. Discussion

3.1. BMP and Pretreatment

The BMP value of unpretreated miscanthus was 153 ± 7 $NmL_{CH4}\cdot g_{VS}^{-1}$. This value lies within the lowest range of published miscanthus BMP values (170 $mL_{CH4}\cdot g_{VS}^{-1}$ [14] to 227 $mL_{CH4}\cdot g_{VS}^{-1}$ [26]). This low methane potential is most certainly linked to the high lignin content of the Floridulus clone [27]. Alkaline pretreatment may therefore be relevant for improving methane production from this clone.

Lime pretreatments have been far less studied than NaOH pretreatments, although sodium has detrimental effects on agricultural soils when digestates are used as biofertilizers. In addition, miscanthus has been scarcely employed in AD studies, which are extensively dedicated to agricultural residues such as rice straw, sugarcane bagasse, corn stover, wheat straw, or other energy crops, such as switchgrass [28]. To the best knowledge of the authors, there has not yet been any study on the impact of lime pretreatment on *Miscanthus x giganteus* BMP. Moreover, what makes this study innovative is that low inputs were set for the pretreatment conditions (low water input with high solid concentration and no heat energy input with ambient temperature conditions). In particular, for humidifying the entire biomass, a minimum amount of water was used. This corresponds to only 13% TS due to the high absorption capacity of miscanthus. Peces et al. (2015) [29] clearly demonstrated that total solids content is a significant parameter for the performance of sonication pretreatment, although it has been rarely considered in pretreatment optimization procedures. However, the doses of lime are consistent with those applied to other types of herbaceous biomass. For example, Jiang et al. (2017) [28] pretreated giant reed biomass with 1, 3, 5, 7, 12, and 20% ($g_{Ca(OH)2}\cdot g_{initialTS}^{-1}$) at 25 °C for 24 h. They observed an increase in methane yields between 7 and 34%. Another study obtained a 23% improvement of BMP with sunflower straw pretreated at 30 °C with 4% ($g_{Ca(OH)2}\cdot 100g_{TS}^{-1}$) for 1 day [18].

Although good performances have been achieved with lime, compared to potassium and sodium hydroxide at the same molarity, it has proven to be significantly less efficient for delignification [30]. Results indicate that with an equivalent molar basis of OH^-, potassium and sodium hydroxide have a performance that is superior to calcium hydroxide [30].

The efficiency of pretreatments also depends upon the substrate. Indeed, a low lignin content is the main factor in promoting enhanced enzymatic saccharification [19] or enhancing anaerobic digestion [16]. Miscanthus is also widely used as animal bedding due to its high absorption capacity [31]. Thus, for an equivalent biomass TS content, less free water would be available with miscanthus than with other types of biomass. These, associated with a low lime solubility (1.65 $g\cdot L^{-1}$ at 20 °C, corresponding to 5.5% in this study), could reduce the amount of lime in contact with the substrate. In addition, Boix et al. (2016) [32] demonstrated that the absorption capacity of miscanthus increases with alkaline treatment. This can be explained by the removal of hydrophobic compounds, due to more exposed OH groups from cellulose or hemicellulose on the stem surface.

3.2. Pretreatment and Co-Digestion

Pretreatments are a promising solution in BMP test series. However, the performances in the LBR could not be confirmed if the methane production was estimated at 59 days. The BMP of manure used for the experiment was 202 ± 30 $NmL_{CH4}\cdot g_{VS}^{-1}$, which was higher than the BMP of raw miscanthus and within a similar range to pretreated miscanthus BMP. The maximum expected methane production with a ratio of 40%$_{VS}$ miscanthus and 60%$_{VS}$ manure is presented in Table 3. While 87% of the expected methane production was reached after 59 days co-digestion with raw miscanthus, 75 and 80% of expected methane production were obtained with miscanthus pretreated at 5 and 10%, respectively. Riggio et al. (2016) [4] carried out the process using an LBR fed with spent cow bedding. They obtained 168 $NmL_{CH4}\cdot g_{VS}^{-1}$ after 60 days, which represents 86% of the BMP value. Thus, the overall performance of the LBR evaluated in this study is satisfactory, although the small increase in methane production at 59 days remains surprising. Dry anaerobic digestion inoculum might require some adaptation to the pretreated biomass. Another explanation could be the high water absorption capacity of miscanthus. As BMP tests were carried out in diluted medium (5 $g_{TSmiscanthus} \cdot L^{-1}$), the high amount of available water can favor contact between lime and biomass. This could enhance the action of lime if it continues during the AD process.

The improvement in methane production after a 10% pretreatment was quite low (6%). This was related to a higher VFA production at the beginning of the AD run (Figure 4a), thus revealing that, unlike a 5% pretreatment, the 10% pretreatment can lead to the release of easily biodegradable matter. Nevertheless, this increase in methane production is not sufficient to justify the application of a full-scale pretreatment.

4. Materials and Methods

4.1. Miscanthus

Miscanthus. x giganteus Floridulus was grown in the North of France (49°53 N, 3°00 E) [27] at the INRA experimental unit of Estrées-Mons and harvested in winter 2015 during its eighth year. The soil is a deep loam soil (Orthic Luvisol, Roma, Italia, FAO classification). The climate is oceanic. The stems were dried at 64 °C for 4 days in a ventilated oven and ground with a crusher (Viking, model GE 220, STIHL, Stuttgart, Germany) to a coarse size (around 6 cm). The TS and volatile solids (VS) content were 94% and 98%$_{TS}$, respectively.

4.2. Experimental Design

To assess the effect of CaO pretreatment on BMP and the first-order kinetics constant k, a two-factor Doehlert-type uniform network was used to define the experimental matrix. The principle and strengths of such a design is described by Goupy et al. (2014) [33] and by Witek-Krowiak et al. (2014) [34]. Briefly, it consists of a two-variable (z = 2) Doehlert design and requires $N = z^2 + z + C$ experiments, with z as the number of variables and C as the number of center points. Here, N was equal to $2^2 + 2 + 1 = 7$. The center point was repeated once. As the experiments were performed in duplicate for each condition, the total number was 16. The two variables of interest (called factors) were defined as the CaO concentration applied for pretreatment (variable A) and the duration of pretreatment (variable B). The ranges to be studied for both factors were selected based on literature, sound reasoning, and preliminary experiments carried out at the laboratory. Thus, for variable (A), the range was between 7.5 and 17.5% (% mean g_{CaO} per 100 g_{TS}). For variable (B), the range was between 1 and 5 days. As no optimum was found within the investigated domain, the range for variable (A) was extended to a second set of experiments and defined between 2.5 and 12.5%. The experimental domains, expressed as coded (±0, 0.5, 0.866, and 1) and real values for each factor, are listed in Table 1 for both designs.

A full second-order polynomial equation was used to model the values obtained for BMP and the first kinetics constant k as a function of the applied lime concentration (A) and the duration of the pretreatment (B). The system can be described by the following equation (Equation (1)):

$$Y = a_0 + a_1 A + a_2 B + a_{12} AB + a_{11} A^2 + a_{22} B^2 \quad (1)$$

where Y is the BMP or the first-order kinetics constant k, a_0 is the constant term, a_1 and a_2 are the linear coefficients associated with each variable, a_{12} is the coefficient associated with the interaction between both variables, and a_{11} and a_{22} are the quadratic coefficients. A detailed calculation of the coefficients is already available in the literature [35].

The model was validated using a Fischer test. The significance of each coefficient in the model was tested using a Student's *t*-test [35]. The results were then compared using standardized effects in a Pareto chart.

4.3. Alkaline Pretreatments

The pretreatments were carried out at ambient temperature, without mixing, and in duplicate in 500 mL flasks using lime (CaO, Akdolit® Q90; purity $\geq$ 92%, Paris, France) and 2 g_{TS} of miscanthus in conditions reported in Table 1. Another originality of this study is the high TS loading (130 $g \cdot L^{-1}$) selected to test conditions with low water input.

4.4. Measure of Methane Potential

All pretreated samples (solid and liquid fractions altogether) were digested in a 500 mL flask with a working volume of 400 mL. Bicarbonate buffer ($NaHCO_3$, 50 $g \cdot L^{-1}$), macroelement and oligoelement solutions, anaerobic sludge at 5 g_{VS} L^{-1}, and the substrate at 5 $g_{TS} \cdot L^{-1}$ were added [36]. Degasification with nitrogen was carried out to obtain anaerobic conditions. Duplicate bottles were incubated at 35 °C for 60 days.

4.5. Methane Production Kinetics

All methane potentials are expressed in $NmL_{CH4} \cdot g_{initialVS}{}^{-1}$. Thus, the eventual losses of organic matter during pretreatments are included in the results. To quantify the impact of pretreatment on the kinetics of methane production, the first-order kinetic constants were calculated by using the least-squares fit of methane production data versus time (t) to the following equation:

$$V = Vmax\left(1 - e^{-kt}\right) \quad (2)$$

where V is the volume of methane ($NmL_{CH4} \cdot g_{VS}{}^{-1}$), V_{max} the maximum producible methane volume ($NmL_{CH4} \cdot g_{VS}{}^{-1}$), k the first-order kinetics constant (d^{-1}), and t is the digestion time (d). V_{max} and k were determined using the Microsoft Excel Solver function.

4.6. Leach Bed Reactors

In order to represent farm batch dry anaerobic digestors used on farms, LBR systems were employed for these experiments. They were previously used and described by Riggio et al. [4]. Experiments were carried out in a 6 L LBR fed with 300 g_{TS} of substrate and inoculum and 1.1 L leachate. The substrate was composed of 85% (in wet weight basis) manure and 15% miscanthus (corresponding to 40%$_{VS}$). Cow manure from wheat straw bedding was collected from a dairy farm in the South of France and stored at −20 °C. Before feeding the reactors, two different lime pretreatments were applied to miscanthus at room temperature, with low water addition (to reach 13%$_{TS}$) and no mixing: 5 and 10% for 5 days. The inoculum used came from a previous experiment. It was composed of a mix of digested manure, miscanthus, and raw sorghum and kept at 35 °C. The leachate originated from a previous experiment and was also maintained in mesophilic conditions. It did not contain VFAs and was diluted with water and buffered with $NaCO_3$ at 1.3 $g \cdot L_{addedwater}{}^{-1}$. The substrate/inoculum ratio was 6 ($g_{VS} \cdot g_{VS}{}^{-1}$) and the TS content of the solid fraction in the reactor (including miscanthus, manure, and inoculum) was 19%. Taking the leachate volume into account, the overall TS content was 12%. The pretreated substrate at 10% CaO for 5 days and the raw substrate (control) were digested in the LBR in duplicate, whereas only one reactor was run for the other pretreatment. Degasification with nitrogen was carried out to obtain anaerobic conditions.

4.7. Analysis

The TS and VS contents were measured according to standard methods [37]. The leachate used was characterized in terms of alkalinity, VFA concentration, and pH. According to the APHA method, alkalinity was performed by 0.1 N hydrochloric acid titration [37]. VFAs were analyzed in a Perkin Elmer Clarus 580 gas chromatographer with helium as the gas vector [38]. The pH was measured with WTW pH-electrode SenTix 41 connected to a WTW inoLab pH 7110 operational manual transmitter.

Biogas flow rate from reactors was recorded every 5 min with the use of a lab-made software connected to a Ritter Milligascounter MGC-1 V3. Biogas volume in BMP tests was monitored using a manometric device (LEO 2, KELLER) and biogas composition was determined by gas chromatography as described in Sambusiti et al. (2012) [39].

4.8. Statistical Analysis

For results obtained from the DOE, Wilcoxon tests were performed using R software (version 3.2, R Development Core Team, R: A Language and Environment for Statistical Computing, R Foundation for Statistical Computing, Austria, Vienna, 2004, ISBN 3-900051-07-0.). Effects were considered to be significant for p-values < 0.05.

Author Contributions: Conceptualization, H.C.; Formal analysis, H.L.T. and J.S.; Investigation, H.L.T.; Methodology, R.E. and H.C.; Writing—original draft, H.L.T.; Writing—review & editing, J.S., R.E. and H.C.

Funding: The authors acknowledge the French National Research Agency for funding the "Biomass For the Future" project (ANR, grant ANR-11-BTBR-0006-BFF).

Acknowledgments: Acknowledgements are addressed to Brancourt and Arnoult (INRA Estrée-Mons and Péronne) who provided miscanthus samples. The authors thank Théo Closet for his involvement in the experiments during his internship.

Conflicts of Interest: The authors declare no conflict of interest.

References

1. Torrijos, M. State of Development of Biogas Production in Europe. *Procedia Environ. Sci.* **2016**, *35*, 881–889. [CrossRef]
2. Deguerce, A.; Capdeville, J.; Perrot, C.; Bioteau, T.; Martinez, J.; Peu, P. Fumiers de bovins, une ressource à fort potentiel pour la filière de méthanisation en France? *Revue Science Eaux & Territoires*, article hors-série, 22 February 2016. Available online: http://www.set-revue.fr/fumiers-de-bovins-une-ressource-fort-potentiel-pour-la-filiere-de-methanisation-en-France (accessed on 30 June 2018).
3. Mata-Alvarez, J.; Macé, S.; Llabrés, P. Anaerobic digestion of organic solid wastes. An overview of research achievements and perspectives. *Bioresour. Technol.* **2000**, *74*, 3–16. [CrossRef]
4. Riggio, S.; Torrijos, M.; Debord, R.; Esposito, G.; van Hullebusch, E.D.; Steyer, J.P.; Escudié, R. Mesophilic anaerobic digestion of farmyard manure of different origins in a batch leach-bed reactor: Substrate characterization and process performance. *Waste Manag.* **2016**, *59*, 129–139. [CrossRef] [PubMed]
5. Tufaner, F.; Avsar, Y. Effects of co-substrate on biogas production from cattle manure: A review. *Environ. Sci. Technol.* **2016**, *13*, 2303–2312. [CrossRef]
6. Botji, T.; Kovàcs, K.L.; Kakuk, B.; Wirth, R.; Ràkhely, G.; Bagi, Z. Pretreatment of poultry manure for efficient biogas production as monosubstrate or co-fermentation with maize silage and corn stover. *Anaerobe* **2017**, *46*, 138–145. [CrossRef]
7. Assemblée Nationale. *Décret n° 2016-929 du 7 Juillet 2016 pris pour L'application de L'article L. 541-39 du code de L'environnement*; Journal officiel de la République Française du 8 juillet 2016 texte n°8; Journal officiel de la République Française: Paris, France, 2016.
8. Perrin, A.; Wohlfahrt, J.; Morandi, F.; Østergård, H.; Flatberg, T.; De La Rua, C.; Bjørkvoll, T.; Gabrielle, B. Integrated design and sustainable assessment of innovative biomass supply chains: A case-study on miscanthus in France. *Appl. Energy* **2017**, *204*, 66–77. [CrossRef]
9. Morandi, F.; Perrin, A.; Østergård, H. Miscanthus as energy crop: Environmental assessment of a miscanthus biomass production case study in France. *J. Clean. Prod.* **2016**, *137*, 313–321. [CrossRef]
10. Muller-Stover, D.S.; Sun, G.; Kroff, P.; Thomsen, S.T.; Hauggaard-Nielsen, H. Anaerobic co-digestion of perennials: Methane potential and digestate nitrogen fertilizer value. *J. Plant Nutr. Soil Sci.* **2016**, *179*, 696–704. [CrossRef]
11. Kiesel, A.; Wagner, M.; Lewandowski, I. Environmental Performance of Miscanthus, Switchgrass and Maize: Can C4 Perennials Increase the Sustainability of Biogas Production? *Sustainability* **2016**, *9*, 5. [CrossRef]

12. Moiceanu, G.; Voicu, G.; Ferdes, M.; Paraschiv, G.; Dincà, M.N.; Vlàdut, V.; Gageanu, I.; Chitoiu, M. Comparative study on biogas production using cow and swine manure mixed with *Miscanthus x giganteus* as substrate. *Rom. Biotechnol. Lett.* **2016**, *21*, 11968–11973.
13. Monlau, F.; Barakat, A.; Trably, E.; Dumas, C.; Steyer, J.P.; Carrère, H. Lignocellulosic Materials into Biohydrogen and Biomethane: Impact of Structural Features and Pretreatment. *Crit. Rev. Environ. Sci. Technol.* **2013**, *43*, 260–322. [CrossRef]
14. Zhou, X.; Li, Q.; Zhang, Y.; Gu, Y. Effect of hydrothermal pretreatment on Miscanthus anaerobic digestion. *Bioresour. Technol.* **2017**, *224*, 721–726. [CrossRef] [PubMed]
15. Nges, I.A.; Chao, L.; Wang, B.; Xiao, L.; Yi, Z.; Liu, J. Physio-chemical pretreatments for improved methane potential of *Miscanthus lutarioriparius*. *Fuel* **2016**, *166*, 29–35. [CrossRef]
16. Monlau, F.; Sambusiti, C.; Barakat, A.; Guo, X.M.; Latrille, E.; Trably, E.; Steyer, J.P.; Carrère, H. Predictive models of biohydrogen and biomethane production based on the compositional and structural features of lignocellulosic materials. *Environ. Sci. Technol.* **2012**, *46*, 12217–12225. [CrossRef] [PubMed]
17. Agbor, V.B.; Cicek, N.; Sparling, R.; Berlin, A.; Levin, D.B. Biomass pretreatment: Fundamentals toward application. *Biotechnol. Adv.* **2011**, *29*, 675–685. [CrossRef] [PubMed]
18. Monlau, F.; Barakat, A.; Steyer, J.P.; Carrere, H. Comparison of seven types of thermo-chemical pretreatments on the structural features and anaerobic digestion of sunflower stalks. *Bioresour. Technol.* **2012**. [CrossRef] [PubMed]
19. Kim, J.S.; Lee, Y.Y.; Kim, T.H. A review on alkaline pretreatment technology for bioconversion of lignocellulosic biomass. *Bioresour. Technol.* **2016**, *199*, 42–48. [CrossRef] [PubMed]
20. Barakat, A.; de Vries, H.; Rouau, X. Dry fractionation process as an important step in current and future lignocellulose biorefineries: A review. *Bioresour. Technol.* **2013**, *134*, 362–373. [CrossRef] [PubMed]
21. Oliveira, J.V.; Alves, M.M.; Costa, J.C. Optimization of biogas production from *Sargassum* sp. using a design of experiments to assess the co-digestion with glycerol and waste frying oil. *Bioresour. Technol.* **2015**, *175*, 480–485. [CrossRef] [PubMed]
22. Joglekar, A.; May, A. Product excellence through design of experiments. *Cereal Foods World* **1987**, *32*, 857–868.
23. Carlsson, M.; Lagerkvist, A.; Morgan-sagastume, F. The effects of substrate pre-treatment on anaerobic digestion systems: A review. *Waste Manag.* **2012**, *32*, 1634–1650. [CrossRef] [PubMed]
24. Sole-Bundo, M.; Carrere, H.; Garfi, M.; Ferrer, I. Enhancement of microalgae anaerobic digestion by thermo-alkaline pretreatment with lime (CaO). *Algal Res.* **2017**, *24*, 199–206. [CrossRef]
25. Rouches, E. *Fungal Pretreatment for Lignocellulosic Biomass Anaerobic Digestion*; Université de Montpellier: Montpellier, France, 2015.
26. Fu, S.; Chen, K.; Zhu, R.; Sun, W.; Zou, H.; Guo, R. Improved anaerobic digestion performance of *Miscanthus floridulus* by different pretreatment methods and preliminary economic analysis. *Energy Convers. Manag.* **2018**, *159*, 121–128. [CrossRef]
27. Arnoult, S.; Obeuf, A.; Béthencourt, L.; Mansard, M.C.; Brancourt-Hulmel, M. *Miscanthus* clones for cellulosic bioethanol production: Relationships between biomass production, biomass production components, and biomass chemical composition. *Ind. Crops Prod.* **2015**, *63*, 316–328. [CrossRef]
28. Jiang, D.; Ge, X.; Zhang, Q.; Zhou, X.; Chan, Z.; Keener, H.; Li, Y. Comparison of sodium hydroxide and calcium hydroxide pretreatments of giant reed for enhanced enzymatic digestibility and methane production. *Bioresour. Technol.* **2017**, *244*, 1150–1157. [CrossRef] [PubMed]
29. Peces, M.; Astals, S.; Mata-alvarez, J. Effect of moisture on pretreatment efficiency for anaerobic digestion of lignocellulosic substrates. *Waste Manag.* **2015**, *46*, 189–196. [CrossRef] [PubMed]
30. Inês, C.; Rodrigues, S.; Ph, C.; Jackson, J.J.; Montross, M.D. A molar basis comparison of calcium hydroxide, sodium hydroxide, and potassium hydroxide on the pretreatment of switchgrass and miscanthus under high solids conditions. *Ind. Crops Prod.* **2016**, *92*, 165–173. [CrossRef]
31. Valle, I.C.; Kandula, D.; Littlejohn, C.; Hill, R.; Walker, M.; Shields, M.; Cummings, N.; Hettiarachchi, D.; Wratten, S. Potential of the beneficial fungus *Trichoderma* to enhance ecosystem-service provision in the biofuel grass *Miscanthus x giganteus* in agriculture. *Sci. Rep.* **2016**, *6*, 25109. [CrossRef]
32. Boix, E.; Georgi, F.; Navard, P. Influence of alkali and Si-based treatments on the physical and chemical characteristics of miscanthus stem fragments. *Ind. Crops Prod.* **2016**, *91*, 6–14. [CrossRef]
33. Goupy, J. *Modelisation Par Les Plans D'Expériences*; Techniques de L'ingénieur: Paris, France, 2014; Volume 33, pp. 1–23.

34. Witek-krowiak, A.; Chojnacka, K.; Podstawczyk, D.; Dawiec, A.; Pokomeda, K. Application of response surface methodology and artificial neural network methods in modelling and optimization of biosorption process. *Bioresour. Technol.* **2014**, *160*, 150–160. [CrossRef] [PubMed]
35. Rossi, C.; Haupt, K. Application of the Doehlert experimental design to molecularly imprinted polymers: Surface response optimization of specific template recognition as a function of the type and degree of cross-linking. *Anal. Bioanal. Chem.* **2007**, *389*, 455–460. [CrossRef] [PubMed]
36. Thomas, H.L.; Pot, D.; Latrille, E.; Trouche, G.; Bonnal, L.; Bastianelli, D.; Carrère, H. Sorghum Biomethane Potential Varies with the Genotype and the Cultivation Site. *Waste Biomass Valoriz.* **2017**, 1–6. [CrossRef]
37. American Public Health Association (APHA). *Standard Methods for the Examination of Water and Wastewater;* American Public Health Association: Washington, DC, USA, 1998.
38. Cazier, E.A.; Trably, E.; Steyer, J.P.; Escudie, R. Biomass hydrolysis inhibition at high hydrogen partial pressure in solid-state anaerobic digestion. *Bioresour. Technol.* **2015**, *190*, 106–113. [CrossRef] [PubMed]
39. Sambusiti, C.; Ficara, E.; Malpei, F.; Steyer, J.P.; Carrère, H. Influence of alkaline pre-treatment conditions on structural features and methane production from ensiled sorghum forage. *Chem. Eng. J.* **2012**, *211–212*, 488–492. [CrossRef]

Sample Availability: Samples of the compounds are not available from the authors.

molecules

Review

An Overview of Current Pretreatment Methods Used to Improve Lipid Extraction from Oleaginous Microorganisms

Alok Patel, Fabio Mikes and Leonidas Matsakas *

Biochemical Process Engineering, Division of Chemical Engineering, Department of Civil, Environmental and Natural Resources Engineering, Luleå University of Technology, 971-87 Luleå, Sweden; alok.kumar.patel@ltu.se (A.P.); fabio.mikes@ltu.se (F.M.)

* Correspondence: leonidas.matsakas@ltu.se; Tel.: +46-(0)920-493043

Received: 7 June 2018; Accepted: 26 June 2018; Published: 28 June 2018

Abstract: Microbial oils, obtained from oleaginous microorganisms are an emerging source of commercially valuable chemicals ranging from pharmaceuticals to the petroleum industry. In petroleum biorefineries, the microbial biomass has become a sustainable source of renewable biofuels. Biodiesel is mainly produced from oils obtained from oleaginous microorganisms involving various upstream and downstream processes, such as cultivation, harvesting, lipid extraction, and transesterification. Among them, lipid extraction is a crucial step for the process and it represents an important bottleneck for the commercial scale production of biodiesel. Lipids are synthesized in the cellular compartment of oleaginous microorganisms in the form of lipid droplets, so it is necessary to disrupt the cells prior to lipid extraction in order to improve the extraction yields. Various mechanical, chemical and physicochemical pretreatment methods are employed to disintegrate the cellular membrane of oleaginous microorganisms. The objective of the present review article is to evaluate the various pretreatment methods for efficient lipid extraction from the oleaginous cellular biomass available to date, as well as to discuss their advantages and disadvantages, including their effect on the lipid yield. The discussed mechanical pretreatment methods are oil expeller, bead milling, ultrasonication, microwave, high-speed and high-pressure homogenizer, laser, autoclaving, pulsed electric field, and non-mechanical methods, such as enzymatic treatment, including various emerging cell disruption techniques.

Keywords: oleaginous microorganisms; lipid extraction; pretreatment; cell disruption

1. Introduction

Over the last decades, the production of biofuels from renewable sources has gained more attention due to critical environmental issues, such as greenhouse gas emission, rapid depletion of fossil fuel supplies, and high energy cost [1]. Microbial oils are found to be a good option to produce biofuels, as many microorganisms, such as microalgae, yeast, bacteria, and fungi have the ability to accumulate oils under special cultivation conditions [2]. Moreover, microbial sources of lipids have many advantages over other sources, including higher lipid productivity in terms of g/L/day, being unaffected by any seasonal climate changes, low labor intensiveness, and easily scale-up [3,4]. The production costs are the major limiting factor to utilize microbial oils for biodiesel production, since the feedstocks to cultivate microorganisms account for 60% to 80% of the overall production cost [5]. To enable the commercial production of microbial lipids, these costs must be reduced by using low-cost feedstocks [5]. Oleaginous microorganisms can utilize various types of organic carbons, regardless of their origin, to accumulate oils in their cytoplasm. They have the specific ability to grow well on inexpensive agricultural waste and industrial by-products [6,7]. Conversion of microbial oil into biodiesel is involving four important upstream and downstream processes, i.e., cultivation, harvesting, lipid extraction, and transesterification [3]. However, the cell disruption process, including lipid extraction from oleaginous microorganisms, is costly and considered as a major bottleneck to produce biodiesel in large scale [8]. The lipids are synthesized intracellularly, which makes downstream processing more problematic for lipid recovery in lab or large-scale [9]. Lipid extraction is usually carried out after the disintegration of cells by pretreatment methods, followed by lipid recovery with organic solvents from the lysed biomass [10]. The cell disruption is an energy-intensive process requiring drying/dewatering of the biomass, which makes the overall process costly [11]. Conventional methods for lipid extraction, such as the Bligh & Dyer as well as the Folch method, involve the use of mixtures of chloroform and methanol which are suitable only for lab-scale [12]. Other problems that are associated with these methods, such as extraction from dry biomass and the use of harmful organic solvents, are also to be taken into account when trying to improve the efficiency of cell disruption [12,13]. Currently, various mechanical, chemical, and enzymatic pretreatment methods are employed to disrupt oleaginous microorganisms on a laboratory scale. These methods include microwave irradiation, ultrasonication, high-speed homogenization, high-pressure homogenization, bead beating, autoclaving, and thermolysis [14]. However, none of these pretreatment methods is effective in higher scale processes [15]. At commercial scale, lipid extraction is usually achieved through a solvent system, where the biomass should be in dry condition, otherwise, the organic solvents cannot establish contact with cells and they remain in the water phase due to their surface charges [16]. However, when considering the expenses involved in drying biomass, the lipid extraction should be done in wet conditions [17]. Disruption of cells by other means also has certain limitations regarding efficiency and lipid yield. For example, extraction of oils from seeds is usually carried out by simple mechanical methods, such as oil press or expeller press, and these methods are also applicable to extract the oils from microalgae. Yet, there is no report for this approach that is used in the extraction of lipids from oleaginous bacteria and yeast [18,19]. For the mentioned techniques, high mechanical pressure is usually applied on the dried biomass to squeeze the oils from cells, but the applied pressure generates excessive heat that can clog the machinery [20]. Although oil press and expeller press are cost-effective methods and they work well with samples of low moisture content, the explored biomass should be moisture-free, otherwise, lipids may pass through the pressed cake [9]. Moreover, the recovery of lipids is not yet sufficient and the drying of biomass again results in high energy and cost demands [21]. Bead beating, another mechanical method, eliminates the drying step, which in turn decreases the overall cost of extraction. In this approach, the wet slurry of biomass spins in a speed rotator that is loaded with fine beads. Since bead beating is only suitable for small amounts of sample, the application on a larger scale is yet again found to be difficult [22].

The problems that are associated with conventional methods can be solved with other physical methods, such as microwave irradiation and ultrasonication. Ultrasonication is one of the most extensively used pretreatment method to disrupt the cellular integrity of oleaginous microorganisms. This technique involves the use of mild pressures and temperatures, which makes the method simple, eco-friendly, and less time-consuming. Moreover, it can be operated without using any beads or chemicals. However, one important weak point of this technique is the generation of free radicals after prolonged treatment, which might have a detrimental effect on the quality of the extracted lipids [23]. Besides ultrasonication, microwave treatment is also a commonly used technique to disrupt cells and extract lipids from oily seeds, and it was already applied in the mid-1980s. Microwaves usually affect dielectric or polar particles within the cells, where a high amount of heat is generated during friction of inter- and intra-molecular movement of particles. The vapors generated due to the presence of water in intracellular compartments exert pressure on the cell wall, therefore leading to cell disruption. In this way, microwave irradiation makes membranes porous and plays a significant role in the lipid extraction process. However, this method requires high electricity expenditures that lead to high cost when being applied on a commercial scale [24].

Hence, in order to find a suitable and cost-effective alternative to mechanical methods, many researchers have been involved in replacing them with biological methods. For example, Jin et al. (2012) used recombinant β-1,3-glucomannanase plMAN5C enzymes to disintegrate the cell wall of oleaginous yeast *Rhodosporidium toruloides* Y4 [25].

Recently published literature is focused mainly on the microalgal biorefinery, including pretreatment that is involved in the lipid extraction process. Hence, in the present review article, we are focusing on the various pretreatment techniques employed to improve the lipid extraction process from different types of oleaginous cellular biomass, such as microalgae, yeast, fungi, and bacteria. The discussed pretreatment methods are mechanical methods, such as expression or expeller press, high-pressure homogenization, high-speed homogenization, bead milling, ultrasonication, microwave, autoclave, acid-catalyzed hot-water, laser, and pulsed electric field treatment. Besides these techniques, some other non-mechanical pretreatments are also discussed here.

2. Microbial Cell Wall and Lipid Composition

Cell disruption is the process of breaking indehiscent bacterial cells and cell wall structures of eukaryotic microorganisms, such as yeast, algae, and fungi [3]. The structure of the cell wall varies with the type of microorganism and the given growth conditions. Knowledge of the cell wall structure of a microorganism helps with the selection of a suitable pretreatment method to disrupt its cellular integrity. Disruption of yeast cell walls is more straightforward when compared to bacterial cells due to their larger cell size and a unique cell wall structure [22]. The cell wall of yeasts contains mainly glucans, mannans, and proteins and the overall structure is thicker than in gram-positive bacteria [26]. Pomraning et al. (2015) suggested that oleaginous yeasts start to synthesize lipid droplets in their compartment after 60 to 72 h of growth and that a significant change in the thickness of cell walls can be observed under a transmission electron microscope. Older cells have thicker cell walls than younger ones [26]. Microalgae are also characterized by a thicker cell wall structure made up of complex carbohydrates and glycoproteins. Jiang et al. (2018) observed the cellular ultrastructure of *Chlorella sorokiniana* SDEC-18 under a transmission electron microscope and revealed that the plasma membrane is surrounded by a thick cell wall [27].

Oleaginous microorganisms synthesize various kinds of lipid classes in their cellular compartment, which, according to the polarity of their head groups, can be classified as neutral lipids that are acting as energy storage (triacylglycerols, free fatty acids, sterols, sterols esters, waxes, and hydrophobic pigments), and polar lipids that are enabling membrane integrity (phospholipids, glycolipids, polysaccharides, and lipoproteins) [28]. The major proportion of total lipids are triacylglycerides with long-chain fatty acids similar to plant oils, making them comparable to conventional vegetable oil [29]. Triacylglycerols (TAG) are fatty acid triesters of glycerol. There are diverse types of TAG with different properties depending on their fatty acid composition [30].

The occurrence of TAG as storage compounds is widespread among eukaryotic organisms, such as microalgae, yeast, fungi, plants, and animals, whereas the occurrence of TAG in bacteria has only rarely been described [31]. However, there are some interesting species of bacteria, such as *Mycobacterium*, *Streptomyces*, *Rhodococcus*, and *Nocardia*, which can synthesize lipids in quantities of up to 70% of the cellular dry weight. Microbial lipid content and composition varies from one species to another and strongly depends on the cultivation conditions.

3. Conventional Methods for Total Lipid Extraction

3.1. *Bligh & Dyer Method*

The Bligh & Dyer method (1959) is a multistep process for lipid extraction, which is used extensively in the literature (more than 47,700 total citations according to Google Scholar) and its use keeps increasing rapidly [32]. It is considered as the standard method for total lipid extraction. Researchers use either the original protocol or a modified version, according to their convenience.

Modifications can be done only at the pretreatment step [9]. Hussain et al. (2014) tried four different methods to extract the lipids from the freeze- and oven-dried oleaginous fungus *Mortierella isabellina* and suggested that the Bligh & Dyer method using methanol:chloroform:water at a ratio of 2:1:0.8 results in the highest lipid yield (41%) from oven-dried fungal biomass [33]. Although the Bligh & Dyer method is widely used, there are some drawbacks. Amongst other limitations, the laborious multistep process is not suitable for large quantities of biomass, and significant amounts of harmful organic solvents are utilized in the process.

3.2. Folch Method

After the Bligh & Dyer method, the Folch method is the second most used method for lipid extraction from oleaginous microbial biomass [34]. It was initially developed to extract and purify the lipids from brain tissue in a two-step process. In the first step, the lipids are extracted from the homogenized tissue with 2:1 chloroform-methanol (*v*/*v*) and in the second step, the non-lipid substances are removed by phase separation after adding at least five-fold volumes of water to the filtrate from step one [34]. Researchers are using this method with minor modification, e.g., Cheirsilp and Kitcha (2015) used the Folch method for the extraction of lipids from the oleaginous fungus *Aspergillus tubingensis* TSIP9. They applied sonication on the mixture of dried biomass and chloroform:methanol (2:1) for 30 min, followed by filtration [35]. Kumar et al. (2015) used a similar method to extract lipids from homogenized biomass of the oleaginous bacteria *Rhodococcus opacus* with chloroform and methanol (2:1, *v*/*v*), followed by 15–20 min of shaking in an orbital shaker at ambient room temperature [36]. A modification of the Folch method for the extraction and purification of lipids from oleaginous yeast *Cryptococcus terricolus* was described by T. A. Pedersen, where the extracted crude lipids were dissolved in chloroform and methanol (2:1, *v*/*v*) and distilled water, followed by a separation step by centrifugation [37]. Hara and Radin (1978) also tried to extract lipids from the tissues with non-toxic solvents, such as hexane:isopropanol, followed by a washing step with aqueous sodium sulfate to remove non-lipid contaminants from the extract [38].

4. Pretreatment of Oleaginous Microbial Biomass to Extract Lipids

The effect of microbial biomass pretreatment on the lipid extraction process has not yet been discussed extensively. The efficiency of lipid extraction varies depending on the pretreatment process that is used to disrupt the cellular integrity as it increases with an increasing degree of cell disruption. However, other parameters, such as residual water content in the case of wet biomass and particulate size in the case of dry biomass, may also affect the pre-treatment process [2,14]. The choice of pretreatment depends on the cellular structure of the microbial biomass. It can be a single-step or a multistep process depending on the physical condition of the biomass (dry or wet). Various pretreatment methods for the disintegration of cellular membranes are currently in use and they can be divided into two main groups of (i) mechanical and (ii) non-mechanical methods (Figure 1).

Some researchers divide the cellular disintegration into thermal treatment methods and non-thermal treatment methods, whereas others explain the pretreatment methods in combination with lipid extraction and categorize them as mechanical methods, such as oil expeller, ultrasonication, and microwave-assisted extraction or chemical methods, such as Soxhlet extraction, supercritical fluid extraction, and accelerated solvent extraction [20]. All of these pretreatment methods have been extensively utilized for the efficient lipid extraction from various oleaginous microorganisms, such as yeast, microalgae, fungi, and bacteria (Table 1).

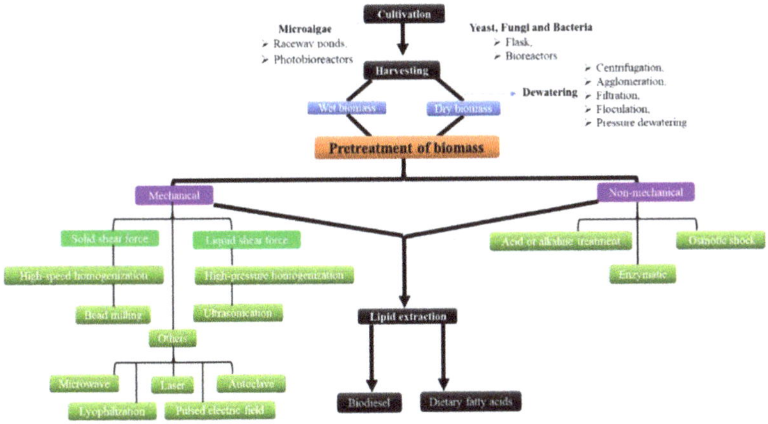

Figure 1. Flowchart showing various pretreatment methods for lipid extraction from oleaginous microbial cells.

Table 1. Comparison of various pretreatment methods for lipid extraction from oleaginous microbial cells.

Oleaginous Micro-Organism	Lipid Extraction Method	Pretreatment of Cells	Lipid Content (%, *w/w*)	References
Oleaginous yeasts				
Rhodosporidium kratochvilovae HIMPA1	Bligh & Dyer method	Ultrasonication at 40 Hz for 5 min	59.7	[39]
	Organic-solvent n-hexane	Acid-catalyzed hot-water treatment	61.9	
	Organic-solvent n-hexane	Microwave irradiation	67.4	
	Organic-solvent n-hexane	Rapid ultrasonication-microwave treatment	70.1	
Cryptococcus curvatus (DSM 70022)	Solvent extraction (chloroform-methanol; 2:1, *v/v*)	Dried biomass, Acid-catalyzed hot-water treatment. (2 mL of 3 M HCl and then digested at 60 °C for 2 h), Sonication for 30 s at 30 kHz	NA	[40]
Rhodotorula glutinis (DSM 10134)			46	
Yarrowia lipolytica (DSM 8218)			48.9	
C. curvatus MUCL 29819	Solvent extraction (chloroform-methanol; 1:1, *v/v*)	Dried yeast cells, Bead milling (glass beads, diameter 0.5 mm)	30.3	[41]
Sporidiobolus pararoseus KM281507	Bligh & Dyer method	Vortexed with glass beads, sonicated at 70 Hz for 30 min	30.7	[42]
S. pararoseus KX709872	Bligh & Dyer method	Vortexed with glass beads for 30 min in the presence of 100 ppm ascorbic acid and sonicated for 30 min in ultrasonication bath	56.6	[43]
Naganishia liquefaciens NITTS2	Solvent extraction (chloroform-methanol; 1:1, *v/v*)	Ultrasonication at 20 kHz for 20 min at 40 °C	55.8	[44]
C. curvatus MTCC 2698	Bligh & Dyer method	Sonication at 40 kHz for 2 min	28.3	[45]
Cryptococcus vishniaccii	Bligh & Dyer method	Sonication at 20 kHz for 5 min	52.3	[46]
Rhodosporidium toruloides and *Lipomyces starkeyi*	Bligh & Dyer method	Acid (2 mol/L of HCl)	25 and 34	[47]
		None	23 and 7	
	Folch method	Acid (2 mol/L of HCl)	34 and 48	
		Enzymatic	31 and 37	
		None	42 and 47	
Oleaginous microalgae				
Schizochytrium sp. ATCC20888	Soxhlet extraction	Enzymatic lysis with alkaline protease	63	[48]
Chlorella vulgaris/Cyanobacteria leptolyngbya	Solvent extraction with hexane or chloroform-methanol (1:1, *v/v*)	Sonicated in an ultrasonic reactor with a clamp-on transducer	16	[48]
Phaeodactylum tricornutum	Solvent extraction (chloroform-methanol; 1:1, *v/v*)	Lyophilization	47	[49]
Scenedesmus sp.	Solvent extraction (chloroform methanol; 1:1, *v/v*)	Enzymatic treatment with cellulase, xylanase and pectinase	86.4 (lipid recovery)	[50]
Tetraselmis sp. KCTC12429BP	Solvent extraction with mixture of hexane and polar solvents (ethanol, isopropanol, methanol, tetrahydrofuran, acetone, acetonitrile)	Lyophilization	5.5 with Chloroform-methanol, 5.2 with hexane-methanol	[51]

Table 1. *Cont.*

Oleaginous Micro-Organism	Lipid Extraction Method	Pretreatment of Cells	Lipid Content (%, *w/w*)	References
Aurantiochytrium sp. KRS101	Solvent extraction with chloroform, chloroform-methanol (2:1, *v/v*), hexane, hexane-isopropanol (3:2, *v/v*), methanol and ethanol	High shear mixer (HSM)	High non-esterifiable lipids with chloroform-methanol and esterifiable lipids with chloroform	[52]
Scenedesmus obliquus	Solvent (chloroform-methanol; 2:1, *v/v*)	Drying of biomass by sun, freeze, and oven followed by microwave, sonication, autoclaving, osmotic shock (10% NaCl)	Highest lipid content of 25.4% was obtained after freeze-drying followed by microwave digestion	[53]
Scendesmus dimorphus	Solvent extraction with ethanol (6 mL/g dry algae), Fractionation with (ethanol: hexane: water; 1:1:1, *v/v/v*)	Extraction autoclave equipped with condenser, mechanical stirring and thermocouple	Oil extraction by fractional method gave neutral lipid (97) with polar lipids (2)	[54]
		Oleaginous fungus		
Mucor circinelloides URM 4182	Solvent extraction With ethanol (96%)	Microwave irradiation at 60 °C for 30 min	31.2	[55]
Cunninghamella echinulata	Soxhlet extraction with diethyl ether anhydrous at 50 °C	Dried biomass ground in a laboratory blender	22.2	[56]
M. circinelloides VI04473 and *Mortierella alpina* UBOCC-A-112046	Folch method, Bligh & Dyer method	Acid hydrolysis with 2 mL 3 N HCl (incubation of the sample at 80 °C for 1 h), bead beating and homogenization (4.0 m/s for 60 s)	NA	[57]
M. circinelloides VI 04473, *Umbelopsis isabellina* UBOCC-A-101350 and *Penicillium glabrum* FRR 419	Lewis extraction	Freeze-dried, biomass, glass beads in high-speed benchtop homogenizer at 6.5 m/s, for 1 min cycle length and 6 cycles	Highest lipid content was obtained from *U. isabellina* at 30 °C	[58]
Alternaria alternata, *Cladosporium cladosporioides*, *Epicoccum nigrum*, *Fusarium oxysporum*, *Aspergillus parasiticus* and *Emericella nidulans* var. lata	Folch method	NA	Highest lipid content (40.8) from *A. alternata*	[59]
Aspergillus tubingensis TSIP9	Folch method	Slurry of biomass and chloroform-methanol sonicated for 30 min	39.5 mg per gram dry substrate (gds)	[35]
		Oleaginous bacteria		
Acinetobacter baylyi ADP1	Bligh & Dyer method	Freeze-dried cells, vortexed	1.6 with wild strain, 12.4 with genetically modified strain	[60]
Rhodococcus opacus	Folch method	Homogenized with chloroform-methanol (2:1, *v/v*), followed by shaking	71 with synthetic medium	[36]
Bacillus subtilis HB1310	Bligh & Dyer method	4 M HCl, incubation at 80 °C for 1 h	39.8	[61]
Rhodopseudomonas palustris (strain 42OL)	Solvent extraction methanol-chloroform (1:2, *v/v*)	Grinding of freeze-dried bacterial cells in a mortar with sand	22 to 39	[62]
Bacillus sp. V10	Bligh & Dyer method	Freeze-drying of the cells	7.4	[63]
R. opacus	Folch method	Homogenized with chloroform-methanol (2:1 *v/v*), followed by shaking	65.8	[64]

NA, not available.

4.1. Mechanical Pretreatment Methods

Mechanical pretreatment for the disintegration of cellular structure is usually carried out by applying mechanical forces or energy transfer through conventional heat, waves, and electric currents. Mechanical forces can be divided into two forms, i.e., solid-shear forces (e.g., bead mill, high-speed homogenization) and liquid-shear forces (e.g., high-pressure homogenization, micro fluidization). The direct energy transfer to the cells can be achieved by waves (laser, ultrasonication, and microwave treatment), conventional heat (autoclave and water bath), or by applying a pulse electric field [2].

4.1.1. Oil or Expeller Pressing

Expeller pressing is the simplest method to extract oils by mechanical crushing. It has many advantages, like smooth and hands-free operational conditions and low needs for maintenance. The lipids are extracted from the dry biomass by applying mechanical pressure to squeeze out the oils from the broken cells. However, this method is relatively slow and requires large amounts of biomass [65]. The applied pressure must be optimal, otherwise it will result in excessive heat generation and blockage problems due to high pressure [66]. Although this method is usually used to extract oils from seeds, some microalgal lipids have also been extracted with this method. For example, filamentous algae were explored for lipid extraction by using the screw expeller press and almost 75% total lipids were extracted from algae by this method [67]. Depending on the type of biomass, various press configurations, such as screw, expeller, piston, etc., can be used. Some researchers suggest that this method is an expensive and slow process when applied on microbial biomass [24,68]. Johnson and Wen (2009) stated that it is a suitable method for feedstocks, like soybean or canola seeds, where lipids can be extracted from the crushed biomass with solvent, while the extraction process may not be suitable for microalgal cells (both mud-like form and dry powder form of algae), where rigid cell walls hinder the extraction process [69]. Topare et al. (2011) suggested that the solvent extraction by Soxhlet apparatus is effective and extracted more than 98% of the lipids from microalgal cells. But, since it is not a cost-effective method, they tried the expeller press method to extract lipids and this method could recover 75% of the oil from algae [67].

4.1.2. Bead Milling

The history of bead milling goes far back to when it was first applied in the manufacturing of cosmetics to reduce the particle size of paint or lacquer and to grind minerals. After proving its effectiveness in the chemical industry, bead milling was successfully applied for the disruption of microbial cells for the downstream processing of intracellular products [20,70–72]. This method has many advantages, such as the need of only single-pass, continuous module of operation, high disruption efficiency, easy biomass loading, mild operating temperature, and applicability to various types of biomass from lab-scale to industrial scale [20,67,71,72]. The operating conditions for the efficient disintegration of cells depend on various factors such as agitator geometry, speed, biomass concentration, slurry flow rate, bead size, bead-to-substrate ratio, etc. Montalescot et al. (2015) reported that disruption of two microalgae, *Nannochloropsis oculata* and *Porphyridium cruentum*, was performed by continuous bead milling where the highest bead filling ratio of >55% v/v was found to be optimal [73]. The type and size of beads also strongly affect the disintegration of microalgal cells. Doucha and Lívanský (2008) suggested that zirconium oxide (ZrO_2) beads are more efficient than glass beads for cellular disintegration because of their higher specific density [74]. Postma et al. (2015) reported that kinetic rate constants can be increased by increasing the speed of the agitator as well as the biomass concentration [72]. They investigated the disintegration of *Chlorella vulgaris* by using zirconium oxide (ZrO_2) beads with a diameter of 1 mm, which gives a lower specific energy consumption, while the agitator speed and biomass loading were 6 m/s and 145 g_{DW}/kg [72]. However, similar specific energy consumptions were also achieved by changing the size of beads with similar flow rate and agitator speed [74]. Balasundaram et al. (2012) investigated the optimal balance between shear forces and impact forces that are required for a differential recovery of intracellular products from the cyanobacteria *C. fritschii* (PCC6912) when a custom-made energy efficient ball mill was used for disintegration [70]. Although bead milling is suitable for disintegration of cells, its high energy consumption during operation and its inefficient energy transfer from rotating shaft to individual cells make it an unfavorable method [20]. Oleaginous yeast *Y. lipolytica* IFP29 (ATCC 20460) was disrupted by ultrasound, microwave irradiation, and bead milling and the lipid extraction efficiency was compared to pretreatment with freezing/defrosting, cold-drying, bead milling, and microwave irradiation before the conventional solvent extractions process [75]. It was suggested that bead milling was efficient for lipid extraction from oleaginous yeast biomass while cold-drying under pressure was the best pretreatment method, giving two times more yield when compared to conventional methods [75].

4.1.3. High-Pressure Homogenization

This method is suitable for the stabilization of emulsification processes in cosmetic, pharmaceutical, and food industries, however, it has also been extensively utilized for the microbial cell disruption of microalgae [76], bacteria [77], and yeast [78]. The cell disruption efficiency varies according to the valve seat configurations of the homogenizer [79]. High disruption efficiency is usually achieved through shear forces of highly pressurized fluids on the stationary valve surface and hydrodynamic cavitation from the shear stress induced by pressure drop [80]. High-pressure homogenization has many advantages as it is a simple continuous operating system and can be applied for wet biomass, where the processing fluid is pressurized in intensifiers and passed through a homogenization chamber. The energy is accumulated in the fluid by the pressure and released into the passage through an orifice valve, where the velocity of the fluid increases to up to 200–400 m/s [81,82]. Increased velocity generates mechanical stress, such as shear and elongational forces, turbulence, and cavitation, which are responsible for disruption of cells [83]. Coccaro et al. (2018) suggested that the most efficient disintegration of *Lactococcus lactis* cells was achieved with small orifice valve size, high operating pressure, and low fluid viscosity [77]. In another study, oleaginous microalgae *Nannochloropsis* sp. were disrupted by prior incubation at 37 °C for 15 h before treatment with high-pressure homogenization at 1200 ± 100 bar, followed by lipid extraction with organic solvents [84]. It was a low solvent, low temperature method for efficient lipid extraction from wet concentrated paste where the recovery was reported to be up to 70% *w/w* of the total lipids and 86% *w/w* of neutral lipids using hexane as solvent [84].

4.1.4. High-Speed Shearing Homogenization

High-speed shearing homogenization (HSH) is usually utilized to prepare foams, emulsions, and suspensions [85]. It is a very effective method to disrupt cells, where a slurry of biomass is stirred in a specific device consisting of a stator–rotor assembly with a small gap (100–3000 μm) [86]. The cells are disintegrated due to hydrodynamic cavitation and the shear forces that are caused by stirring at high rpm, which creates high shear rates (20,000–100,000 s^{-1}) [20]. High-speed shearing homogenization was used to prepare the extracts of *Agaricus blazei murill* for the extraction of α-glucan with a final carbohydrate content of 96% [86]. Kwak et al. (2018) used a high-shear mixer to disrupt the cells and extract the lipids from the wet biomass of the oleaginous microalgae *Aurantiochytrium* sp. KRS101 [52]. They suggested that the performance of the high-shear mixer was quite similar between wet and dry biomass of microalgae when extracted with different solvents, such as hexane, hexane-isopropanol, and ethanol. The mixtures developed a strong shear stress and cavitation effect when stirred at 15,000 rpm for 10 min, which was enough to extract all the esterifiable lipids from the microalgae [52]. The most important feature of this method is that it can be directly used for high moisture containing samples, thus reducing the water footprint and downstream process costs [86]. However, extensive heat generation and high energy consumption during the operation are the major drawbacks when it comes to scale-up processes [14].

4.1.5. Ultrasonication

The ultrasonication method was found to be the most applicable and efficient method for lipid extraction from oleaginous microbial biomass. Cavitation and acoustic streaming are two different phenomena that are created during the application of ultrasound to the cells. Cavitation creates pressure on the cells in the form of microbubbles, leading to the disruption of cell walls and membranes. Ultrasonication for the disruption of microalgal cells has been tested with various types of solvents, such as chloroform-methanol [87–89], n-hexane [90,91], diethyl ether [92], and other solvents [87].

Three freshwater-isolated microalgal species *Chlorella* sp., *Nostoc* sp., and *Tolypothrix* sp. were disrupted by different methods, such as bead beating, autoclave, microwave, sonication, and 10% sodium chloride solution treatment [93]. Among the tested methods, sonication was found to be the most effective method to disrupt the microalgal cells and the highest lipid content was obtained by disruption of *Chlorella* sp. [93]. The three microalgae species *Phaeodactylum tricornutum*, *Nannochloropsis gaditana*, and *Chaetoceros calcitrans* were tested for lipid extraction with the conventional Bligh & Dyer extraction method (1959) with prior ultrasonication treatment [94]. *Trichosporon oleaginosus* and an oleaginous fungal strain were treated with ultrasonication (520 kHz, 40 W, and 50 Hz, 2800 W) using various solvents, including water, hexane, methanol, and chloroform-methanol (1:1, v/v), followed by lipid extraction and comparison of the process efficiency to the conventional chloroform-methanol (2:1, v/v) extraction method [95]. The results suggested that almost all of the lipids (100%) were extracted from *T. oleaginosus* and the SKF-5 strain after a very short incubation (15 min) at a relatively low temperature (25 °C) with chloroform-methanol, followed by pretreatment with ultrasonication at 50 Hz and 2800 W [95]. Mecozzi et al. (2002) performed an experiment for lipid extraction from marine mucilage samples using an ultrasonic cleaning bath at 35 kHz [92]. They used two different solvent systems and suggested that diethyl ether was more suitable than methanol to assist the ultrasonication for lipid extraction. Moreover, the disruption due to the acoustic cavitation phenomena minimized the oxidative damage on the lipids [92]. Wu et al. (2012) investigated the ultrasonication treatment at low frequency (20 kHz) with high intensity (0.0403 W/cm^3) and found it to be effective for the disruption of *Microcystis aeruginosa* [96]. They suggested that the acoustic cavitation phenomenon at a low ultrasonic frequency is mainly responsible for the damage of cells due to sufficient shear forces being directly applied to cells. But, while the mechanical energy of cavitation is lower at high ultrasonic frequencies, the ultrasonic degradation of water generates free radicals that weaken the cell wall of cyanobacteria [96]. In another study, the lipids were extracted from microalgae *Scenedesmus obliquus* by using sun-, freeze-, and oven-dried biomass. The cells were disrupted by microwave-, sonication-, autoclaving-, and osmotic shock treatment [53]. The results suggested that the lipid yield of dried samples that were subjected to microwave treatment ($20.73 \pm 4.16\%$) was higher than for autoclaving and osmotic shock, while the results were comparable with sonication ($19.49 \pm 3.30\%$) [53]. Wang et al. (2014) treated two microalgae *Scenedesmus dimorphus* and *Nannochloropsis oculate* with a high frequency focused ultrasound (3.2 MHz, 40 W) and a low frequency non-focused ultrasound (20 kHz, 100 W). The results revealed that high frequency focused ultrasound was a more energy efficient process for maximal cell disintegration [97].

4.1.6. Microwave Irradiation

Microwave irradiation is another extensively used method for lipid extraction from oleaginous microorganisms, where electromagnetic waves are applied to the suspension of cells in an organic solvent. During the microwave treatment of polar compounds, the applied alternative current is converted into electromagnetic energy, and finally, in heating energy [98], as the polar compounds align themselves in the direction of the applied electric field and rotate at high speed when the microwave field alters. The process is accelerated when ions are present in the working system [75]. High heat is generated during the frictional movement of polar compounds or ions, not involving conventional radiant heat [99]. Microwave heating consumes almost two to three times less energy than that involves in the conventional heating [100]. Guerra et al. (2014) treated the oleaginous microalgae *Chlorella* sp.

with microwaves and enhanced lipid yield was recorded as compared to the conventional Bligh & Dyer method, therefore it further boosts chemical and energy savings. Moreover, lipid extraction using a single-step microwave-assisted extraction was more convenient and effective than the multistep, time consuming traditional Bligh & Dyer method [101]. Furthermore, Lee et al. (2010) stated that the lipid extraction yield was higher for the microwave method when compared to autoclaving, bead beating, ultrasonication, and 10% NaCl solution extraction methods [89]. Microwaves were reported to be a useful tool for the extraction of plant oils and animal fats, and, in addition, their implementation has the advantage of easy scale-up [24]. Teo et al. (2014) performed trials on the extraction of lipids from the marine microalgae *Nannochloropsis* sp. and *Tetraselmis* sp. using four different solvent extraction methods (Hara & Radin, Folch, Chen, and Bligh & Dyer) along with conventional heating and microwave irradiation. The highest lipid yield was obtained when they used the Hara & Radin (8.19%) and the Folch (8.47%) method following the microwave irradiation [102]. Boldor et al. (2010) used microwave treatment for the extraction of oils from Chinese tallow tree in batch and continuous flow mode [24]. They suggested that the application of microwave-assisted solvent extraction to extract the lipids from seeds has many advantages over conventional methods, including short operating time and reduced energy consumption [24]. The cells of three freshwater-isolated microalgal species, *Botryococcus* sp., *Chlorella vulgaris* and *Scenedesmus* sp., were disrupted by different methods, such as bead beating, autoclave, microwave, sonication, and 10% NaCl treatment [89]. Among all of the tested methods, the microwave oven was found to be the simplest, easiest, and most effective method to disrupt the microalgal cells. The highest lipid content was obtained by the disruption of *Botryococcus* sp. [89]. Although microwave treatment is a suitable technique to extract lipids in a short amount of time, it has some drawbacks. Its use is limited to polar solvents and the method is unsuitable for volatile compounds. Moreover, the formation of free radicals and the increased temperature make microwave treatment less favorable [103].

4.1.7. Autoclaving

Autoclaving is usually utilized for the sterilization of laboratory equipment and media prior to the growth of microorganisms. Various microalgal species, such as *Haematococcus pluvialis* [104], *Botryococcus* sp., *C. vulgaris*, and *Scenedesmus* sp. [89] were disrupted by autoclaving at 121 °C and 1.5 MPa for 5 or 30 min [89,104]. Rakesh et al. (2015) treated four oleaginous microalgae, *Chlorococcum* sp. MCC30, *Botryococcus* sp. MCC31, *Botryococcus* sp. MCC32, and *Chlorella sorokiniana* MICG5, with various methods, such as autoclaving, microwave irradiation, osmotic shock treatment, and pasteurization, and reported that the highest amount of nutraceutically important unsaturated fatty acids was obtained when *Botryococcus* sp. was treated with autoclaving [105]. Similarly, Florentino de Souza Silva et al. (2014) suggested that autoclaving is a more efficient technique than ultrasonication but not as efficient as microwaving and electroflotation by alternating current (EFAC), when mixed cultures of microalgae were treated with different methods [106].

4.1.8. Pulsed Electric Field

Pulsed electric field (PEF) treatment works based on electroporation phenomena, including electromechanical compression and electric field-induced tension, where an external electric field is used to induce the critical electrical potential across the cell membrane [107,108]. The increase in membrane porosity is directly proportional to the strength of the applied electric field and pulses and the pore formation in the membrane can be reversible or irreversible, depending on the size and number of pores in comparison to the total surface area of the membrane or cell wall [107]. Eing et al. (2013) treated the oleaginous microalgae *Auxenochlorella protothecoides* with a PEF at 35 kV/cm and the pulse duration was set to 1 μs. They suggested that the lipid yield after PEF treatment and extraction with ethanol was four times higher than it was for untreated cells [109]. Similarly, the oleaginous microalga *Synechocystis* PCC 6803 was treated with a pulsed electric field (intensity > 35 kWh/m^3) and isopropanol as solvent [110].

In another study, oleaginous microalgae *Ankistrodesmus falcatus* wet biomass was treated with PEF using the green solvent ethyl acetate and the results demonstrated that the lipid yield was 83–88% higher when compared to the untreated cells [111].

4.1.9. Laser

Laser treatment is a well-known technique to disintegrate the cellular membrane without damaging the compartments of the cell factory or other interior compounds. Most importantly, laser treatment is free of the use of any organic solvent, fast, and requires no laborious effort [112,113]. Previously, researchers have studied the efficiency and mechanism of this method of cell lysis in static mode for different microorganisms, like *Escherichia coli*, *Saccharomyces cerevisiae*, and microalgae at various wavelengths and energy inputs. In a study, oleaginous microalgae *N. oculata* cells were disrupted by using various pretreatment methods, such as microwave, water bath, blender, ultrasonic, and laser treatment, and it was revealed that the highest disruption efficiency was achieved with laser treatment (96.53%), followed by microwave treatment (94.92%) [114]. However, the number and scope of these studies are limited and further investigations, especially in the continuous system, are required in order to examine the potential applications of this cell disruption method.

4.1.10. Acid-Catalyzed Hot-Water

Hot water treatment is a well-known technique to disintegrate the crystalline nature of cellulosic biomass [115]. Hot water requires high pressure at an elevated temperature in order to remain in liquid form. It is to be noted that this pretreatment under acidic conditions is applicable for the extraction of lipids from biomass in wet condition ensuring the cost-effectiveness since no extra energy input is required for dewatering processes. Lipids with high free fatty acid content were extracted from *C. vulgaris* by using acid-catalyzed hot-water treatment and the anionic surfactant sodium dodecyl benzene sulfonate (SDBS) [116]. The lipid extraction yield was 266.0 mg/g of cell weight from a total fatty acid content of 296.0 mg/g of cell weight, when the concentration of sulfuric acid and SDBS were 2.0% and 0.2%, respectively [116]. *C. vulgaris* cells were disrupted for efficient lipid extraction using acid-catalyzed hot-water treatment [117]. The lipid extraction yield was 337.4 mg/g of cell weight from a total fatty acid content of 381.6 mg/g of cell weight, given a 1% sulphuric acid concentration and heating at 120 °C for 60 min, when compared to 83.2 mg/g of cell weight lipid yield with no heating and no catalyst [117]. This method is also suitable to extract lipids rich in docosahexaenoic acid (DHA) from *Aurantiochytrium* sp. [118]. During acid-catalyzed hot-water treatment, cells are disrupted along with the degradation of other cellular components, leading to excess acid loading and devaluation of co-products.

4.2. *Non-Mechanical Pretreatment Methods*

Conventional mechanical techniques have several drawbacks, including insufficient extraction yields, the use of toxic solvents, and long processing time. Hence, there is a need for rapid, less energy intensive methods for the lipid extraction from wet biomass. Non-mechanical disruption methods, such as enzymatic and chemical cell lysis, are mainly used in lab-scale processes for bioanalytical purposes. The energy consumption of mechanical methods is always higher when compared to non-mechanical methods. Lee et al. (2012) compared the energy consumed by lipid extraction from microalgae with the energy levels that are required for other methods [119]. The advantages and disadvantages along with process parameters of various cell disruption methods are summarized and compared in Table 2.

Table 2. Summary and comparison of various mechanical and non-mechanical pretreatment methods for cellular degradation.

Pretreatment Methods	Mode of Action	Energy Consumption	Scale-Up Possibility	Advantages	Disadvantages	References
Ultrasonication	Cavitation, acoustic streaming and liquid shear stress	Medium/low	Yes/no	Less processing time, lower solvent consumption, greater penetration of solvent into cellular compartment	High power consumption, difficult to scale up	[23,91,92,96,97, 120,121]
Oil/expeller press	Mechanical compaction and shear forces	High	Yes	Easy process, no solvent	Large amount of sample required, slow process, unsuitable for samples with high moisture content	[67,122]
High-speed homogenization	Cavitation and shear forces	High/medium	Yes	Simple process, effective, short contact time	High energy consumption, increased temperature during operation	[20,85,86]
High-pressure homogenization	Cavitation and shear forces	High/medium	Yes	Solvent-free, simple process, effective, short contact time	High maintenance cost, less efficient with filamentous microorganisms, no residual effect	[22,123–125]
Bead milling	Mechanical compaction and shear forces	High/medium	Yes	Solvent-free, suitable for samples with high moisture content	Low efficiency with rigid cells, depending on various parameters such as bead size and agitation, no residual effect	[19,72,73,75,103, 124,126–128]
Microwave irradiation	Temperature increase, molecular energy increase	High/medium	Yes/no	Eco-friendly, reduced processing time and solvent consumption	Filtration or centrifugation is necessary to remove the solid residue, unsuitable for non-polar or volatile compounds	[78,95,106,107, 128–130]
Pulsed electric field treatment	Pore formation due to electric waves	High	Yes/no	Relatively simple, high energetic efficiency, relatively fast	High maintenance costs, high temperature, dependence on medium composition, decomposition of fragile compounds	[20,56,107–111]
Enzymatic treatment	Specific enzyme-substrate interaction	Low	Yes	Simple, high energetic efficiency	Long processing time and high capital cost	[22,25,71,123, 125,130–133]

4.2.1. Enzymatic Pretreatment

The extraction of lipids using enzymatic pretreatment completely depends on the cell wall characteristics of the subjected oleaginous microorganism [130]. This technique includes various cell wall degrading enzymes, such as xylanase, cellulase, amylase, papain, pectinase, and hemicellulase [131]. Enzymatic pretreatment is a well-known technique in the vegetable oil industry to degrade the structural polysaccharides of the cell wall of oily seeds [134,135]. It constitutes a favourable cell disintegration method due to its specificity and mild operating temperature, as well as its low time and energy requirements. Furthermore, the method is devoid of harmful solvents and harsh physical conditions, such as shear forces [132]. It has been reported that enzymatic pretreatment is suitable for extracting lipids from the oleaginous yeast. For example, the oleaginous yeast *Rhodosporidium toruloides* was treated with the recombinant β-1,3-glucomannanase plMAN5C, and almost 96.6% of the total lipid content was extracted directly from the culture with ethyl acetate at room temperature and atmospheric pressure without dewatering [25]. Moreover, this method has also been applied to extract lipids from oleaginous microalgae. The microalga *C. vulgaris* was treated with cellulases for 72 h and the hydrolysis efficiency of the cell wall carbohydrates was 85.3%. After enzymatic hydrolysis, the lipid extraction efficiency by solvent extraction was higher than without hydrolysis [136]. Bonturi et al. (2015) extracted lipids from intact and pretreated cells of oleaginous yeasts *R. toruloides* and *L. starkeyi* using various methods such as Folch, Pedersen, hexane, and Bligh & Dyer methods involving acid and enzyme pretreatment. They suggested that enzymatic pretreatment is not an efficient technique for *L. starkeyi* due to the sulfide bonds in its cell wall, which increase the strength and rigidity of its organelles [47].

In another study, the oleaginous microalga *Scenedesmus* sp. was treated with various enzymes, such as cellulase, xylanase, and pectinase under varying conditions, including enzyme concentration, temperature, pH, and incubation time [50]. The results demonstrated that the combination of cellulase, xylanase, and pectinase for 190 min improved the lipid extraction yields by 96.4% when compared to the untreated microalga [50]. Another oleaginous marine microalga, *Nannochloropsis* sp., was treated with cellulase and mannanase and the results revealed the improvement of lipid extraction yields from 40.8% to over 73% [137]. Treatment of the same microalga with similar enzymes under different conditions significantly improved the recovery of lipids from *Nannochloropsis* sp. biomass [133]. The enzymatic hydrolysis of *Chloroccum* sp. by using cellulase obtained from *Trichoderma reesei*, ATCC 26921, was an effective method to enhance the saccharification process of microalgal biomass for bioethanol production [138]. Hence, enzymatic treatment can improve the lipid extraction from various oleaginous microorganisms and scaling up the process is relatively easy. However, long processing times and high capital costs hinder the scale-up of enzymatic pretreatment for lipid extraction in the biorefineries [138].

4.2.2. Other Emerging Methods for the Extraction of Lipids from Oleaginous Microorganisms

Researchers have used a limited number of other pretreatment methods, including chemical treatments to disrupt the microbial cells. Supercritical fluid extraction is an extensively used method to extract lipids from oleaginous microbial biomass [9,127,130,139–142]. Bai et al. (2014) used free nitrous acid (FNA) as an effective and low cost pretreatment method to extract lipids from microalgae [143]. Boyd et al. (2012) used switchable hydrophilicity solvents, such as *N*,*N*-dimethylcyclohexylamine for the lipid extraction from freeze-dried samples of *Botryococcus braunii* microalgae [144]. Two microalgae species, *N. oculata* and *Dunaliella salina*, were treated with a novel simultaneous distillation and extraction process (SDEP) for lipid extraction under wet conditions by using d-limonene, a-pinene and p-cymene as solvents [145]. Du et al. (2013) tried to use switchable solvents such as secondary amines for the extraction of lipids from wet and non-broken algae [146]. Kim et al. (2012) used a mixture of ionic liquid [Bmim][CF_3SO_3] and methanol for the extraction of lipids from *C. vulgaris* [147]. Lee et al. (2013) treated *Chlorella* sp. biomass with organic nanoclays, such as Mg–APTES clay, Al–APTES clay, Ca–APTES clay, and Mg–N3 clay [148]. In other reports, researchers used H_2O_2 with or without $FeSO_4$ to disrupt the cell walls of *C. vulgaris* [149]. Jo et al. (2014) reported a quick method for dimethyl carbonate-mediated lipid extraction from *Chlorella* sp. [150]. Hydrothermal liquefaction (HTL) is the most emerging technique to convert high moisture algal biomass to crude biooil, producing more lipids per mass of microalgae than the other extraction method [151]. This technique is more advantageous than the traditional thermochemical conversion processes, such as pyrolysis, where high energy is required for drying the biomass [152]. In this process, biomolecules are decomposed in hot compressed water via the combined action of elevated temperature, elevated pressure, and hydrolytic attack. Reaction temperature and catalysts are two important factors in hydrothermal liquefaction that decide the fractionation of water-soluble and water-insoluble biocrude from algae [153]. Sheehan and Savage (2017) developed a kinetic model for hydrothermal liquefaction to predict the biocrude yield from protein, lipid, and carbohydrate rich microalgae, and suggested that feedstocks containing more proteins or lipids give higher biocrude yields than those that are abundant in carbohydrates [154]. For example, Hietala et al. (2016) performed isothermal and non- isothermal hydrothermal liquefaction with microalgae *Nannochloropsis* sp. and predicted that up to 46% *w*/*w* biocrude yields are achievable with short holding time of 1 min at 400 °C [155].

5. Conclusions and Recommendations

This review article summarizes the various techniques that are available for total lipid extraction from the biomass of various oleaginous microorganisms using different mechanical and non-mechanical pretreatment methods. Since all mechanical pretreatment methods are followed by the solvent extraction of lipids, overall costs of downstream processing can be increased. Some researcher use green solvents instead of potentially harmful chloroform and methanol to make the extraction process more feasible. Supercritical fluid extraction techniques appear to be a good option to avoid toxic solvents, but the initial capital cost for equipment is high. Another alternative, involving the use of solvent-free extractions, is offering an environment-friendly and cost-effective option on a laboratory scale. However, more research is required in order to scale up the process. Drying of biomass prior to pretreatment is another energy-intensive step, so the target should be to extract lipids from wet cellular biomass to avoid extra cost for drying. Several pretreatment methods allow for the extraction process in wet conditions. For example, hydrothermal liquefaction is an emerging method in which wet microalgae are converted into crude biooil and this technique is also suitable for scale-up of process. However, using only one single pretreatment method may not be sufficient to reach a maximal lipid extraction yield from lysed biomass, hence it may be advantageous to apply multiple different pretreatment methods on both lab- and large-scale. For example, if ultrasonication and microwave irradiation are used in combination for the cell disruption under wet conditions, four different physical phenomena would work together to easily break the cells and release the lipids to the external environment. During ultrasonication, the transmission of sonic waves causes cavitation where microbubbles form during the rarefaction phase of the sound wave and collapse during the compression phase. Disintegrated microbubbles release shock waves in the form of mechanical energy, which causes irreparable shearing in the cell wall of oleaginous microorganisms. After the ultrasonication step, microwave treatment causes the alignment of polar compounds in the cellular compartments in the direction of the applied electric field followed high-speed rotation when the microwave field alters. With this method combination, both the duration of the process as well as the solvent consumption could be decreased.

Author Contributions: All the authors participated in the collection of data from the literature, analysis of data and drafting the manuscript. The final publication was prepared with contributions from all authors.

Funding: The authors will like to thank Bio4Energy, a strategic research environment appointed by the Swedish government, for supporting this work.

Conflicts of Interest: The authors declare no conflict of interest.

References

1. Patel, A.; Arora, N.; Sartaj, K.; Pruthi, V.; Pruthi, P.A. Sustainable biodiesel production from oleaginous yeasts utilizing hydrolysates of various non-edible lignocellulosic biomasses. *Renew. Sustain. Energy Rev.* **2016**, *62*, 836–855. [CrossRef]
2. Ma, Y.; Gao, Z.; Wang, Q.; Liu, Y. Biodiesels from Microbial Oils: Opportunity and Challenges. *Bioresour. Technol.* **2018**. [CrossRef] [PubMed]
3. Yellapu, S.K.; Kaur, R.; Kumar, L.R.; Tiwari, B.; Zhang, X.; Tyagi, R.D. Recent developments of downstream processing for microbial lipids and conversion to biodiesel. *Bioresour. Technol.* **2018**, *256*, 515–528. [CrossRef] [PubMed]
4. Huang, C.; Chen, X.X.; Xiong, L.; Yang, X.; Chen, X.X.; Ma, L.; Chen, Y. Microbial oil production from corncob acid hydrolysate by oleaginous yeast Trichosporon coremiiforme. *Biomass Bioenergy* **2013**, *49*, 273–278. [CrossRef]
5. Koutinas, A.A.; Chatzifragkou, A.; Kopsahelis, N.; Papanikolaou, S.; Kookos, I.K. Design and techno-economic evaluation of microbial oil production as a renewable resource for biodiesel and oleochemical production. *Fuel* **2014**, *116*, 566–577. [CrossRef]
6. Dourou, M.; Aggeli, D.; Papanikolaou, S.; Aggelis, G. Critical steps in carbon metabolism affecting lipid accumulation and their regulation in oleaginous microorganisms. *Appl. Microbiol. Biotechnol.* **2018**, *102*, 2509–2523. [CrossRef] [PubMed]
7. Angerbauer, C.; Siebenhofer, M.; Mittelbach, M.; Guebitz, G.M.M. Conversion of sewage sludge into lipids by *Lipomyces starkeyi* for biodiesel production. *Bioresour. Technol.* **2008**, *99*, 3051–3056. [CrossRef] [PubMed]
8. Lee, S.Y.; Cho, J.M.; Chang, Y.K.; Oh, Y.K. Cell disruption and lipid extraction for microalgal biorefineries: A review. *Bioresour. Technol.* **2017**, *244*, 1317–1328. [CrossRef] [PubMed]
9. Ranjith Kumar, R.; Hanumantha Rao, P.; Arumugam, M. Lipid Extraction Methods from Microalgae: A Comprehensive Review. *Front. Energy Res.* **2015**, *2*, 1–9. [CrossRef]

10. Dong, T.; Knoshaug, E.P.; Pienkos, P.T.; Laurens, L.M.L. Lipid recovery from wet oleaginous microbial biomass for biofuel production: A critical review. *Appl. Energy* **2016**, *177*, 879–895. [CrossRef]
11. Soccol, C.R.; Dalmas Neto, C.J.; Soccol, V.T.; Sydney, E.B.; da Costa, E.S.F.; Medeiros, A.B.P.; de Souza Vandenberghe, L.P. Pilot scale biodiesel production from microbial oil of Rhodosporidium toruloides DEBB 5533 using sugarcane juice: Performance in diesel engine and preliminary economic study. *Bioresour. Technol.* **2017**, *223*, 259–268. [CrossRef] [PubMed]
12. Halim, R.; Danquah, M.K.; Webley, P.A. Extraction of oil from microalgae for biodiesel production: A review. *Biotechnol. Adv.* **2012**, *30*, 709–732. [CrossRef] [PubMed]
13. Ranjan, A.; Patil, C.; Moholkar, V.S. Mechanistic assessment of microalgal lipid extraction. *Ind. Eng. Chem. Res.* **2010**, *49*, 2979–2985. [CrossRef]
14. D'Hondt, E.; Martín-Juárez, J.; Bolado, S.; Kasperoviciene, J.; Koreiviene, J.; Sulcius, S.; Elst, K.; Bastiaens, L. Cell disruption technologies. In *Microalgae-Based Biofuels and Bioproducts: From Feedstock Cultivation to End-Products*; Elsevier: New York, NY, USA, 2017; pp. 133–154, ISBN 9780081010273.
15. Chisti, Y. Biodiesel from microalgae. *Biotechnol. Adv.* **2007**, *25*, 294–306. [CrossRef] [PubMed]
16. Kim, J.; Yoo, G.; Lee, H.; Lim, J.; Kim, K.; Kim, C.W.; Park, M.S.; Yang, J.W. Methods of downstream processing for the production of biodiesel from microalgae. *Biotechnol. Adv.* **2013**, *31*, 862–876. [CrossRef] [PubMed]
17. Chen, Y.H.; Walker, T.H. Fed-batch fermentation and supercritical fluid extraction of heterotrophic microalgal *Chlorella protothecoides* lipids. *Bioresour. Technol.* **2012**, *114*, 512–517. [CrossRef] [PubMed]
18. Jeevan Kumar, S.P.; Vijay Kumar, G.; Dash, A.; Scholz, P.; Banerjee, R. Sustainable green solvents and techniques for lipid extraction from microalgae: A review. *Algal Res.* **2017**, *21*, 138–147. [CrossRef]
19. Postma, P.R.; Suarez-Garcia, E.; Safi, C.; Olivieri, G.; Olivieri, G.; Wijffels, R.H.; Wijffels, R.H. Energy efficient bead milling of microalgae: Effect of bead size on disintegration and release of proteins and carbohydrates. *Bioresour. Technol.* **2017**, *224*, 670–679. [CrossRef] [PubMed]
20. Günerken, E.; D'Hondt, E.; Eppink, M.H.M.; Garcia-Gonzalez, L.; Elst, K.; Wijffels, R.H. Cell disruption for microalgae biorefineries. *Biotechnol. Adv.* **2015**, *33*, 243–260. [CrossRef] [PubMed]
21. Demirbaş, A. Production of Biodiesel from Algae Oils. *Energy Sources Part A Recover. Util. Environ. Eff.* **2008**, *31*, 163–168. [CrossRef]
22. Geciova, J.; Bury, D.; Jelen, P. Methods for disruption of microbial cells for potential use in the dairy industry—A review. *Int. Dairy J.* **2002**, *12*, 541–553. [CrossRef]
23. Chemat, F.; Khan, M.K. Applications of ultrasound in food technology: Processing, preservation and extraction. *Ultrason. Sonochem.* **2011**, *18*, 813–835. [CrossRef] [PubMed]
24. Boldor, D.; Kanitkar, A.; Terigar, B.G.; Leonardi, C.; Lima, M.; Breitenbeck, G.A. Microwave assisted extraction of biodiesel feedstock from the seeds of invasive chinese tallow tree. *Environ. Sci. Technol.* **2010**, *44*, 4019–4025. [CrossRef] [PubMed]
25. Jin, G.; Yang, F.; Hu, C.; Shen, H.; Zhao, Z.K. Enzyme-assisted extraction of lipids directly from the culture of the oleaginous yeast Rhodosporidium toruloides. *Bioresour. Technol.* **2012**, *111*, 378–382. [CrossRef] [PubMed]
26. Pomraning, K.R.; Wei, S.; Karagiosis, S.A.; Kim, Y.M.; Dohnalkova, A.C.; Arey, B.W.; Bredeweg, E.L.; Orr, G.; Metz, T.O.; Baker, S.E. Comprehensive metabolomic, lipidomic and microscopic profiling of *Yarrowia lipolytica* during lipid accumulation identifies targets for increased lipogenesis. *PLoS ONE* **2015**, *10*, e0123188. [CrossRef] [PubMed]
27. Jiang, L.; Zhang, L.; Nie, C.; Pei, H. Lipid productivity in limnetic *Chlorella* is doubled by seawater added with anaerobically digested effluent from kitchen waste. *Biotechnol. Biofuels* **2018**, *11*, 1–14. [CrossRef] [PubMed]
28. D'ESPAUX, L.; Mendez-Perez, D.; Li, R.; Keasling, J.D. Synthetic biology for microbial production of lipid-based biofuels. *Curr. Opin. Chem. Biol.* **2015**, *29*, 58–65. [CrossRef] [PubMed]
29. Patel, A.; Pravez, M.; Deeba, F.; Pruthi, V.; Singh, R.P.; Pruthi, P.A. Boosting accumulation of neutral lipids in *Rhodosporidium kratochvilovae* HIMPA1 grown on hemp (*Cannabis sativa* Linn) seed aqueous extract as feedstock for biodiesel production. *Bioresour. Technol.* **2014**, *165*, 214–222. [CrossRef] [PubMed]
30. Patel, A.; Pruthi, V.; Pruthi, P.A. Synchronized nutrient stress conditions trigger the diversion of CDP-DG pathway of phospholipids synthesis towards de novo TAG synthesis in oleaginous yeast escalating biodiesel production. *Energy* **2017**, *139*, 962–974. [CrossRef]

31. Ambat, I.; Srivastava, V.; Sillanpää, M. Recent advancement in biodiesel production methodologies using various feedstock: A review. *Renew. Sustain. Energy Rev.* **2018**, *90*, 356–369. [CrossRef]
32. Bligh, E.; Dyler, W.J. A rapid Method of Total Lipid Extraction and Purification. *Can. J. Biochem. Physiol.* **1959**, *37*, 911–917. [CrossRef] [PubMed]
33. Hussain, J.; Ruan, Z.; Nascimento, I.A.; Liu, Y.; Liao, W. Lipid profiling and corresponding biodiesel quality of *Mortierella isabellina* using different drying and extraction methods. *Bioresour. Technol.* **2014**, *169*, 768–772. [CrossRef] [PubMed]
34. Folch, J.; Lees, M.; Sloane Stanley, G. A simple method of isolation and purification of total lipids from animal tissues. *J. Biol. Chem.* **1957**, *226*, 497–509. [PubMed]
35. Cheirsilp, B.; Kitcha, S. Solid state fermentation by cellulolytic oleaginous fungi for direct conversion of lignocellulosic biomass into lipids: Fed-batch and repeated-batch fermentations. *Ind. Crops Prod.* **2015**, *66*, 73–80. [CrossRef]
36. Kumar, S.; Gupta, N.; Pakshirajan, K. Simultaneous lipid production and dairy wastewater treatment using Rhodococcus opacus in a batch bioreactor for potential biodiesel application. *J. Environ. Chem. Eng.* **2015**, *3*, 1630–1636. [CrossRef]
37. Pedersen, T.A. Lipid Formation in Cryptococcus terricolus. III. Extraction and Purification of Lipids. *Acta Chem. Scand.* **1962**, *16*, 374–382. [CrossRef]
38. Hara, A.; Radin, N.S. Lipid extraction of tissues with a low toxicity solvent. *Anal. Biochem.* **1978**, *90*, 420–426. [CrossRef]
39. Patel, A.; Arora, N.; Pruthi, V.; Pruthi, P.A. A novel rapid ultrasonication-microwave treatment for total lipid extraction from wet oleaginous yeast biomass for sustainable biodiesel production. *Ultrason. Sonochem.* **2018**, in press. [CrossRef]
40. Johnravindar, D.; Karthikeyan, O.P.; Selvam, A.; Murugesan, K.; Wong, J.W.C. Lipid accumulation potential of oleaginous yeasts: A comparative evaluation using food waste leachate as a substrate. *Bioresour. Technol.* **2018**, *248*, 221–228. [CrossRef] [PubMed]
41. Huang, X.; Luo, H.; Mu, T.; Shen, Y.; Yuan, M.; Liu, J. Enhancement of lipid accumulation by oleaginous yeast through phosphorus limitation under high content of ammonia. *Bioresour. Technol.* **2018**, *262*, 9–14. [CrossRef] [PubMed]
42. Manowattana, A.; Techapun, C.; Watanabe, M.; Chaiyaso, T. Bioconversion of biodiesel-derived crude glycerol into lipids and carotenoids by an oleaginous red yeast *Sporidiobolus pararoseus* KM281507 in an airlift bioreactor. *J. Biosci. Bioeng.* **2018**, *125*, 59–66. [CrossRef] [PubMed]
43. Chaiyaso, T.; Srisuwan, W.; Techapun, C.; Watanabe, M.; Takenaka, S. Direct bioconversion of rice residue from canteen waste into lipids by new amylolytic oleaginous yeast *Sporidiobolus pararoseus* KX709872. *Prep. Biochem. Biotechnol.* **2018**, *6068*, 1–11. [CrossRef] [PubMed]
44. Selvakumar, P.; Sivashanmugam, P. Study on Lipid Accumulation in Novel Oleaginous Yeast *Naganishia liquefaciens* NITTS2 Utilizing Pre-digested Municipal Waste Activated Sludge: A Low-cost Feedstock for Biodiesel Production. *Appl. Biochem. Biotechnol.* **2018**. [CrossRef] [PubMed]
45. Chatterjee, S.; Mohan, S.V. Microbial lipid production by *Cryptococcus curvatus* from vegetable waste hydrolysate. *Bioresour. Technol.* **2018**, *254*, 284–289. [CrossRef] [PubMed]
46. Deeba, F.; Patel, A.; Arora, N.; Pruthi, V.; Pruthi, P.A.; Negi, Y.S. Amaranth seeds (*Amaranthus palmeri* L.) as novel feedstock for biodiesel production by oleaginous yeast. *Environ. Sci. Pollut. Res.* **2018**, *25*, 353–362. [CrossRef] [PubMed]
47. Bonturi, N.; Matsakas, L.; Nilsson, R.; Christakopoulos, P.; Miranda, E.A.; Berglund, K.A.; Rova, U. Single cell oil producing yeasts *Lipomyces starkeyi* and *Rhodosporidium toruloides*: Selection of extraction strategies and biodiesel property prediction. *Energies* **2015**, *8*, 5040–5052. [CrossRef]
48. Lin, Y.; Xie, X.; Yuan, B.; Fu, J.; Liu, L.; Tian, H.; Chen, T.; He, D. Optimization of Enzymatic Cell Disruption for Improving Lipid Extraction from *Schizochytrium* sp. through Response Surface Methodology. *J. Oleo Sci.* **2018**, *67*, 215–224. [CrossRef] [PubMed]
49. Vandamme, D.; Gheysen, L.; Muylaert, K.; Foubert, I. Impact of harvesting method on total lipid content and extraction efficiency for *Phaeodactylum tricornutum*. *Sep. Purif. Technol.* **2018**, *194*, 362–367. [CrossRef]
50. Zhang, Y.; Kong, X.; Wang, Z.; Sun, Y.; Zhu, S.; Li, L.; Lv, P. Optimization of enzymatic hydrolysis for effective lipid extraction from microalgae *Scenedesmus* sp. *Renew. Energy* **2018**, *125*, 1049–1057. [CrossRef]

51. Ellison, C.R.; Overa, S.; Boldor, D. Central composite design parameterization of microalgae/cyanobacteria co-culture pretreatment for enhanced lipid extraction using an external clamp-on ultrasonic transducer. *Ultrason. Sonochem.* **2018**, in press. [CrossRef] [PubMed]
52. Kwak, M.; Kang, S.G.; Hong, W.K.; Han, J.I.; Chang, Y.K. Simultaneous cell disruption and lipid extraction of wet *Aurantiochytrium* sp. KRS101 using a high shear mixer. *Bioprocess Biosyst. Eng.* **2018**, 1–8. [CrossRef] [PubMed]
53. Ansari, F.A.; Gupta, S.K.; Nasr, M.; Rawat, I.; Bux, F. Evaluation of various cell drying and disruption techniques for sustainable metabolite extractions from microalgae grown in wastewater: A multivariate approach. *J. Clean. Prod.* **2018**, *182*, 634–643. [CrossRef]
54. Ju, C.; Wang, F.; Huang, Y.; Fang, Y. Selective extraction of neutral lipid from wet algae paste and subsequently hydroconversion into renewable jet fuel. *Renew. Energy* **2018**, *118*, 521–526. [CrossRef]
55. Carvalho, A.K.F.; Bento, H.B.S.; Izário Filho, H.J.; de Castro, H.F. Approaches to convert *Mucor circinelloides* lipid into biodiesel by enzymatic synthesis assisted by microwave irradiations. *Renew. Energy* **2018**, *125*. [CrossRef]
56. Al-Hawash, A.B.; Li, S.; Zhang, X.; Zhang, X.; Ma, F. Productivity of γ-Linoleic acid by oleaginous fungus *Cunninghamella echinulata* using a pulsed high magnetic field. *Food Biosci.* **2018**, *21*, 1–7. [CrossRef]
57. Forfang, K.; Zimmermann, B.; Kosa, G.; Kohler, A.; Shapaval, V. FTIR spectroscopy for evaluation and monitoring of lipid extraction efficiency for oleaginous fungi. *PLoS ONE* **2017**, *12*, e0170611. [CrossRef] [PubMed]
58. Kosa, G.; Kohler, A.; Tafintseva, V.; Zimmermann, B.; Forfang, K.; Afseth, N.K.; Tzimorotas, D.; Vuoristo, K.S.; Horn, S.J.; Mounier, J.; et al. Microtiter plate cultivation of oleaginous fungi and monitoring of lipogenesis by high-throughput FTIR spectroscopy. *Microb. Cell Fact.* **2017**, *16*, 1–12. [CrossRef] [PubMed]
59. Bagy, M.M.K.; Abd-Alla, M.H.; Morsy, F.M.; Hassan, E.A. Two stage biodiesel and hydrogen production from molasses by oleaginous fungi and *Clostridium acetobutylicum* ATCC 824. *Int. J. Hydrogen Energy* **2014**, *39*, 3185–3197. [CrossRef]
60. Santala, S.; Efimova, E.; Kivinen, V.; Larjo, A.; Aho, T.; Karp, M.; Santala, V. Improved Triacylglycerol Production in *Acinetobacter baylyi* ADP1 by Metabolic Engineering. *Microb. Cell Fact.* **2011**, *10*, 1–10. [CrossRef] [PubMed]
61. Zhang, Q.; Li, Y.; Xia, L. An oleaginous endophyte *Bacillus subtilis* HB1310 isolated from thin-shelled walnut and its utilization of cotton stalk hydrolysate for lipid production. *Biotechnol. Biofuels* **2014**, *7*, 152. [CrossRef] [PubMed]
62. Carlozzi, P.; Buccioni, A.; Minieri, S.; Pushparaj, B.; Piccardi, R.; Ena, A.; Pintucci, C. Production of bio-fuels (hydrogen and lipids) through a photofermentation process. *Bioresour. Technol.* **2010**, *101*, 3115–3120. [CrossRef] [PubMed]
63. Cea, M.; Sangaletti-Gerhard, N.; Acuña, P.; Fuentes, I.; Jorquera, M.; Godoy, K.; Osses, F.; Navia, R. Screening transesterifiable lipid accumulating bacteria from sewage sludge for biodiesel production. *Biotechnol. Rep.* **2015**, *8*, 116–123. [CrossRef] [PubMed]
64. Goswami, L.; Tejas Namboodiri, M.M.; Vinoth Kumar, R.; Pakshirajan, K.; Pugazhenthi, G. Biodiesel production potential of oleaginous *Rhodococcus opacus* grown on biomass gasification wastewater. *Renew. Energy* **2017**, *105*, 400–406. [CrossRef]
65. Harun, R.; Singh, M.; Forde, G.M.; Danquah, M.K. Bioprocess engineering of microalgae to produce a variety of consumer products. *Renew. Sustain. Energy Rev.* **2010**, *14*, 1037–1047. [CrossRef]
66. Ramesh, D. Lipid identification and extraction techniques. In *Biotechnological Applications of Microalgae: Biodiesel and Value-Added Products*; Bux, F., Ed.; CRC Press: Boca Raton, FL, USA, 2013; pp. 89–97.
67. Topare, N.S.; Raut, S.J.; Renge, V.C.; Khedkar, S.V.; Chavan, Y.P.; Bhagat, S.L. Extraction of oil from algae by solvent extraction and oil expeller method. *Int. J. Chem. Sci.* **2011**, *9*, 1746–1750.
68. Mubarak, M.; Shaija, A.; Suchithra, T.V. A review on the extraction of lipid from microalgae for biodiesel production. *Algal Res.* **2015**, *7*, 117–123. [CrossRef]
69. Johnson, M.B.; Wen, Z. Production of biodiesel fuel from the microalga schizochytrium limacinum by direct transesterification of algal biomass. *Energy Fuels* **2009**, *23*, 5179–5183. [CrossRef]
70. Balasundaram, B.; Skill, S.C.; Llewellyn, C.A. A low energy process for the recovery of bioproducts from cyanobacteria using a ball mill. *Biochem. Eng. J.* **2012**, *69*, 48–56. [CrossRef]

71. Bunge, F.; Pietzsch, M.; Müller, R.; Syldatk, C. Mechanical disruption of *Arthrobacter* sp. DSM 3747 in stirred ball mills for the release of hydantoin-cleaving enzymes. *Chem. Eng. Sci.* **1992**, *47*, 225–232. [CrossRef]
72. Postma, P.R.; Miron, T.L.; Olivieri, G.; Barbosa, M.J.; Wijffels, R.H.; Eppink, M.H.M. Mild disintegration of the green microalgae *Chlorella vulgaris* using bead milling. *Bioresour. Technol.* **2015**, *184*, 297–304. [CrossRef] [PubMed]
73. Montalescot, V.; Rinaldi, T.; Touchard, R.; Jubeau, S.; Frappart, M.; Jaouen, P.; Bourseau, P.; Marchal, L. Optimization of bead milling parameters for the cell disruption of microalgae: Process modeling and application to *Porphyridium cruentum* and *Nannochloropsis oculata*. *Bioresour. Technol.* **2015**, *196*, 339–346. [CrossRef] [PubMed]
74. Doucha, J.; Lívanský, K. Influence of processing parameters on disintegration of *Chlorella* cells in various types of homogenizers. *Appl. Microbiol. Biotechnol.* **2008**, *81*, 431–440. [CrossRef] [PubMed]
75. Meullemiestre, A.; Breil, C.; Abert-Vian, M.; Chemat, F. Microwave, ultrasound, thermal treatments, and bead milling as intensification techniques for extraction of lipids from oleaginous *Yarrowia lipolytica* yeast for a biojetfuel application. *Bioresour. Technol.* **2016**, *211*, 190–199. [CrossRef] [PubMed]
76. Samarasinghe, N. Effect of High Pressure Homogenization on Aqueous Phase Solvent Extraction of Lipids from Nannochloris Oculata Microalgae. *J. Energy Nat. Resour.* **2012**, *1*. [CrossRef]
77. Coccaro, N.; Ferrari, G.; Donsì, F. Understanding the break-up phenomena in an orifice-valve high pressure homogenizer using spherical bacterial cells (*Lactococcus lactis*) as a model disruption indicator. *J. Food Eng.* **2018**. [CrossRef]
78. Ekpeni, L.E.N.; Benyounis, K.Y.; Nkem-Ekpeni, F.F.; Stokes, J.; Olabi, A.G. Underlying factors to consider in improving energy yield from biomass source through yeast use on high-pressure homogenizer (hph). *Energy* **2015**, *81*, 74–83. [CrossRef]
79. Shene, C.; Monsalve, M.T.; Vergara, D.; Lienqueo, M.E.; Rubilar, M. High pressure homogenization of *Nannochloropsis oculata* for the extraction of intracellular components: Effect of process conditions and culture age. *Eur. J. Lipid Sci. Technol.* **2016**, *118*, 631–639. [CrossRef]
80. Halim, R.; Rupasinghe, T.W.T.; Tull, D.L.; Webley, P.A. Mechanical cell disruption for lipid extraction from microalgal biomass. *Bioresour. Technol.* **2013**, *140*, 53–63. [CrossRef] [PubMed]
81. Donsì, F.; Annunziata, M.; Ferrari, G. Microbial inactivation by high pressure homogenization: Effect of the disruption valve geometry. *J. Food Eng.* **2013**, *115*, 362–370. [CrossRef]
82. Floury, J.; Legrand, J.; Desrumaux, A. Analysis of a new type of high pressure homogeniser. Part B. study of droplet break-up and recoalescence phenomena. *Chem. Eng. Sci.* **2004**, *59*, 1285–1294. [CrossRef]
83. Lee, L.; Norton, I.T. Comparing droplet breakup for a high-pressure valve homogeniser and a Microfluidizer for the potential production of food-grade nanoemulsions. *J. Food Eng.* **2013**, *114*, 158–163. [CrossRef]
84. Olmstead, I.L.D.; Kentish, S.E.; Scales, P.J.; Martin, G.J.O. Low solvent, low temperature method for extracting biodiesel lipids from concentrated microalgal biomass. *Bioresour. Technol.* **2013**, *148*, 615–619. [CrossRef] [PubMed]
85. Hua, X.; Xu, S.; Wang, M.; Chen, Y.; Yang, H.; Yang, R. Effects of high-speed homogenization and high-pressure homogenization on structure of tomato residue fibers. *Food Chem.* **2017**, *232*, 443–449. [CrossRef] [PubMed]
86. Zhang, A.; Deng, J.; Liu, X.; He, P.; He, L.; Zhang, F.; Linhardt, R.J.; Sun, P. Structure and conformation of α-glucan extracted from *Agaricus blazei* Murill by high-speed shearing homogenization. *Int. J. Biol. Macromol.* **2018**, *113*, 558–564. [CrossRef] [PubMed]
87. Araujo, G.S.; Matos, L.J.B.L.; Fernandes, J.O.; Cartaxo, S.J.M.; Gonçalves, L.R.B.; Fernandes, F.A.N.; Farias, W.R.L. Extraction of lipids from microalgae by ultrasound application: Prospection of the optimal extraction method. *Ultrason. Sonochem.* **2013**, *20*, 95–98. [CrossRef] [PubMed]
88. Šoštarič, M.; Klinar, D.; Bricelj, M.; Golob, J.; Berovič, M.; Likozar, B. Growth, lipid extraction and thermal degradation of the microalga *Chlorella vulgaris*. *New Biotechnol.* **2012**, *29*, 325–331. [CrossRef] [PubMed]
89. Lee, J.-Y.; Yoo, C.; Jun, S.-Y.; Ahn, C.-Y.; Oh, H.-M. Comparison of several methods for effective lipid extraction from microalgae. *Bioresour. Technol.* **2010**, *101*, S75–S77. [CrossRef] [PubMed]
90. Wiyarno, B.; Yunus, R.M.; Mel, M. Extraction of Algae Oil from *Nannocloropsis* sp.: A Study of Soxhlet and Ultrasonic-Assisted Extractions. *J. Appl. Sci.* **2011**, *11*, 3607–3612. [CrossRef]
91. Cravotto, G.; Boffa, L.; Mantegna, S.; Perego, P.; Avogadro, M.; Cintas, P. Improved extraction of vegetable oils under high-intensity ultrasound and/or microwaves. *Ultrason. Sonochem.* **2008**, *15*, 898–902. [CrossRef] [PubMed]

92. Mecozzi, M.; Amici, M.; Romanelli, G.; Pietrantonio, E.; Deluca, A. Ultrasound extraction and thin layer chromatography-flame ionization detection analysis of the lipid fraction in marine mucilage samples. *J. Chromatogr. A* **2002**, *963*, 363–373. [CrossRef]
93. Prabakaran, P.; Ravindran, A.D. A comparative study on effective cell disruption methods for lipid extraction from microalgae. *Lett. Appl. Microbiol.* **2011**, *53*, 150–154. [CrossRef] [PubMed]
94. Ríos, S.D.; Castañeda, J.; Torras, C.; Farriol, X.; Salvadó, J. Lipid extraction methods from microalgal biomass harvested by two different paths: Screening studies toward biodiesel production. *Bioresour. Technol.* **2013**, *133*, 378–388. [CrossRef] [PubMed]
95. Zhang, X.; Yan, S.; Tyagi, R.D.; Drogui, P.; Surampalli, R.Y. Ultrasonication assisted lipid extraction from oleaginous microorganisms. *Bioresour. Technol.* **2014**, *158*, 253–261. [CrossRef] [PubMed]
96. Wu, X.; Joyce, E.M.; Mason, T.J. Evaluation of the mechanisms of the effect of ultrasound on *Microcystis aeruginosa* at different ultrasonic frequencies. *Water Res.* **2012**, *46*, 2851–2858. [CrossRef] [PubMed]
97. Wang, M.; Yuan, W.; Jiang, X.; Jing, Y.; Wang, Z. Disruption of microalgal cells using high-frequency focused ultrasound. *Bioresour. Technol.* **2014**, *153*, 315–321. [CrossRef] [PubMed]
98. Martínez-Palou, R. Microwave-assisted synthesis using ionic liquids. *Mol. Divers.* **2010**, *14*, 3–25. [CrossRef] [PubMed]
99. Budarin, V.L.; Shuttleworth, P.S.; De Bruyn, M.; Farmer, T.J.; Gronnow, M.J.; Pfaltzgraff, L.; Macquarrie, D.J.; Clark, J.H. The potential of microwave technology for the recovery, synthesis and manufacturing of chemicals from bio-wastes. *Catal. Today* **2015**, *239*, 80–89. [CrossRef]
100. Wahlen, B.D.; Willis, R.M.; Seefeldt, L.C. Biodiesel production by simultaneous extraction and conversion of total lipids from microalgae, cyanobacteria, and wild mixed-cultures. *Bioresour. Technol.* **2011**, *102*, 2724–2730. [CrossRef] [PubMed]
101. Martinez-Guerra, E.; Gude, V.G.; Mondala, A.; Holmes, W.; Hernandez, R. Extractive-transesterification of algal lipids under microwave irradiation with hexane as solvent. *Bioresour. Technol.* **2014**, *156*, 240–247. [CrossRef] [PubMed]
102. Teo, C.L.; Idris, A. Enhancing the various solvent extraction method via microwave irradiation for extraction of lipids from marine microalgae in biodiesel production. *Bioresour. Technol.* **2014**, *171*, 477–481, Ahead of Print. [CrossRef] [PubMed]
103. Zheng, H.; Yin, J.; Gao, Z.; Huang, H.; Ji, X.; Dou, C. Disruption of chlorella vulgaris cells for the release of biodiesel-producing lipids: A comparison of grinding, ultrasonication, bead milling, enzymatic lysis, and microwaves. *Appl. Biochem. Biotechnol.* **2011**, *164*, 1215–1224. [CrossRef] [PubMed]
104. Mendes-Pinto, M.M.; Raposo, M.F.J.; Bowen, J.; Young, A.J.; Morais, R. Evaluation of different cell disruption process on encysted cells of *Haematococcus pluvialis*. *J. Appl. Phycol.* **2001**, *13*, 19–24. [CrossRef]
105. Rakesh, S.; Dhar, D.W.; Prasanna, R.; Saxena, A.K.; Saha, S.; Shukla, M.; Sharma, K. Cell disruption methods for improving lipid extraction efficiency in unicellular microalgae. *Eng. Life Sci.* **2015**, *15*, 443–447. [CrossRef]
106. Florentino de Souza Silva, A.P.; Costa, M.C.; Colzi Lopes, A.; Fares Abdala Neto, E.; Carrhá Leitão, R.; Mota, C.R.; Bezerra dos Santos, A. Comparison of pretreatment methods for total lipids extraction from mixed microalgae. *Renew. Energy* **2014**, *63*, 762–766. [CrossRef]
107. Tsong, T.Y. Electroporation of cell membranes. Minireview. *Biophys. J.* **1991**, *60*, 297–306. [CrossRef]
108. Weaver, J.C.; Harrison, G.I.; Bliss, J.G.; Mourant, J.R.; Powell, K.T. Electroporation: High frequency of occurrence of a transient high-permeability state in erythrocytes and intact yeast. *FEBS Lett.* **1988**, *229*, 30–34. [CrossRef]
109. Eing, C.; Goettel, M.; Straessner, R.; Gusbeth, C.; Frey, W. Pulsed electric field treatment of microalgae—Benefits for microalgae biomass processing. *IEEE Trans. Plasma Sci.* **2013**, *41*, 2901–2907. [CrossRef]
110. Sheng, J.; Vannela, R.; Rittmann, B.E. Evaluation of cell-disruption effects of pulsed-electric-field treatment of *Synechocystis* PCC 6803. *Environ. Sci. Technol.* **2011**, *45*, 3795–3802. [CrossRef] [PubMed]
111. Zbinden, M.D.A.; Sturm, B.S.M.; Nord, R.D.; Carey, W.J.; Moore, D.; Shinogle, H.; Stagg-Williams, S.M. Pulsed electric field (PEF) as an intensification pretreatment for greener solvent lipid extraction from microalgae. *Biotechnol. Bioeng.* **2013**, *110*, 1605–1615. [CrossRef] [PubMed]
112. Byreddy, A.R.; Gupta, A.; Barrow, C.J.; Puri, M. Comparison of cell disruption methods for improving lipid extraction from thraustochytrid strains. *Mar. Drugs* **2015**, *13*, 5111–5127. [CrossRef] [PubMed]

113. Cheng, J.; Sun, J.; Huang, Y.; Feng, J.; Zhou, J.; Cen, K. Dynamic microstructures and fractal characterization of cell wall disruption for microwave irradiation-assisted lipid extraction from wet microalgae. *Bioresour. Technol.* **2013**, *150*, 67–72. [CrossRef] [PubMed]
114. McMillan, J.R.; Watson, I.A.; Ali, M.; Jaafar, W. Evaluation and comparison of algal cell disruption methods: Microwave, waterbath, blender, ultrasonic and laser treatment. *Appl. Energy* **2013**, *103*, 128–134. [CrossRef]
115. Yu, Q.; Zhuang, X.; Lv, S.; He, M.; Zhang, Y.; Yuan, Z.; Qi, W.; Wang, Q.; Wang, W.; Tan, X. Liquid hot water pretreatment of sugarcane bagasse and its comparison with chemical pretreatment methods for the sugar recovery and structural changes. *Bioresour. Technol.* **2013**, *129*, 592–598. [CrossRef] [PubMed]
116. Park, J.Y.; Nam, B.; Choi, S.A.; Oh, Y.K.; Lee, J.S. Effects of anionic surfactant on extraction of free fatty acid from *Chlorella vulgaris*. *Bioresour. Technol.* **2014**, *166*, 620–624. [CrossRef] [PubMed]
117. Park, J.-Y.; Oh, Y.-K.; Lee, J.-S.; Lee, K.; Jeong, M.-J.; Choi, S.-A. Acid-catalyzed hot-water extraction of lipids from *Chlorella vulgaris*. *Bioresour. Technol.* **2014**, *153*, 408–412. [CrossRef] [PubMed]
118. Choi, S.A.; Jung, J.Y.; Kim, K.; Lee, J.S.; Kwon, J.H.; Kim, S.W.; Yang, J.W.; Park, J.Y. Acid-catalyzed hot-water extraction of docosahexaenoic acid (DHA)-rich lipids from *Aurantiochytrium* sp. KRS101. *Bioresour. Technol.* **2014**, *161*, 469–472. [CrossRef] [PubMed]
119. Lee, A.K.; Lewis, D.M.; Ashman, P.J. Disruption of microalgal cells for the extraction of lipids for biofuels: Processes and specific energy requirements. *Biomass Bioenergy* **2012**, *46*, 89–101. [CrossRef]
120. Ferreira, A.F.; Dias, A.P.S.; Silva, C.M.; Costa, M. Effect of low frequency ultrasound on microalgae solvent extraction: Analysis of products, energy consumption and emissions. *Algal Res.* **2016**, *14*, 9–16. [CrossRef]
121. Keris-Sen, U.D.; Sen, U.; Soydemir, G.; Gurol, M.D. An investigation of ultrasound effect on microalgal cell integrity and lipid extraction efficiency. *Bioresour. Technol.* **2014**, *152*, 407–413. [CrossRef] [PubMed]
122. Popoola, T.O.S.; Yangomodou, O.D. Extraction, properties and utilization potentials of cassava seed oil. *Biotechnology* **2006**, *5*, 38–41. [CrossRef]
123. Gogate, P.R.; Kabadi, A.M. A review of applications of cavitation in biochemical engineering/biotechnology. *Biochem. Eng. J.* **2009**, *44*, 60–72. [CrossRef]
124. Yusaf, T.; Al-Juboori, R.A. Alternative methods of microorganism disruption for agricultural applications. *Appl. Energy* **2014**, *114*, 909–923. [CrossRef]
125. Follows, M.; Hetherington, P.J.; Dunnill, P.; Lilly, M.D. Release of enzymes from bakers' yeast by disruption in an industrial homogenizer. *Biotechnol. Bioeng.* **1971**, *13*, 549–560. [CrossRef] [PubMed]
126. Currie, J.A.; Dunnill, P.; Lilly, M.D. Release of protein from Bakers' yeast (*Saccharomyces cerevisiae*) by disruption in an industrial agitator mill. *Biotechnol. Bioeng.* **1972**, *14*, 725–736. [CrossRef]
127. Safi, C.; Camy, S.; Frances, C.; Varela, M.M.; Badia, E.C.; Pontalier, P.Y.; Vaca-Garcia, C. Extraction of lipids and pigments of *Chlorella vulgaris* by supercritical carbon dioxide: Influence of bead milling on extraction performance. *J. Appl. Phycol.* **2014**, *26*, 1711–1718. [CrossRef]
128. Clavijo Rivera, E.; Montalescot, V.; Viau, M.; Drouin, D.; Bourseau, P.; Frappart, M.; Monteux, C.; Couallier, E. Mechanical cell disruption of *Parachlorella kessleri* microalgae: Impact on lipid fraction composition. *Bioresour. Technol.* **2018**, *256*, 77–85. [CrossRef] [PubMed]
129. Cheng, J.; Huang, R.; Li, T.; Zhou, J.; Cen, K. Biodiesel from wet microalgae: Extraction with hexane after the microwave-assisted transesterification of lipids. *Bioresour. Technol.* **2014**, *170*, 69–75. [CrossRef] [PubMed]
130. Nadar, S.S.; Rao, P.; Rathod, V.K. Enzyme assisted extraction of biomolecules as an approach to novel extraction technology: A review. *Food Res. Int.* **2018**, *108*, 309–330. [CrossRef] [PubMed]
131. Sowbhagya, H.B.; Srinivas, P.; Krishnamurthy, N. Effect of enzymes on extraction of volatiles from celery seeds. *Food Chem.* **2010**, *120*, 230–234. [CrossRef]
132. Puri, M.; Sharma, D.; Barrow, C.J. Enzyme-assisted extraction of bioactives from plants. *Trends Biotechnol.* **2012**, *30*, 37–44. [CrossRef] [PubMed]
133. Zuorro, A.; Maffei, G.; Lavecchia, R. Optimization of enzyme-assisted lipid extraction from *Nannochloropsis microalgae*. *J. Taiwan Inst. Chem. Eng.* **2016**, *67*, 106–114. [CrossRef]
134. Latif, S.; Anwar, F. Physicochemical studies of hemp (*Cannabis sativa*) seed oil using enzyme-assisted cold-pressing. *Eur. J. Lipid Sci. Technol.* **2009**, *111*, 1042–1048. [CrossRef]
135. Shankar, D.; Agrawal, Y.C.; Sarkar, B.C.; Singh, B.P.N. Enzymatic hydrolysis in conjunction with conventional pretreatments to soybean for enhanced oil availability and recovery. *JAOCS J. Am. Oil Chem. Soc.* **1997**, *74*, 1543–1547. [CrossRef]

136. Cho, H.S.; Oh, Y.K.; Park, S.C.; Lee, J.W.; Park, J.Y. Effects of enzymatic hydrolysis on lipid extraction from *Chlorella vulgaris*. *Renew. Energy* **2013**, *54*, 156–160. [CrossRef]
137. Maffei, G.; Bracciale, M.P.; Broggi, A.; Zuorro, A.; Santarelli, M.L.; Lavecchia, R. Effect of an enzymatic treatment with cellulase and mannanase on the structural properties of *Nannochloropsis microalgae*. *Bioresour. Technol.* **2018**, *249*, 592–598. [CrossRef] [PubMed]
138. Harun, R.; Danquah, M.K. Enzymatic hydrolysis of microalgal biomass for bioethanol production. *Chem. Eng. J.* **2011**, *168*, 1079–1084. [CrossRef]
139. Sathish, A.; Sims, R.C. Biodiesel from mixed culture algae via a wet lipid extraction procedure. *Bioresour. Technol.* **2012**, *118*, 643–647. [CrossRef] [PubMed]
140. Kostić, M.D.; Joković, N.M.; Stamenković, O.S.; Rajković, K.M.; Milić, P.S.; Veljković, V.B. Optimization of hempseed oil extraction by n-hexane. *Ind. Crops Prod.* **2013**, *48*, 133–143. [CrossRef]
141. Sawangkeaw, R.; Ngamprasertsith, S. A review of lipid-based biomasses as feedstocks for biofuels production. *Renew. Sustain. Energy Rev.* **2013**, *25*, 97–108. [CrossRef]
142. Kumar, D.; Singh, B.; Korstad, J. Utilization of lignocellulosic biomass by oleaginous yeast and bacteria for production of biodiesel and renewable diesel. *Renew. Sustain. Energy Rev.* **2017**, *73*, 654–671. [CrossRef]
143. Bai, X.; Ghasemi Naghdi, F.; Ye, L.; Lant, P.; Pratt, S. Enhanced lipid extraction from algae using free nitrous acid pretreatment. *Bioresour. Technol.* **2014**, *159*, 36–40. [CrossRef] [PubMed]
144. Boyd, A.R.; Champagne, P.; McGinn, P.J.; MacDougall, K.M.; Melanson, J.E.; Jessop, P.G. Switchable hydrophilicity solvents for lipid extraction from microalgae for biofuel production. *Bioresour. Technol.* **2012**, *118*, 628–632. [CrossRef] [PubMed]
145. Dejoye Tanzi, C.; Abert Vian, M.; Chemat, F. New procedure for extraction of algal lipids from wet biomass: A green clean and scalable process. *Bioresour. Technol.* **2013**, *134*, 271–275. [CrossRef] [PubMed]
146. Du, Y.; Schuur, B.; Samorì, C.; Tagliavini, E.; Brilman, D.W.F. Secondary amines as switchable solvents for lipid extraction from non-broken microalgae. *Bioresour. Technol.* **2013**, *149*, 253–260. [CrossRef] [PubMed]
147. Kim, Y.H.; Choi, Y.K.; Park, J.; Lee, S.; Yang, Y.H.; Kim, H.J.; Park, T.J.; Hwan Kim, Y.; Lee, S.H. Ionic liquid-mediated extraction of lipids from algal biomass. *Bioresour. Technol.* **2012**, *109*, 312–315. [CrossRef] [PubMed]
148. Lee, Y.C.; Huh, Y.S.; Farooq, W.; Chung, J.; Han, J.I.; Shin, H.J.; Jeong, S.H.; Lee, J.S.; Oh, Y.K.; Park, J.Y. Lipid extractions from docosahexaenoic acid (DHA)-rich and oleaginous *Chlorella* sp. biomasses by organic-nanoclays. *Bioresour. Technol.* **2013**, *137*, 74–81. [CrossRef] [PubMed]
149. Steriti, A.; Rossi, R.; Concas, A.; Cao, G. A novel cell disruption technique to enhance lipid extraction from microalgae. *Bioresour. Technol.* **2014**, *164*, 70–77. [CrossRef] [PubMed]
150. Jo, Y.J.; Lee, O.K.; Lee, E.Y. Dimethyl carbonate-mediated lipid extraction and lipase-catalyzed in situ transesterification for simultaneous preparation of fatty acid methyl esters and glycerol carbonate from *Chlorella* sp. KR-1 biomass. *Bioresour. Technol.* **2014**, *158*, 105–110. [CrossRef] [PubMed]
151. Godwin, C.M.; Hietala, D.C.; Lashaway, A.R.; Narwani, A.; Savage, P.E.; Cardinale, B.J. Algal polycultures enhance coproduct recycling from hydrothermal liquefaction. *Bioresour. Technol.* **2017**, *224*, 630–638. [CrossRef] [PubMed]
152. Hietala, D.C.; Koss, C.K.; Narwani, A.; Lashaway, A.R.; Godwin, C.M.; Cardinale, B.J.; Savage, P.E. Influence of biodiversity, biochemical composition, and species identity on the quality of biomass and biocrude oil produced via hydrothermal liquefaction. *Algal Res.* **2017**, *26*, 203–214. [CrossRef]
153. Xu, D.; Savage, P.E. Effect of temperature, water loading, and Ru/C catalyst on water-insoluble and water-soluble biocrude fractions from hydrothermal liquefaction of algae. *Bioresour. Technol.* **2017**, *239*, 1–6. [CrossRef] [PubMed]
154. Sheehan, J.D.; Savage, P.E. Modeling the effects of microalga biochemical content on the kinetics and biocrude yields from hydrothermal liquefaction. *Bioresour. Technol.* **2017**, *239*, 144–150. [CrossRef] [PubMed]
155. Hietala, D.C.; Faeth, J.L.; Savage, P.E. A quantitative kinetic model for the fast and isothermal hydrothermal liquefaction of *Nannochloropsis* sp. *Bioresour. Technol.* **2016**, *214*, 102–111. [CrossRef] [PubMed]

molecules

Review

Pre-treatment of Oil Palm Biomass for Fermentable Sugars Production

Nur Fatin Athirah Ahmad Rizal [1], Mohamad Faizal Ibrahim [1,2,*], Mohd Rafein Zakaria [1,2], Suraini Abd-Aziz [1], Phang Lai Yee [1] and Mohd Ali Hassan [1]

1 Department of Bioprocess Technology, Faculty of Biotechnology and Biomolecular Sciences, Universiti Putra Malaysia, 43400 UPM Serdang, Selangor, Malaysia; nurfatinrizal@gmail.com (N.F.A.A.R.); mohdrafein@upm.edu.my (M.R.Z.); suraini@upm.edu.my (S.A.-A.); phanglaiyee@upm.edu.my (P.L.Y.); alihas@upm.edu.my (M.A.H.)

2 Laboratory of Biopolymer and Derivatives, Institute of Tropical Forestry and Forest Products, Universiti Putra Malaysia, 43400 UPM Serdang, Selangor, Malaysia

* Correspondence: faizal_ibrahim@upm.edu.my; Tel.: +603-8947-1936

Received: 24 April 2018; Accepted: 31 May 2018; Published: 7 June 2018

Abstract: Malaysia is the second largest palm oil producer in the world and this industry generates more than 80 million tonnes of biomass every year. When considering the potential of this biomass to be used as a fermentation feedstock, many studies have been conducted to develop a complete process for sugar production. One of the essential processes is the pre-treatment to modify the lignocellulosic components by altering the structural arrangement and/or removing lignin component to expose the internal structure of cellulose and hemicellulose for cellulases to digest it into sugars. Each of the pre-treatment processes that were developed has their own advantages and disadvantages, which are reviewed in this study.

Keywords: lignocellulosic biomass; fermentation; pre-treatment; palm oil process flow

1. Introduction

Oil palm tree (*Elaeis guineensis jacq.*) was introduced to Malaysia by British in early 1870′s as an ornamental or decorative plant. The first commercial planting of oil palm was in 1917, in Tennamaran Estate in Selangor. The demand for oil during the industrial revolution in the 19th century as a lubricant in steam engines and machinery, and soap has led to the production of palm oil, which is also introduced to reduce the country's economic dependence on rubber and tin [1]. Now, Malaysia is the second largest of world palm oil producers and exporters that accounts for more than 30% of world palm oil production and 37% of world exports in 2016 [2]. Being one of the biggest producers and exporters of crude palm oil (CPO) and palm oil products, Malaysia has an important role to fulfil the global need for oils and fats.

In line with the increase of the palm oil production capacity in Malaysia, a large amount of waste is being generated from this industry. Processing fresh fruit bunch (FFB) in the mill generates oil palm empty fruit bunch (OPEFB), oil palm mesocarp fiber (OPMF), oil palm kernel shell (OPKS), and a large amount of palm oil mill effluent (POME). Some biomass is also generated in the plantation area, such as oil palm frond (OPF) and oil palm trunk (OPT). In total, the palm oil industry generates more than 80 million tonnes of oil palm biomass in 2016, and the value will keep on increasing to fulfil the demand [3]. In order to maintain the sustainability of palm oil industry, a proper waste management has been developed and continuously improved to meet the economic and environmental challenges.

At the moment, the palm oil industries are still practicing the traditional waste management, with only little improvement. The oil palm biomasses, such as OPKS and OPMF, are still burned in a boiler to generate steam and electricity for mill operation [4]. The OPEFB is dumped at the mill without a proper treatment or brought to the plantation for mulching [5]. POME is still treated in ponds, whereby the final discharge still pollutes the environment, including changing the biodiversity of a nearby river [6]. Only a few factories have already implemented anaerobic digesters for treating POME and collecting methane [7]. Some factories, such as Global Green Synergy (http://www.ggs.my/) and Bionik Fertilizer (http://www.kulimnursery.com/), have started producing biocompost, biocharcoal, dried fiber, and pellets from oil palm biomass. Nevertheless, there is large potential for this biomass to be converted into various value-added products that will generate additional income to the industry, and at the same time, reduce the impact on the environment.

2. Oil Palm Biomass

Palm oil is the most important product for Malaysia that has helped to change the scenario of Malaysia agriculture and economy perspective. Palm oil industry provides a high economic return for Malaysia. The growth national income (GNI) is RM80 billion, which place this industry as the fourth GNI contributor in the country [8]. This industry becomes larger from year to year due to the world requirement on palm oil products. The plantation area for oil palm has gradually increased from 0.5 million hectares in early 1975 to 4.5 million hectares in late 2006 [9]. In 2015, Malaysia had 5.64 million hectares of oil palm plantations [10]. With the growth of palm oil industry, the amount of biomass residues generated also shows a significant increase. As a leading industry in the world's oil production, the palm oil industry has left behind a large amount of biomass from its plantation and milling activities as compared to other types of agricultural biomass. Palm oil industry in Malaysia generates approximately 83 million tonnes of oil palm biomass in 2012 and it is expected to grow to be more than 100 million tonnes by 2020 [8]. Production of biomass from the Malaysia's palm oil industry is shown in Figure 1.

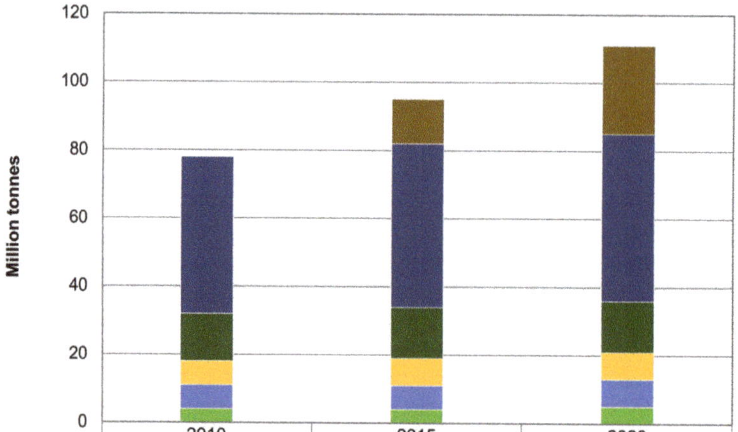

	2010	2015	2020
Forecast	0	13	26
OPF	46	48	49
OPT	14	15	15
OPMF	7	8	8
OPEFB	7	7	8
OPKS	4	4	5

Figure 1. Production of oil palm biomass, i.e.,: oil palm frond (OPF), oil palm trunk (OPT), oil palm mesocarp fiber (OPMF), oil palm empty fruit bunch (OPEFB), and oil palm kernel shell (OPKS). Data obtained from Malaysia Innovation Agency [8].

The process flow of palm oil production at the mill with the production of oil palm biomass is illustrated in Figure 2. It shows that in every input of 100,000 tonnes of FFB for processing palm oil, a total of 43,700 tonnes of oil palm biomass residuals were generated. This value accounted for 23,000 tonnes of OPEFB, 5000 tonnes of OPKS, and 15,700 tonnes of OPMF. It should be noted that the OPEFB is the most abundant biomass produced from the palm oil mill. This value does not include the amount of biomass in wastewater, which is known as palm oil mill effluent (POME), which generates 69,000 tonnes per 100,000 tonnes of FFB input. In total, the palm oil mills in Malaysia generate 7.34 million tonnes of OPEFB, 7.72 million tonnes of OPMF, 4.46 million tonnes of OPKS, and 64 million tonnes of POME per year [3]. In the palm oil mill, only 10% of biomass is reused for electricity generation, while the remaining 90% are disposed of as wastes [11]. However, the occurrence of these oil palm waste has created a major disposal problem. The fundamental principles of waste management are to minimize and recycle the waste, recover the energy, and finally dispose of the waste [12].

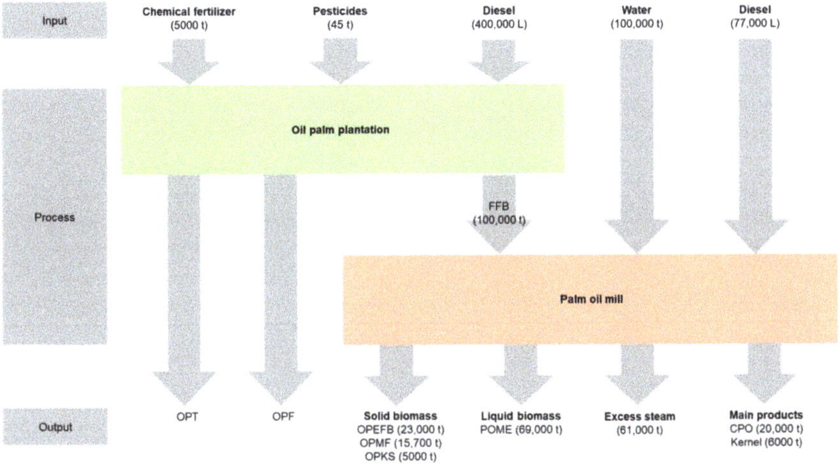

Figure 2. Material flow in the palm oil mill showing the production of oil palm empty fruit bunch (OPEFB), oil palm mesocarp fiber (OPMF), oil palm kernel shell (OPKS) and palm oil mill effluent (POME) from fresh fruit bunch (FFB). Units represent t: tonnes and L: litre. Data adapted from Yoshizaki et al. [13] and Hayashi [14].

Similar to other lignocellulosic biomass, oil palm biomass also consists of cellulose, hemicellulose, and lignin as major components of its cell wall, forming a fibre-like structure that makes the oil palm biomass recalcitrant. The cellulose and hemicellulose are polysaccharides that can be converted into sugar monomers, which can be used as fermentation substrates for various products. However, these components are protected by lignin, which is a complex and large complex structure containing cross-linked phenolic polymers that cover the internal layer of hemicellulose and cellulose [15]. It confers a rigid, impermeable resistance to microbial attack and oxidative stress [16]. The composition of lignin plays an important role in the selection of suitable lignocellulosic biomass as a substrate for sugar production prior to fermentation. A higher lignin content makes the structural arrangement rigid and highly ordered, therefore, it increases the biomass recalcitrance [17]. High lignin content will also contribute to the use of a vigorous pre-treatment process and/or require high amounts of enzymes or chemicals for delignification, subsequently increasing the total pre-treatment cost. Besides, the hydrolysis yield of sugar over the total biomass weight will also be less than for those substrates with a lower lignin content.

In the case of oil palm biomass, OPF fibre showed the lowest composition of lignin. This fibre generated after being mechanically pressed to obtain sugar juice, leaving over the soft fibrous structure with lignin composition of less than 20%, as shown in Table 1. This biomass has been successfully pre-treated into sugar for various fermentation processes, such as biobutanol [18] and bioethanol [19,20]. However, the major challenge for utilizing this biomass as a fermentation substrate is the logistic approach to transfer the raw OPF to the processing plant since it is generated in the plantation area [21]. Another potential oil palm biomass is OPEFB. This biomass is the most abundant oil palm biomass that is produced in the mill that has been widely reported as a promising feedstock for sugar production. This is because this biomass is comprised of sugar of above 70% (cellulose + hemicellulose) and the lignin content is less than 25%, as shown in Table 1. Another abundant oil palm biomass that is produced in the mill is OPMF. However, a part of this biomass is burned in the boiler for steam and electricity generation in the mill. Besides, this biomass composed of 25–28% of lignin, higher than the lignin content presence in OPEFB. Studies have been conducted to pretreat OPMF to produce sugar [17,22,23].

Besides OPEFB and OPMF, palm oil mill also produces oil palm decanter cake (OPDC) and OPKS as waste. It should be noted that OPKS is fully utilized to generate steam and electricity by burning it in a boiler. However, several studies have been conducted to utilize this biomass in order to produce various value-added products. Most of OPKS is studied for biochar and activated carbon production, since this biomass is very compact, high density, and low moisture content, which are the criteria for biochar and activated carbon production [24–28]. It is also composed of very high lignin content of more than 50%, thus making this biomass unsuitable for sugar production. Another valuable oil palm biomass is OPDC, which is obtained from the three-phase decanter system. This oil separator functions to further extract the remaining oil present in POME before it is discharged as an effluent. The three-phase decanter extracts oil from sludge leaving over the remaining liquid (POME) and solid residue (OPDC) [29]. OPDC contains a significantly low amount of potential sugar of ~26% [29]. The major advantage of OPDC is that this biomass naturally has a small fibre size due to mechanical pressing during oil extraction [30]. Small particle size is also a key factor for the effective pre-treatment and hydrolysis by enzymes, since enzyme action is highly affected by the surface area [17]. Unfortunately, OPDC is not produced in all the mills in Malaysia since the three-phase oil separator is installed only in several palm oil mills. Besides, OPDC contains very high ash amount (22%) as it comes from waste mixed with water after the washing procedures. The lignin content is also considerably high at approximately 31%, which requires a harsh pre-treatment process to remove lignin.

Table 1. Chemical composition of various raw oil palm biomasses

Oil Palm Biomass	Cellulose (%)	Hemicellulose (%)	Lignin (%)	Reference
Oil palm empty fruit bunch (OPEFB)	28–41	21–37	18–23	[17,31–34]
Oil palm mesocarp fibre (OPMF)	25–28	21–24	25–28	[17,35]
Oil palm kernel shell (OPKS)	28	22	44	[26]
Oil palm frond (OPF)	33	23	15	[36]
Oil palm trunk (OPT)	56	16	19	[37]
Oil palm decanter cake (OPDC)	22	4	31	[29]

3. Pre-treatment

The structural arrangement of oil palm biomass and its composition play an important role for the conversion efficiency into fermentable sugars. In general, pre-treatment can be categorized into chemical, physical, physico-chemical, and biological pre-treatment [38], whereby each category has their own advantages and disadvantages, as shown in Table 2. Pre-treatment at upstream operation includes a physical pre-treatment, such as size reduction and thermo-chemical process that involve the disruption of the recalcitrant biomass. It increases substrate porosity by delignifying the lignin, hence, enables the maximal exposure of cellulose for cellulase action, which subsequently improve the hydrolysis process, minimize the energy consumption, and maximize the sugar recovery [39]. According to Taherzadeh and Karimi [40], an effective and economical pre-treatment should meet the following requirements: (1) avoiding destruction of hemicelluloses and cellulose, (2) avoiding formation of possible inhibitors, (3) minimizing the energy demand, (4) reducing the cost of size reduction for feedstock and cost of material for the construction of pre-treatment reactors, (5) consumption of little or no chemicals, and (6) using a compatible chemical. Effective pre-treatment is fundamental for optimal successful hydrolysis and downstream operations.

Table 2. Advantages and disadvantages of various pre-treatment methods.

Pre-treatment Category	Pre-treatment Methods	Advantages	Disadvantages	References
Physical	Milling, grinding, chipping, shredding	• Easily reduces the substrate size and increases the surface area • Short process duration	• High energy requirement • High cost for equipment and machinery setup • Required additional pre-treatment as it does not remove lignin	[41,42]
Chemical	Acid	• Short process duration • Possible to fully hydrolyse the entire solid	• Corrosive on the equipment • Not economically feasible	[43,44]
	Alkali	• Short process duration • Selectively attacks the lignin	• Effective if the biomass has a low lignin content • Less effective if the lignin content is high in biomass • Usually requires high temperature to dissolve lignin	[44]
	Ionic liquids (Lis)	• Inexpensive • Not toxic to enzymes and fermentation • Can be recovered and recycled	• Inefficient on the recovery of ILs	[45,46]
Physico-chemical	Steam explosion	• No chemicals used • Low energy input • Environment-friendly	• Incomplete destruction of the lignin-carbohydrate matrix • Risk of condensation and precipitation of soluble lignin components • Destruction of a portion of xylan in hemicellulose • Inhibitor formed at higher temperatures	[44,47]
	Liquid hot water	• Hydrolysis of hemicellulose and removal of lignin • Lower temperature used • Fewer inhibitors produced at high temperature • Effective for large-scale application	• Amount of solubilised product is higher	[48]
	Superheated steam	• Improved energy efficiency • Low environmental impact when condensate is reused • Time saving • Environment friendly • Cost effective for large-scale	• Partial hemicelluloses degradation • Risk of condensation and precipitation of soluble lignin components • Not effective in removing lignin.	[22,23,49]
Biological	Fungi	• Selectively degrades lignin • Environment-friendly • Requires mild operating conditions	• Slow process and therefore it needs long process duration • Requires large space • Some carbohydrate fraction is consumed by the microorganism • Needs to be conducted in sterile conditions	[50,51]
	Ligninolytic enzymes	• Selectively degrades lignin and does not digest the carbohydrate structure • Environment-friendly • Requires mild operating conditions • Can be conducted in non-sterile conditions	• Enzyme cost is very high	[52,53]

3.1. Physical Pre-treatment

Physical pre-treatment, also known as mechanical pre-treatment, is a process that uses mechanical methods, such as milling, chipping, grinding, and shredding to reduce the particle size and to increase the surface area of the biomass. Large biomass surface area increases the enzyme accessible area, and therefore increases the degree of depolymerization of biomass [44]. This pre-treatment is also able to partially modify the structure of biomass, reduce cellulose crystallinity, and disrupt the chemical bonding [36]. For example, the crystallinity index (CrI) of OPEFB reduced from 56% to 9% after grinding using ball milling for 120 min. A shorter milling duration is required for OPF, since this substrate has a lower lignin content than OPEFB [36]. In many practices, physical pre-treatment is used as an initial pre-treatment before the substrate is processed using other kinds of pre-treatment methods [40]. Sun and Cheng [42] reported that the selection of the pre-treatment method for chipping, grinding, and milling is depending on the final particle size of the biomass, usually in the range of 10–30 mm after chipping, and 0.2–2 mm after milling or grinding. Biomass size smaller than 400 mesh (0.04 mm) has little effect on the rate and yield of sugar production. In a study by Rizal et al. [17], the particle size was shown to play an important role in enhancing the hydrolysis of OPEFB and OPMF into glucose. The substrate with a size of 0.25 mm improved the hydrolysis yield by 4.6–4.8-fold.

3.2. Chemical Pre-treatment

Chemical pre-treatment can be divided into two categories, which are acid and alkaline pre-treatments. The acid pre-treatment acts by solubilizing the hemicellulose fraction of biomass and exposes the cellulose to be converted to sugars [54]. It can be performed while using either concentrated or diluted acid. However, the utilization of concentrated acid is less desirable for the subsequent process due to the formation of inhibitors that inhibit enzyme action during hydrolysis and/or microorganisms during fermentation. Besides, its major disadvantages include serious corrosion problems and the use of sophisticated equipment for acid recovery, which leads to high operational and maintenance costs [55]. These limitations reduce the interest in applying this method on a commercial scale [56]. Examples of acid reagents used for pre-treatment process are hydrochloric acid (HCl), phosphoric acid (H_3PO_4), sulfuric acid (H_2SO_4), and nitric acid (HNO_3) [12,25,26].

The alkaline pre-treatment involves the use of bases, such as sodium, potassium, calcium, and ammonium hydroxides [54,57]. This pre-treatment causes the swelling of lignocellulosic biomass, dissolution of lignin and hemicellulose, and de-esterification of intermolecular ester bonds, which subsequently reduces the extent of polymerization and CrI. Since alkaline pre-treatment dissolves mostly lignin, this pre-treatment is preferable than acids that dissolve carbohydrates. The disruption of lignin structure increases the exposure of the internal surface, and makes it accessible to enzyme digestion, which improves the hydrolysis efficiency [58]. The alkaline pre-treatment of OPEFB using 2% NaOH by Ibrahim et al. [59] produced approximately 32 g/L of sugar. Alkaline pre-treatment also has been tested on OPDC, which improved the sugar production from <1 g/L (untreated) to ~6 g/L after pre-treatment using 1% of NaOH [29]. A similar observation was also reported by Barlianti et al. [20] for the NaOH pre-treatment of OPEFB and OPF.

Besides acids and alkaline chemicals, ionic liquids (ILs) and deep eutectic solvents (DES) are some of the chemicals that have been tested for the pre-treatment of lignocellulosic biomass. ILs are salts occurring in the liquid form that was employed to dissolve the lignocellulosic biomass [60]. Meanwhile, DES can be formed between a variety of quaternary ammonium salts and carboxylic acids capable of self-association, often through hydrogen bond interactions. This interaction forms a eutectic mixture with a melting point that was lower than that of each individual component [61,62]. ILs and DES is an inexpensive process, which can be conducted in low/mild temperature, non-toxic to enzymes or cells during saccharification and fermentation, and selectively depolymerize the lignin, thus making this pre-treatment process attractive to the industrial scale [60,63]. Besides, ILs and DES can also be recovered and recycled [60]. However, the recovery process requires a step to remove the inhibitor, in which the additional cost should be considered [45,46]. It was observed that approximately 11% of lignin composition was removed from OPF after being treated using ILs, which was higher than that of enzymatic delignification by laccase [63].

3.3. Physico-Chemical Pre-treatment

Physico-chemical pre-treatment involves both chemical and physical interactions in the pre-treatment process [64]. Steam explosion (SE), hydrothermal pre-treatment, ammonia fibre explosion (AFEX), liquid hot water (LHW), and supercritical CO_2 (SC-CO_2) are among the most widely used pre-treatment methods [55,65,66]. These pre-treatments are considered as the most effective, environment-friendly, and the process has been optimized with a variety of feedstocks on a pilot scale for industrial applications [31,44,67]. It has the capability of changing the structure of biomass, increasing the enzyme accessible surface area, and reducing the degree of biomass polymerization [38,44,68]. In addition, the modification of biomass structure using physico-chemical pre-treatment could enhance enzymatic hydrolysis [17,69].

Most of the physico-chemical pre-treatments are conducted at a high temperature and pressure, in aqueous solutions, and in a closed system [65]. The steam penetrates into the biomass and breaks the structural components, shearing the cell walls and partially hydrolyses the glycosidic bonds of hemicellulose. Physico-chemical pre-treatments use water to avoid the negative effects on the environment [70]. The effective pre-treatment mechanism occurs when autoionization of water at high temperature generates hydronium (H_3O^+) ions and reduces the pH, in which the solution formed acts, like an acid [67,70,71]. This acid is able to further solubilize the hemicellulose component and acetyl residues from xylan, which is then liberated in the form of acetic acid. This acetic acid will further catalyze the hydrolysis and this process is qualified as autohydrolysis [67]. Besides, it was reported that the concentration of hydronium ions from disassociation of acetic acid is higher than that from water autoionization [72].

Steam-assisted fractionation of lignocellulosic biomass improves enzymatic hydrolysis after an exposure to high pressure and temperature. After a certain pre-treatment duration, the biomass is rapidly decompressed, resulting in the degradation of lignocellulosic biomass and reducing its recalcitrance [31,73]. Typically, the oil palm biomass is treated at the temperature range of 160–240 °C and pressures of 20–50 bars for 1–120 min [22,49,73,74]. However, it should be noted that a higher temperature will cause severe degradation of cellulose, lowering the cellulose compositional percentage that will reduce the total glucose yield that was obtained after enzymatic hydrolysis. The selection of suitable temperature and pre-treatment duration is based on the types of oil palm biomass. It was reported that lower temperature is more efficient than a pre-treatment with higher temperature, even though at longer pre-treatment duration [57].

3.4. Biological Pre-treatment

Recently, this environmentally friendly approach has received renewed attention as a pre-treatment method for enhancing the enzymatic hydrolysis of lignocellulosic biomass into sugars. The major advantages of this pre-treatment are low capital costs, low energy usage, little or no chemicals involved, and can be conducted in mild environmental conditions [75,76]. Biological pre-treatment can be divided into two major categories, i.e.,: (1) microbial and (2) enzymatic pre-treatment. Microbial pre-treatment employs microorganism, including white fungi, brown fungi, soft rot fungi and bacteria to modify the lignocellulosic composition [52]. White rot fungi are most widely used for biological pre-treatment of lignocellulosic biomass as they effectively destroy the cell wall and lignin [77]. There are several types of white rot fungi such as *Phanerochaete chrysosporium, Clostridium butyricum, Trichoderma viride, Pycnoporus cinnarbarinus, Dichomitus squalens, Phlebia radiate, Trametes versicolor, Aspergillus oryza,* and *Pleurotus ostreaus* that have been investigated to pretreat different kinds of lignocellulosic biomass [66,67,78]. Lignin degradation by white rot fungi occurs through the action of lignin degrading peroxidases and laccase [78]. The major disadvantage of this pre-treatment is that it requires a long process duration to grow the fungi on lignocellulosic biomass.

Enzymatic pre-treatment employs enzymes to delignify the lignin component of oil palm biomass. The process is faster than microbial pre-treatment, as the enzyme will act directly on the biomass. The process takes about 72 h [79] as compared to a microbial pre-treatment that could take up to 40 days [80]. Enzymatic pre-treatment is also selective as it only attacks the lignin component, leaving the cellulose and hemicellulose intact. However, in order to effectively pretreat the biomass using enzymes, the biomass must first be mechanically pretreated to reduce the particle size as the substrate surface area affects the pre-treatment efficiency [17]. OPT was ground using wet disk milling to a size of 80 μm before being pretreated while using extracted enzymes that were produced by local isolates. They found that the mixture of enzymes rich of xylanase improves the hydrolysis efficiency of OPT [81]. The common ligninolytic enzymes are peroxidases, such as lignin peroxidase (LiP EC 1.11.1.14) and manganese peroxidase (MnP EC 1.11.1.13), as well as laccase (EC 1.10.3.2; benzenediol: oxygen oxidoreductase) [78]. All of these enzymes play a role in lignin degradation, and due to their dependence on molecular oxygen as opposed to hydrogen peroxide, they are becoming an item of interest in the industries related to enzyme [82].

3.5. Combination Pre-treatment

Combinations of different pre-treatments are always put into consideration in order to obtain an optimal fractionation of different components and achieve high yields of fermentable sugars. Selection of the pre-treatments to be combined and the sequence of the pre-treatment processes depend on the substrate characteristics, the availability of equipment, and the operational cost. The operational costs of pre-treatment are influenced by the pre-treatment duration, the amount of energy used, and the quantity of chemicals/enzymes that are applied in the process. For example, OPEFB and OPMF were ground using a hammer mill to a size of 0.25 mm, before being pretreated with laccase [17]. As an example, the OPF was ground to a size of 0.25–0.5 mm, before being pretreated using ILs, followed by an enzymatic delignification by laccase. This combination of pre-treatments resulted in a higher lignin removal as compared to single pre-treatment using either ILs or laccase [63]. OPEFB and OPF pretreated with hot compressed water (HCW) and wet disk (WD) milling has provided higher hydrolysis yield as compared to single pre-treatment [22]. A phosphoric acid pre-treatment, followed by fungal pre-treatment on OPEFB, reduced the CrI, which then improved the ethanol production in a simultaneous saccharification and fermentation (SSF) process [83].

A summary of the pre-treatment method from various studies is listed in Table 3. Different pre-treatment methods and conditions showed different yields of fermentable sugars obtained, which also contributed to different types of oil palm biomass used and the amount/composition of enzymes during enzymatic hydrolysis.

Table 3. Pre-treatment performances on various oil palm biomass.

Pre-treatment Methods	Oil Palm Biomass	Pre-treatment Conditions	Yield	Reference
Physical Ball milling	Oil palm empty fruit bunch (OPEFB)	120 min	79% of glucose	[36]
Physical Ball milling	Oil palm frond (OPF)	60 min	84% of glucose	[36]
Physico-chemical Hydrothermal	Oil palm empty fruit bunch (OPEFB)	170–250 °C 10–20 min	100% of glucose	[73]
Physico-chemical Hot compresses water	Oil palm frond (OPF)	10 bar 178 °C 11.1 min 9.6 liquid-solid ratio	97% of glucose	[84]
Chemical Aqueous ammonia	Oil palm empty fruit bunch (OPEFB)	60 °C 12 h 21% of aqueous ammonia	41% of glucose	[85]

Table 3. *Cont.*

Pre-treatment Methods	Oil Palm Biomass	Pre-treatment Conditions	Yield	Reference
Chemical Solvent-ionic liquid	Oil palm frond (OPF)	80 °C 15 min 10% of solid loading	100% of glucose	[86]
Chemical Dilute acid pre-treatment at high temperature	Oil palm trunk (OPT)	3% H_2SO_4 180 °C 40 min	80% of glucose	[87]
Combination Hammer mill Superheated steam (SHS) Laccase	Oil palm empty fruit bunch (OPEFB)	Size 0.25 mm SHS 160 °C, 20 min Laccase 100 U/g	72% of glucose	[17]
Combination Hammer mill Superheated steam (SHS) Laccase	Oil palm mesocarp fibre (OPMF)	Size 0.25 mm SHS 180 °C, 20 min Laccase 400 U/g	63% of glucose	[17]
Combination Alkaline hydrothermal and wet disk milling	Oil palm mesocarp fiber (OPMF)	1.5% NaOH	97% of glucose	[35]

4. Conclusions

Although some pre-treatments could achieve up to 100% of glucose recovery, the efficiency and the suitability of the pre-treatment should be considered based on energy and time consumption, cost for chemicals and/or enzymes, initial capital for setting up the plant, the inhibitors released after the pre-treatment process, the waste generated from the pre-treatment process, the environmental impact, and the conversion of oil palm biomass into sugar. The mechanical pre-treatment to reduce substrate's size is important since enzymatic hydrolysis by cellulase is highly affected by the accessible surface area to digest cellulose into glucose. Therefore, many mechanical pre-treatments, such as wet disk milling, hammer mill, and ball milling resulted in a high glucose recovery. Besides, even though a single pre-treatment could save the energy and time, a combination of more than two pre-treatments efficiently enhances the glucose or sugar recovery. However, the compatibility of combining the pre-treatments is limited, hence more research on the combination pre-treatment should be conducted.

Author Contributions: Writing-Original Draft Preparation, N.F.A.A.R and M.F.I.; Writing-Review & Editing, M.F.I., M.R.Z., S.A.A., P.L.Y and M.A.H.; Supervision, M.F.I.

Funding: This research was funded by Science and Technology Research Partnership for Sustainable Development (SATREPS) grant number 6300156.

Acknowledgments: The authors gratefully acknowledged the financial support from Science and Technology Research Partnership for Sustainable Development (SATREPS).

Conflicts of Interest: The authors declare no conflict of interest.

References

1. MPOC The Oil Palm Tree. Available online: http://www.mpoc.org.my/The_Oil_Palm_Tree.aspx (accessed on 24 April 2018).
2. MPOC MPOC Annual Report 2016. Available online: http://www.mpoc.org.my/pubs_view.aspx?id=5ab6aafb-7ce5-406a-b48e-64d1022b736b (accessed on 1 May 2018).
3. Loh, S.K. The potential of the Malaysian oil palm biomass as a renewable energy source. *Energy Convers. Manag.* **2017**, *141*, 285–298. [CrossRef]
4. Yoshizaki, T.; Shirai, Y.; Hassan, M.A.; Baharuddin, A.S.; Raja Abdullah, N.M.; Sulaiman, A.; Busu, Z. Improved economic viability of integrated biogas energy and compost production for sustainable palm oil mill management. *J. Clean. Prod.* **2013**, *44*, 1–7. [CrossRef]

5. Ibrahim, M.F.; Abd-Aziz, S.; Razak, M.N.A.; Phang, L.Y.; Hassan, M.A. Oil palm empty fruit bunch as alternative substrate for acetone-butanol-ethanol production by *Clostridium butyricum* EB6. *Appl. Biochem. Biotechnol.* **2012**, *166*, 1615–1625. [CrossRef] [PubMed]
6. Sharuddin, S.S.; Ramli, N.; Hassan, M.A.; Mustapha, N.A.; Amran, A.; Mohd-Nor, D.; Sakai, K.; Tashiro, Y.; Shirai, Y.; Maeda, T. Bacterial community shift revealed Chromatiaceae and Alcaligenaceae as potential bioindicators in the receiving river due to palm oil mill effluent final discharge. *Ecol. Indic.* **2017**, *82*, 526–529. [CrossRef]
7. Sumathi, S.; Chai, S.P.P.; Mohamed, A.R.R. Utilization of oil palm as a source of renewable energy in Malaysia. *Renew. Sustain. Energy Rev.* **2008**, *12*, 2404–2421. [CrossRef]
8. Agensi Inovasi Malaysia. *National Biomass Strategy 2020: New Wealth Creation for Malaysia's Biomass Industry*; Agensi Inovasi Malaysia: Cyberjaya, Malaysia, 2013.
9. MPOB The First Palm Oil Plantation in Malaysia (1917). Available online: http://palmoilis.mpob.gov.my/index.php/gs/about-galeri-sawit?id=89 (accessed on 13 February 2018).
10. Aikanathan, S.; Basiron, Y.; Sundram, K.; Chenayah, S.; Sasekumar, A. Sustainable management of oil palm plantation industry and the perception implications. *J. Oil Palm Environ. Health* **2015**, *6*, 10–24. [CrossRef]
11. Sulaiman, F.; Abdullah, N.; Gerhauser, H.; Shariff, A. A perspective of oil palm and its wastes. *J. Phys. Sci.* **2010**, *21*, 67–77. [CrossRef]
12. Awalludin, M.F.; Sulaiman, O.; Hashim, R.; Nadhari, W.N.A. W. An overview of the oil palm industry in Malaysia and its waste utilization through thermochemical conversion, specifically via liquefaction. *Renew. Sustain. Energy Rev.* **2015**, *50*, 1469–1484. [CrossRef]
13. Yoshizaki, T.; Shirai, Y.; Hassan, M.A.; Baharuddin, A.S.; Abdullah, N.M.R.; Sulaiman, A.; Busu, Z. Economic analysis of biogas and compost projects in a palm oil mill with clean development mechanism in Malaysia. *Environ. Dev. Sustain.* **2012**, *14*, 1065–1079. [CrossRef]
14. Hayashi, K. Environmental impact of palm oil industry in Indonesia. *Proc. Int. Symp. EcoTopia Sci.* **2007**, *7*, 646–651.
15. Rubin, E.M. Genomics of cellulosic biofuels. *Nature* **2008**, *454*, 841–845. [CrossRef] [PubMed]
16. Bajpai, P. *Pretreatment of Lignocellulosic Biomass for Biofuel Production*; Springer Briefs in Molecular Science; Springer: Singapore, 2016; ISBN 978-981-10-0686-9.
17. Rizal, N.A.; Ibrahim, M.F.; Zakaria, M.; Bahrin, E.K.; Abd-Aziz, S.; Hassan, M. Combination of superheated steam with laccase pretreatment together with size reduction to enhance enzymatic hydrolysis of oil palm biomass. *Molecules* **2018**, *23*, 811.
18. Aliyu, A.S.; Aziz, A.A.; Yahya, A.; Lattiff, Z.A. Potential of oil palm frond liquid extract and fibre as feedstock for bio-butanol production. *J. Teknol.* **2015**, *74*, 63–67. [CrossRef]
19. Kumneadklang, S.; Larpkiattaworn, S.; Niyasom, C.; O-Thong, S. Bioethanol production from oil palm frond by simultaneous saccharification and fermentation. *Energy Procedia* **2015**, *79*, 784–790. [CrossRef]
20. Barlianti, V.; Dahnum, D.; Hendarsyah, H.; Abimanyu, H. Effect of alkaline pretreatment on properties of lignocellulosic oil palm waste. *Procedia Chem.* **2015**, *16*, 195–201. [CrossRef]
21. Abdullah, S.S.S.; Shirai, Y.; Ali, A.A.M.; Mustapha, M.; Hassan, M.A. Case study: Preliminary assessment of integrated palm biomass biorefinery for bioethanol production utilizing non-food sugars from oil palm frond petiole. *Energy Convers. Manag.* **2016**, *108*, 233–242. [CrossRef]
22. Zakaria, M.R.; Norrrahim, M.N.F.; Hirata, S.; Hassan, M.A. Hydrothermal and wet disk milling pretreatment for high conversion of biosugars from oil palm mesocarp fibre. *Bioresour. Technol.* **2015**, *181*, 263–269. [CrossRef] [PubMed]
23. Nordin, N.; Ariffin, H.; Andou, Y.; Hassan, M.; Shirai, Y.; Nishida, H.; Yunus, W.; Karuppuchamy, S.; Ibrahim, N. Modification of oil palm mesocarp fibre characteristics using superheated steam treatment. *Molecules* **2013**, *18*, 9132–9146. [CrossRef] [PubMed]
24. Zainal, N.H.; Aziz, A.A.; Idris, J.; Jalani, N.F.; Mamat, R.; Ibrahim, M.F.; Hassan, M.A.; Abd-Aziz, S. Reduction of POME final discharge residual using activated bioadsorbent from oil palm kernel shell. *J. Clean. Prod.* **2018**, *182*. [CrossRef]
25. Rugayah, A.F.; Astimar, A.A.; Norzita, N. Preparation and characterisation of activated carbon from palm kernel shell by physical activation with steam. *J. Oil Palm Res.* **2014**, *26*, 251–264.

26. Zainal, N.H.; Aziz, A.A.; Idris, J.; Mamat, R.; Hassan, M.A.; Bahrin, E.K.; Abd-Aziz, S. Microwave-assisted pre-carbonisation of palm kernel shell produced charcoal with high heating value and low gaseous emission. *J. Clean. Prod.* **2017**, *142*, 2945–2949. [CrossRef]
27. Ruiz, H.A.; Zambrano, M.A; Giraldo, L. Production and characterisation of activated carbon from oil-palm shell for carboxylic acid. *Orient. J. Chem.* **2015**, *31*, 753–762. [CrossRef]
28. Kong, S.H.; Loh, S.K.; Bachmann, R.T.; Rahim, S.A.; Salimon, J. Biochar from oil palm biomass: A review of its potential and challenges. *Renew. Sustain. Energy Rev.* **2014**, *39*, 729–739. [CrossRef]
29. Razak, M.N.A.; Ibrahim, M.F.; Yee, P.L.; Hassan, M.A.; Abd-Aziz, S. Utilization of oil palm decanter cake for cellulase and polyoses production. *Biotechnol. Bioprocess Eng.* **2012**, *17*, 547–555. [CrossRef]
30. Sahad, N.; Som, A.M.; Baharuddin, A.S.; Mokhtar, N.; Busu, Z.; Sulaiman, A. Physicochemical characterization of oil palm decanter cake (OPDC) for residual oil recovery. *BioResources* **2014**, *9*, 6361–6372. [CrossRef]
31. Medina, J.D.C.; Woiciechowski, A.; Filho, A.Z.; Nigam, P.S.; Ramos, L.P.; Soccol, C.R. Steam explosion pretreatment of oil palm empty fruit bunches (EFB) using autocatalytic hydrolysis: A biorefinery approach. *Bioresour. Technol.* **2016**, *199*, 173–180. [CrossRef] [PubMed]
32. Akhtar, J.; Idris, A. Oil palm empty fruit bunches a promising substrate for succinic acid production via simultaneous saccharification and fermentation. *Renew. Energy* **2017**, *114*, 917–923. [CrossRef]
33. Abdul, P.M.; Jahim, J.M.; Harun, S.; Markom, M.; Lutpi, N.A.; Hassan, O.; Balan, V.; Dale, B.E.; Mohd Nor, M.T. Effects of changes in chemical and structural characteristic of ammonia fibre expansion (AFEX) pretreated oil palm empty fruit bunch fibre on enzymatic saccharification and fermentability for biohydrogen. *Bioresour. Technol.* **2016**, *211*, 200–208. [CrossRef] [PubMed]
34. Ibrahim, M.F.; Razak, M.N.A.; Phang, L.Y.; Hassan, M.A.; Abd-Aziz, S. Crude cellulase from oil palm empty fruit bunch by *Trichoderma asperellum* UPM1 and *Aspergillus fumigatus* UPM2 for fermentable sugars production. *Appl. Biochem. Biotechnol.* **2013**, *170*, 1320–1335. [CrossRef] [PubMed]
35. Zakaria, M.R.; Hirata, S.; Hassan, M.A. Combined pretreatment using alkaline hydrothermal and ball milling to enhance enzymatic hydrolysis of oil palm mesocarp fibre. *Bioresour. Technol.* **2014**, *169*, 236–243. [CrossRef] [PubMed]
36. Zakaria, M.R.; Fujimoto, S.; Hirata, S.; Hassan, M.A. Ball milling pretreatment of oil palm biomass for enhancing enzymatic hydrolysis. *Appl. Biochem. Biotechnol.* **2014**, *173*, 1778–1789. [CrossRef] [PubMed]
37. Lai, L.-W.; Idris, A. Disruption of oil palm trunks and fronds by microwave-alkali pretreatment. *BioResources* **2013**, *8*, 2792–2804. [CrossRef]
38. Kumar, A.K.; Sharma, S. Recent updates on different methods of pretreatment of lignocellulosic feedstocks: A review. *Bioresour. Bioprocess.* **2017**, *4*, 7. [CrossRef] [PubMed]
39. Zhu, J.; Wan, C.; Li, Y. Enhanced solid-state anaerobic digestion of corn stover by alkaline pretreatment. *Bioresour. Technol.* **2010**, *101*, 7523–7528. [CrossRef] [PubMed]
40. Taherzadeh, M.J.; Karimi, K. Pretreatment of lignocellulosic wastes to improve ethanol and biogas production: A review. *Int. J. Mol. Sci.* **2008**, *9*, 1621–1651. [CrossRef] [PubMed]
41. Hendriks, A.T.; Zeeman, G. Pretreatments to enhance the digestibility of lignocellulosic biomass. *Bioresour. Technol.* **2009**, *100*, 10–18. [CrossRef] [PubMed]
42. Sun, Y.; Cheng, J. Hydrolysis of lignocellulosic materials for ethanol production: A review. *Bioresour. Technol.* **2002**, *83*, 1–11. [CrossRef]
43. Binder, J.B.; Raines, R.T. Fermentable sugars by chemical hydrolysis of biomass. *Proc. Natl. Acad. Sci. USA* **2010**, *107*, 4516–4521. [CrossRef] [PubMed]
44. Agbor, V.B.; Cicek, N.; Sparling, R.; Berlin, A.; Levin, D.B. Biomass pretreatment: Fundamentals toward application. *Biotechnol. Adv.* **2011**, *29*, 675–685. [CrossRef] [PubMed]
45. Sen, S.M.; Binder, J.B.; Raines, R.T.; Maravelias, C.T. Conversion of biomass to sugars via ionic liquid hydrolysis: Process synthesis and economic evaluation. *Biofuels Bioprod. Biorefin.* **2012**, *6*, 444–452. [CrossRef]
46. Domínguez de María, P. Recent trends in (ligno)cellulose dissolution using neoteric solvents: Switchable, distillable and bio-based ionic liquids. *J. Chem. Technol. Biotechnol.* **2014**, *89*, 11–18. [CrossRef]
47. Mosier, N.S.; Hendrickson, R.; Brewer, M.; Ho, N.; Sedlak, M.; Dreshel, R.; Welch, G.; Dien, B.S.; Aden, A.; Ladisch, M.R. Industrial scale-up of pH-controlled liquid hot water pretreatment of corn fibre for fuel ethanol production. *Appl. Biochem. Biotechnol.* **2005**, *125*, 77–97. [CrossRef]

48. Yang, B.; Wyman, C.E. Effect of xylan and lignin removal by batch and flowthrough pretreatment on the enzymatic digestibility of corn stover cellulose. *Biotechnol. Bioeng.* **2004**, *86*, 88–95. [CrossRef] [PubMed]
49. Bahrin, E.K.; Baharuddin, A.S.; Ibrahim, M.F.; Razak, M.N.A.; Sulaiman, A.; Abd-Aziz, S.; Hassan, M.A.; Shirai, Y.; Nishida, H. Physicochemical property changes and enzymatic hydrolysis enhancement of oil palm empty fruit bunches treated with superheated steam. *BioResources* **2012**, *7*, 1784–1801.
50. Sánchez, C. Lignocellulosic residues: Biodegradation and bioconversion by fungi. *Biotechnol. Adv.* **2009**, *27*, 185–194. [CrossRef] [PubMed]
51. Kunamneni, A.; Plou, F.J.; Ballesteros, A.; Alcalde, M. Laccases and their applications: A patent review. *Recent Pat. Biotechnol.* **2008**, *2*, 10–24. [CrossRef] [PubMed]
52. Wan, C.; Li, Y. Fungal pretreatment of lignocellulosic biomass. *Biotechnol. Adv.* **2012**, *30*, 1447–1457. [CrossRef] [PubMed]
53. Pinto, P.A.; Dias, A.A.; Fraga, I.; Marques, G.; Rodrigues, M.A.M.; Colaço, J.; Sampaio, A.; Bezerra, R.M.F. Influence of ligninolytic enzymes on straw saccharification during fungal pretreatment. *Bioresour. Technol.* **2012**, *111*, 261–267. [CrossRef] [PubMed]
54. Alvira, P.; Tomás-Pejó, E.; Ballesteros, M.; Negro, M.J. Pretreatment technologies for an efficient bioethanol production process based on enzymatic hydrolysis: A review. *Bioresour. Technol.* **2010**, *101*, 4851–4861. [CrossRef] [PubMed]
55. Ren, N.Q.; Zhao, L.; Chen, C.; Guo, W.Q.; Cao, G.L. A review on bioconversion of lignocellulosic biomass to H2: Key challenges and new insights. *Bioresour. Technol.* **2016**, *215*, 92–99. [CrossRef] [PubMed]
56. Mosier, N.; Wyman, C.; Dale, B.; Elander, R.; Lee, Y.Y.; Holtzapple, M.; Ladisch, M. Features of promising technologies for pretreatment of lignocellulosic biomass. *Bioresour. Technol.* **2005**, *96*, 673–686. [CrossRef] [PubMed]
57. Chaturvedi, V.; Verma, P. An overview of key pretreatment processes employed for bioconversion of lignocellulosic biomass into biofuels and value added products. *3 Biotech* **2013**, *3*, 415–431. [CrossRef] [PubMed]
58. Park, Y.C.; Kim, J.S. Comparison of various alkaline pretreatment methods of lignocellulosic biomass. *Energy* **2012**, *47*, 31–35. [CrossRef]
59. Ibrahim, M.F.; Abd-Aziz, S.; Yusoff, M.E.M.; Phang, L.Y.; Hassan, M.A. Simultaneous enzymatic saccharification and ABE fermentation using pretreated oil palm empty fruit bunch as substrate to produce butanol and hydrogen as biofuel. *Renew. Energy* **2015**, *77*, 447–455. [CrossRef]
60. Elgharbawy, A.A.; Alam, M.Z.; Moniruzzaman, M.; Goto, M. Ionic liquid pretreatment as emerging approaches for enhanced enzymatic hydrolysis of lignocellulosic biomass. *Biochem. Eng. J.* **2016**, *109*, 252–267. [CrossRef]
61. Abbott, A.P.; Boothby, D.; Capper, G.; Davies, D.L.; Rasheed, R.K. Deep Eutectic Solvents formed between choline chloride and carboxylic acids: Versatile alternatives to ionic liquids. *J. Am. Chem. Soc.* **2004**, *126*, 9142–9147. [CrossRef] [PubMed]
62. Zhang, Q.; De Oliveira Vigier, K.; Royer, S.; Jérôme, F. Deep eutectic solvents: Syntheses, properties and applications. *Chem. Soc. Rev.* **2012**, *41*, 7108. [CrossRef] [PubMed]
63. Financie, R.; Moniruzzaman, M.; Uemura, Y. Enhanced enzymatic delignification of oil palm biomass with ionic liquid pretreatment. *Biochem. Eng. J.* **2016**, *110*, 1–7. [CrossRef]
64. Galbe, M.; Zacchi, G. Pretreatment: The key to efficient utilization of lignocellulosic materials. *Biomass Bioenergy* **2012**, *46*, 70–78. [CrossRef]
65. Rabemanolontsoa, H.; Saka, S. Various pretreatments of lignocellulosics. *Bioresour. Technol.* **2016**, *199*, 83–91. [CrossRef] [PubMed]
66. Sun, S.; Sun, S.; Cao, X.; Sun, R. The role of pretreatment in improving the enzymatic hydrolysis of lignocellulosic materials. *Bioresour. Technol.* **2016**, *199*, 49–58. [CrossRef] [PubMed]
67. Capolupo, L.; Faraco, V. Green methods of lignocellulose pretreatment for biorefinery development. *Appl. Microbiol. Biotechnol.* **2016**, *100*, 9451–9467. [CrossRef] [PubMed]
68. Chen, H.; Liu, J.; Chang, X.; Chen, D.; Xue, Y.; Liu, P.; Lin, H.; Han, S. A review on the pretreatment of lignocellulose for high-value chemicals. *Fuel Process. Technol.* **2017**, *160*, 196–206. [CrossRef]
69. Singh, R.; Hu, J.; Regner, M.R.; Round, J.W.; Ralph, J.; Saddler, J.N.; Eltis, L.D. Enhanced delignification of steam-pretreated poplar by a bacterial laccase. *Sci. Rep.* **2017**, *7*, 42121. [CrossRef] [PubMed]

70. Brodeur, G.; Yau, E.; Badal, K.; Collier, J.; Ramachandran, K.B.; Ramakrishnan, S. Chemical and physicochemical pretreatment of lignocellulosic biomass: A review. *Enzyme Res.* **2011**, *2011*, 787532. [CrossRef] [PubMed]
71. Yan, L.; Ma, R.; Li, L.; Fu, J. Hot water pretreatment of lignocellulosic biomass: An effective and environmentally friendly approach to enhance biofuel production. *Chem. Eng. Technol.* **2016**, *39*, 1759–1770. [CrossRef]
72. Heitz, M.; Carrasco, F.; Rubio, M.; Chauvette, G.; Chornet, E.; Jaulin, L.; Overend, R.P. Generalized correlations for the aqueous liquefaction of lignocellulosics. *Can. J. Chem. Eng.* **1986**, *64*, 647–650. [CrossRef]
73. Zakaria, M.R.; Hirata, S.; Hassan, M.A. Hydrothermal pretreatment enhanced enzymatic hydrolysis and glucose production from oil palm biomass. *Bioresour. Technol.* **2015**, *176*, 142–148. [CrossRef] [PubMed]
74. Zakaria, M.R.; Hirata, S.; Fujimoto, S.; Hassan, M.A. Combined pretreatment with hot compressed water and wet disk milling opened up oil palm biomass structure resulting in enhanced enzymatic digestibility. *Bioresour. Technol.* **2015**, *193*, 128–134. [CrossRef] [PubMed]
75. Isroi; Millati, R.; Syamsiah, S.; Niklasson, C.; Cahyanto, M.N.; Lundquist, K.; Taherzadeh, M.J. Biological pretreatment of lignocelluloses with white-rot fungi and its applications: A review. *BioResources* **2011**, *6*, 5224–5259. [CrossRef]
76. Asgher, M.; Ahmad, Z.; Iqbal, H.M.N. Alkali and enzymatic delignification of sugarcane bagasse to expose cellulose polymers for saccharification and bio-ethanol production. *Ind. Crops Prod.* **2013**, *44*, 488–495. [CrossRef]
77. Potumarthi, R.; Baadhe, R.R.; Nayak, P.; Jetty, A. Simultaneous pretreatment and sacchariffication of rice husk by *Phanerochete chrysosporium* for improved production of reducing sugars. *Bioresour. Technol.* **2013**, *128*, 113–117. [CrossRef] [PubMed]
78. Masran, R.; Zanirun, Z.; Kamal Bahrin, E.; Ibrahim, M.F.; Lai Yee, P.; Abd-Aziz, S. Harnessing the potential of ligninolytic enzymes for lignocellulosic biomass pretreatment. *Appl. Microbiol. Biotechnol.* **2016**, *100*, 5231–5246. [CrossRef] [PubMed]
79. Kusi, O.A.; Premjet, D.; Premjet, S. A review article of biological pre-treatment of agricultural biomass. *Pertanika J. Trop. Agric. Sci.* **2018**, *41*, 19–40.
80. Sindhu, R.; Binod, P.; Pandey, A. Biological pretreatment of lignocellulosic biomass – an overview. *Bioresour. Technol.* **2015**, *199*, 76–82. [CrossRef] [PubMed]
81. Widyasti, E.; Shikata, A.; Hashim, R.; Sulaiman, O.; Sudesh, K.; Wahjono, E.; Kosugi, A. Biodegradation of fibrillated oil palm trunk fibre by a novel thermophilic, anaerobic, xylanolytic bacterium Caldicoprobacter sp. CL-2 isolated from compost. *Enzyme Microb. Technol.* **2018**, *111*, 21–28. [CrossRef] [PubMed]
82. Heap, L.; Green, A.; Brown, D.; van Dongen, B.; Turner, N. Role of laccase as an enzymatic pretreatment method to improve lignocellulosic saccharification. *Catal. Sci. Technol.* **2014**, *9*. [CrossRef]
83. Ishola, M.M.; Isroi; Taherzadeh, M.J. Effect of fungal and phosphoric acid pretreatment on ethanol production from oil palm empty fruit bunches (OPEFB). *Bioresour. Technol.* **2014**, *165*, 9–12. [CrossRef] [PubMed]
84. Goh, C.S.; Lee, K.T.; Bhatia, S. Hot compressed water pretreatment of oil palm fronds to enhance glucose recovery for production of second generation bio-ethanol. *Bioresour. Technol.* **2010**, *101*, 7362–7367. [CrossRef] [PubMed]
85. Jung, Y.H.; Kim, I.J.; Han, J.-I.; Choi, I.-G.; Kim, K.H. Aqueous ammonia pretreatment of oil palm empty fruit bunches for ethanol production. *Bioresour. Technol.* **2011**, *102*, 9806–9809. [CrossRef] [PubMed]
86. Tan, H.T.; Lee, K.T.; Mohamed, A.R. Pretreatment of lignocellulosic palm biomass using a solvent-ionic liquid [BMIM]Cl for glucose recovery: An optimisation study using response surface methodology. *Carbohydr. Polym.* **2011**, *83*, 1862–1868. [CrossRef]
87. Noparat, P.; Prasertsan, P.; O-Thong, S.; Pan, X. Dilute acid pretreatment of oil palm trunk biomass at high temperature for enzymatic hydrolysis. *Energy Procedia* **2015**, *79*, 924–929. [CrossRef]

Sample Availability: Samples of the compounds are not available from the authors.

molecules

MDPI

Article

Subcritical Water-Carbon Dioxide Pretreatment of Oil Palm Mesocarp Fiber for Xylooligosaccharide and Glucose Production

Norlailiza Ahmad [1], Mohd Rafein Zakaria [1,2,*], Mohd Zulkhairi Mohd Yusoff [1,2], Shinji Fujimoto [3], Hiroyuki Inoue [3], Hidayah Ariffin [1,2], Mohd Ali Hassan [1] and Yoshihoto Shirai [4]

1 Department of Bioprocess Technology, Faculty of Biotechnology and Biomolecular Sciences, Universiti Putra Malaysia, Serdang, Selangor 43400 UPM, Malaysia; lailizaahmad@gmail.com (N.A.); mzulkhairi@upm.edu.my (M.Z.M.Y.); hidayah@upm.edu.my (H.A.); alihas@upm.edu.my (M.A.H.)
2 Laboratory of Biopolymer and Derivatives, Institute of Tropical Forestry and Forest Products, Universiti Putra Malaysia, Serdang, Selangor 43400 UPM, Malaysia
3 Research Institute for Sustainable Chemistry, National Institute of Advanced Industrial Science and Technology (AIST), 3-11-32 Kagamiyama, Higashi-Hiroshima, Hiroshima 739-0046, Japan; s.fujimoto@aist.go.jp (S.F.); inoue-h@aist.go.jp (H.I.)
4 Department of Biological Functions and Engineering, Graduate School of Life Science and Systems Engineering, Kyushu Institute of Technology, Fukuoka 804-8550, Japan; shirai@life.kyutech.ac.jp
* Correspondence: mohdrafein@upm.edu.my; Tel.: +60-3-8947-1946, Fax: +60-3-8946-7590

check for updates

Received: 24 April 2018; Accepted: 29 May 2018; Published: 30 May 2018

Abstract: The present work aimed to investigate the pretreatment of oil palm mesocarp fiber (OPMF) in subcritical H_2O-CO_2 at a temperature range from 150–200 °C and 20–180 min with CO_2 pressure from 3–5 MPa. The pretreated solids and liquids from this process were separated by filtration and characterized. Xylooligosaccharides (XOs), sugar monomers, acids, furans and phenols in the pretreated liquids were analyzed by using HPLC. XOs with a degree of polymerization X2–X4 comprising xylobiose, xylotriose, xylotetraose were analyzed by using HPAEC-PAD. Enzymatic hydrolysis was performed on cellulose-rich pretreated solids to observe xylose and glucose production. An optimal condition for XOs production was achieved at 180 °C, 60 min, 3 MPa and the highest XOs obtained was 81.60 mg/g which corresponded to 36.59% of XOs yield from total xylan of OPMF. The highest xylose and glucose yields obtained from pretreated solids were 29.96% and 84.65%, respectively at cellulase loading of 10 FPU/g-substrate.

Keywords: oil palm mesocarp fiber; subcritical H_2O-CO_2; pretreatment; xylooligosaccharides; glucose

1. Introduction

Malaysia is the second largest oil palm producer, with more than 15 million tonnes of palm oil produced, along with the production of biomass such as oil palm empty fruit bunch (OPEFB), oil palm mesocarp fiber (OPMF) and oil palm frond fiber (OPFF) [1]. OPMF is one of the potential and attractive biomass which can be used as a biomaterial to produce many bio-products such as biosugar, biogas, biochar and biocomposite which can be further used by various industries. Generally, OPMF consists of cellulose (23–29%), hemicellulose (21–34%), lignin (21–32%), extractives and ash [2–4]. Due to the complex structure of OPMF, different pretreatments have been performed to disrupt the lignocellulose structure to give maximum access of enzymes to hemicellulose and cellulose [2,5]. Autohydrolysis is one of the preferable pretreatment methods as it uses a green approach such as compressed hot water with various reaction temperatures and times to hydrolyze xylan into shorter oligosaccharides such as xylooligosaccharides (XOs) and xylose [6,7]. Recently, subcritical H_2O-CO_2 pretreatment has become more attractive as it offers benefits such as the mild conditions used, less formation of undesirable by-products and the use of non-toxic gases [8]. Subcritical H_2O-CO_2 pretreatment produces carbonic acid that facilitates the hydrolysis of hemicellulose in biomass with no negative impact on the environment since when the pressure is released, the gas will be neutralized [9,10]. It was reported that high-pressure CO_2 penetrates the small pores in the biomass and helps in disrupting the biomass structure, thus improving the hydrolysis rate of hemicellulose in the biomass [8,11]. The combined severity factor (CS_{PCO2}) is used to evaluate the influence of temperature, time and high-pressure CO_2 on the hydrolysis of xylan [8–12].

The structure, degree of polymerization (DP) and yield of XOs depend on the type of biomass and methods used in the production stage [12]. XOs can be obtained abundantly in pretreatment liquids, together with undesired by-products such as acetic acid, furfural, 5-hydroxymethylfurfural (5-HMF) and tannic acid [11,13,14]. Purification steps are necessary to remove these undesired by-products to obtain high purity XOs. Different types of XOs can be produced from xylan such as xylobiose (X2), xylotriose (X3), xylotetraose (X4) and xylopentaose (X5) [15,16]. XOs are produced from corncobs [6], cotton stalks, tobacco stalks, sunflower stalks, wheat straw [17], sugarcane bagasse [13], OPFF [18] and OPEFB [16].

Oil palm biomass such as OPEFB and OPFF were reported to produce XOs by an autohydrolysis process [16,18]. Under optimal pretreatment conditions, 17.6 g/L of XOs was produced from OPEFB at pretreatment severity log *Ro* = 3.91 (210 °C) with DP X5–X40 and 6.15 g/L with DP X5–X10 was recorded [16]. Autohydrolysis pretreatment of OPFF at 121 °C for 60 min and subsequent enzymatic hydrolysis with xylanase at 8 U/100 mg of autohydrolyzate resulted in the production of 17.5% and 13.9% of XOs and xylobiose, respectively [16]. It was reported that in hydrothermal pretreatment of OPMF monomeric xylose and XOs were produced at pretreatment severity log, *Ro* = 3.25–3.94, and XOs concentrations were detected in the range from 5.0 to 7.0 g/L. It was suggested that OPMF is a suitable biomass to produce XOs [19]. It is worth noting that in all previously reported oil palm biomass experiments autohydrolysis processes alone were conducted and the formation of XOs was obtained at high pretreatment severities and with subsequent enzymatic hydrolysis by xylanase. Addition of CO_2 in the autohydrolysis process and operation under subcritical and supercritical conditions offered several advantages over autohydrolysis alone [7–9]. An attempt to obtain a higher XO yield from OPMF under mild operational conditions with the application of initial pressurized CO_2 (0–5 MPa) was performed in this study.

Due to the high demand and potential uses of XOs in the industry as well as abundant sources of OPMF from the oil palm industry, the present study aimed to evaluate the production of XOs and glucose from OPMF using subcritical H_2O-CO_2 pretreatment under isothermal and non-isothermal conditions. The present work was also conducted to prove that impregnation of CO_2 in the subcritical H_2O reaction could reduce the formation of inhibitory by-products resulting in the improved production of XOs compared to subcritical H_2O without CO_2 assistance. The efficiency of the pretreatment was evaluated based on the types and concentration of XOS produced and glucose yield from cellulose conversion by enzymatic hydrolysis. To the best of our knowledge, this is the first study on OPMF for XOs and glucose production under subcritical H_2O-CO_2 pretreatment process conditions.

2. Results and Discussion

2.1. Compositional Analysis

The chemical compositions of the biomass were complex and varied according to its structure and origin. Table 1 shows that OPMF used in this study mainly comprised cellulose (23.6%), hemicellulose (22.3%), lignin (28.2%), and solvent extractives (8.3%) and was within the range of earlier reports [2,3]. Different pretreatments were performed first to find the best pretreatment of OPMF as it contains a high lignin content compared to other oil palm biomass [4,19]. The determined hemicellulose content was relatively similar to that of other lignocellulosic materials such as tobacco stalks, wheat straws, corn stover and olive stones, with contents of 20.0%, 20.9%, 22.0% and 23.3%, respectively [17,20,21]. All of these biomasses were potentially used as substrates to produce XOs. The present finding showed that OPMF was among the biomasses that could potentially be used as raw materials for XOs production.

Table 1. Chemical composition of OPMF used in this study in comparison to previous reports.

Chemical Component	Content (wt %)			
Solvent extractives	8.3 ± 0.4 [a]	11.4 ± 0.2 [b]	6.3 ± 0.51 [a]	-
Cellulose	23.6 ± 0.9	25.0 ± 1.7	28.8 ± 0.48	42.8 ± 0.69
Hemicellulose	22.3 ± 0.5	25.7 ± 3.3	25.3 ± 0.65	33.1 ± 2.01
Klason Lignin	28.2 ± 1.4	25.5 ± 0.5	28.9 ± 2.07	20.5 ± 3.44
Ash	5.8 ± 0.7	5.8 ± 0.2	2.6 ± 0.34	3.6 ± 0.74
Reference	This study	Zakaria et al. [5]	Iberahim et al. [2]	Nordin et al. [3]

'-' Not determined; [a] Ethanol extractives; [b] Acetone extractives.

2.2. *Physico-Chemical Properties of Untreated and Pretreated Samples*

2.2.1. Solids Recovery

Subcritical H_2O-CO_2 pretreatment is one of the thermochemical methods that aim to disrupt the hemicellulose structure and break down xylan into a smaller chain of XOs and xylose. In this study, CS_{PCO2} was used to evaluate the effect of temperature, time and pressure of CO_2 on the hydrothermal process of OPMF. The CS_{PCO2} was used to monitor hydrothermal reaction by pH value obtained from Henry's law equation and to facilitate the efficiency of pretreatment [9]. As shown in Table 2, subcritical H_2O-CO_2 were performed at 150–200 °C for 20–60 min at 0, 3, 5 MPa which corresponds to CS_{PCO2} = −0.93 to −0.06 and the physicochemical properties were compared with subcritical H_2O. It was observed that pretreated solids recovery was in the range from 62.6–84.1% and decreased towards increasing CS_{PCO2} and reached 62.6% at the final CS_{PCO2}= −0.06. The reduction of solid recovery yields towards higher CS_{PCO2} can be explained from xylan solubilization into pretreated liquids [6].

2.2.2. Xylooligosaccharide Content in the Pretreatment Liquids

In pretreatment liquids, hemicellulose was observed to depolymerize into xylan-derived products such as xylose, XOs, arabinose and furfural during hydrothermal pretreatment under the conditions tested (Table 2). XOs represented the major compound present in the pretreatment conditions examined and the increased of XOs production were observed from CS_{PCO2}= −0.93 to −0.19 due to the higher solubilization of hemicellulose components concomitant with higher pretreatment severities. As pretreatment severity increased, the XOs yield started to decrease at CS_{PCO2} = −0.06 and this corresponded to a sharp increase of xylose monomer concentration up to 16.40 mg/g at CS_{PCO2} = −0.06. The highest XOs was recorded at CS_{PCO2} = −0.19 (180 °C, 60 min, 3 MPa), with 8.16 g/L and this value was equivalent to 36.6% of XOs yield from xylan and corresponded to 81.60 mg/g of raw OPMF. At this condition, xylose and furfural were recorded with concentrations of 1.85 g/L and 14.13 g/L, respectively.

Approximately 49.15% of the total xylan was degraded to major compound XOs, followed by xylose and furfural. The XOs concentration obtained in this study was slightly lower compared to that reported by Morais et al. [8] using a wheat straw with XOs production of 11.4 g/L which corresponded to 61.7% of XOs from the total xylan at CS_{PCO2} = −0.33 (215 °C, 30 bar CO_2). Ho et al. [16] found the highest XOs concentration with 17.6 g/L was obtained from OPEFB by autohydrolysis pretreatment at log, *Ro* = 3.91. Therefore XOs was estimated based on xylose and arabinose and 6.15 g/L of XOs with DP X5–X10 was obtained. Interestingly, in this study, by comparing XOs yield from subcritical H_2O-CO_2 at CS_{PCO2} = −0.93 with subcritical H_2O treatment, the XOs value was increased from

1.12 g/L to 1.66 g/L which corresponded to a 48.2% increment. In another study, Zakaria et al. [19] reported XOs were found to be a major compound with the highest value, 7.0 g/L at severity factor log, *R*o = 3.94 from OPMF after the hydrothermal process.

The present study has shown that the impregnation of CO_2 in the hydrothermal process has improved XOs production. Sabiha-Hanim et al. [18] reported a maximum of 48% of the hemicellulose was hydrolyzed using an autoclave system (121 °C, 20–80 min). In term of competitive XOs yields from different biomass, Otieno and Ahring [22] has performed autohydrolysis pretreatment at 145 °C for 60 min on *Miscanthus sinensis*, *Panicum virgatum*, *Calamagroustis acutiflora* and bagasse and found that XOs yields were 65.0%, 84.2%, 87.9% and 92.3%, respectively even though the initial dry mass of xylan was >20%. Lower xylan conversion to XOs probably due to aggregation of xylan with lignin during repolymerization that formed precipitates upon cooling process [23]. It can be concluded that the production of XOs heavily dependent on the types of biomass and selection of pretreatment conditions tested such as temperature, reaction time, initial CO_2 pressure and solid to liquid ratio [13,14].

2.2.3. Monomeric Sugars, Acids, Furans and Tannic Acids Content in the Pretreatment Liquids

Other monomeric sugars like glucose and arabinose were detected in low concentrations in the pretreated liquids. Glucose amount at all conditions was recorded low from 0.12 g/L to 0.30 g/L indicated that this treatment only caused small solubilization of cellulose into the pretreated liquids [14]. Garrote et al. [6] reported a maximum value of glucose in the pretreated liquid of only 0.8 g/L which indicated that hydrothermal treatment at 160–220 °C did not affect the cellulose structure of the biomass. Other by-products produced from hydrothermal process heavily depending on the types of materials and the pretreatment conditions applied. The acidic condition created during hydrothermal process released by-products such as acetic acid, 5-HMF, furfural, formic acid and tannic acid [24]. As shown in Table 2 acetic acid concentration increases as the CS_{PCO2} increases and achieved a maximum value at $CS_{PCO2} = -0.06$ with 381.60 mg/g of raw OPMF. The increasing trend of acetic acid showed that the breakdown of hemicellulose components and xylan side-chains occurred in this pretreatment process and acetic acid can act as a catalyst in carbohydrate degradation [6,14].

Furfural and 5-HMF were formed from degradation of pentose and hexose sugars, respectively and further degradation of furfural and 5-HMF produced formic acid [7,21,24]. The trend of XOs, xylose and furfural over CS_{PCO2} showed a correlation of degradation of xylan-derived product from OPMF. As the severity increased, the XOs concentrations decreased and xylose concentration increased which indicated the sugar degradation occurred caused by the severe pretreatment conditions [14]. Tannic acids were soluble degradation by-product from lignin component formed during the hydrothermal process. This finding was in agreement with a previous study by Zakaria et al. [19], whereby tannic acid was affected by the treatment severities.

2.2.4. pH of the Pretreatment Liquids

The pH of pretreatment liquids presented in Table 2 was calculated using the van Walsum equation [9] and measured pHs were recorded in the range from 4.16–4.32 across all conditions tested. It was obvious that pH of the pretreated liquids in subcritical H_2O-CO_2 was more acidic in comparison to subcritical H_2O case, probably due to the presence of carbonic acid formed from the reaction of H_2O and CO_2 in the reactor during pretreatment process [9] together with higher concentrations of acetic acid and formic acids. Lower pH values obtained from subcritical H_2O pretreatment at higher pretreatment severity were probably due to acetic acid accumulation caused by cleavage of acetyl groups during the hemicellulose degradation [25]. On the other hand, the presence of high dense CO_2 and hot water promoted gas diffusion into the biomass and caused more hemicellulose disruption [26].

Table 2. Physical and chemical properties of pretreated samples under various pretreatment conditions.

Reaction Conditions	Subcritical H_2O								Subcritical CO_2-H_2O							
T (°C)	150		150		160		170		170		180		180		200	
Time (min)	60		60		40		20		40		40		60		20	
Pressure (MPa)	0		3		3		5		3		3		3		5	
Log (*Ro*)	3.25		3.25		3.37		3.36		3.66		3.96		4.13		4.25	
pH (pretreated liquid)	4.41		4.18		4.22		4.16		4.27		4.32		4.32		4.31	
CS_{pCO2}	-		−0.93		−0.85		−0.80		−0.61		−0.36		−0.19		−0.06	
Solid recovery (*w*/*w* %)	83.29		84.14		82.41		80.37		70.22		63.25		68.45		62.56	
Composition/yields	g/L	mg/g	g/L	mg/g	g/L	mg/g	g/L	mg/g	g/L	mg/g	g/L	mg/g	g/L	mg/g	g/L	mg/g
XOs *	1.12	11.23	1.66	16.60	2.14	21.40	1.33	13.30	4.84	48.40	6.62	66.20	8.16	81.60	3.45	34.52
Xylose	0	0.05	0.03	0.26	0.05	0.50	0.15	1.50	0.33	3.30	1.17	11.70	1.85	18.50	1.64	16.40
Glucose	0.19	1.92	0.14	1.40	0.12	1.20	0.19	1.90	0.15	1.50	0.30	3.00	0.20	2.00	0.23	2.30
Arabinose	0.54	5.41	0.51	5.10	0.57	5.70	0.68	6.80	0.73	7.30	0.39	3.90	0.31	3.10	0.16	1.60
Acetic acid	4.07	40.70	3.36	33.60	3.78	37.80	4.94	49.40	14.73	147.30	22.59	225.90	32.33	323.30	38.16	381.60
Furfural	0	0	0	0	0	0	0.88	8.80	2.47	24.70	7.30	73.00	14.13	141.30	22.53	225.30
5-HMF	0	0	0	0	0	0	0	0	0	0	0.37	3.70	0.60	6.00	1.00	10.00
Formic acid	8.06	80.64	8.07	80.70	8.12	81.20	8.40	84.00	11.66	116.60	13.27	132.70	17.06	170.60	18.03	180.30
Tannic acid	0.51	-	0.37	-	0.54	-	0.21	-	0.77	-	1.09	-	1.07	-	1.72	-

* The XOs was calculated by subtracting total xylose monomeric sugars obtained after hydrolyzing pretreatment liquid with 4% H_2SO_4 and monomeric sugars in the pretreatment liquid as suggested by Sluiter et al. [27].

2.3. *Types of XOs Produced*

Figure 1 shows a yield of XOs over combined severity factor, CS_{PCO2} and the characteristics of XOs from pretreatment liquid samples were determined based on their degree of polymerization by using Dionex ICS 3000. Xylobiose (DP X2), xylotriose (DP X3) and xylotetraose (DP X4) were XOs obtained from pretreatment liquid samples and the highest XOs yield was obtained at $CS_{PCO2} = -0.19$ (180 °C, 60 min, 3 MPa) with 8.16 g/L. This value was equivalent to DP X2–X4 of the total xylan and 81.60 mg/g of raw OPMF. Under these conditions, xylobiose, xylotriose and xylotetraose were recorded with concentrations of 24.11 mg/g, 23.18 mg/g and 25.19 mg/g, respectively. From this value, 88.82% of total XOs obtained have DP X2–X4 and the rest of XOs was probably in higher DP form. Higher concentration of XOs yields for DP X2–X4 was recorded at higher pretreatment severity and this can be explained that at a higher temperature and longer reaction time, most of the longer chain XOs were degraded to shorter chain length oligosaccharides and other by-products such as furfural, therefore lower amount of XOs was recovered [28].

Sabiha-Hanim et al. [18] reported that XOs from OPFF after hydrothermal treatment at 121 °C for 60 min contain mainly xylobiose and xylotriose and after subjection to enzymatic hydrolysis using xylanase from *Trichoderma viride*. The XOs that mainly comprise DP X2–X6 were also observed in other agricultural wastes such as tobacco stalks, cotton stalks, sunflower stalks and wheat straw from acid hydrolysis processed [17]. In other study using different oil palm biomass, Ho and co-workers [16] found that XOs obtained from empty fruit bunch (EFB) was mainly with DP X5–X40 with XOs concentration 17.64 g/L after underwent autohydrolysis process at log *Ro* = 3.91. Similarly, XOs generated from xylan of natural grass using enzymatic hydrolysis with *Trichoderma viride* was recorded containing major xylobiose (11.0%) and a small amount of xylotriose [29].

It is worth noting that most of the previous studies on types of XOs detected from xylan involved an additional biological treatment such as enzymatic hydrolysis and acid hydrolysis. In contrast, the type of XOs detected in the present work were only from the subcritical H_2O-CO_2 pretreatment process without further treatment by any other hydrolysis process. Hence, the type and DP of XOs exhibited mainly depend on the hydrolysis treatment and condition used [13]. XOs have novel applications in many industries such as the food, pharmaceutical and health industries. The XOs with short DP range from DP X1–X6 have a beneficial and advantageous function as prebiotics in food-related products [30]. XOs are potential compounds that can behave as prebiotics when ingested as it can stimulate beneficial bacteria inside the colon [31].

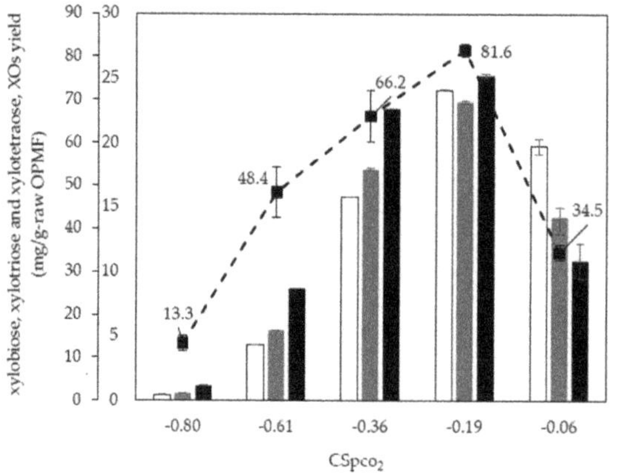

Figure 1. XOs characterization (with respective error bar) for (□) xylobiose, (▒) xylotriose, (■) xylotetraose, (- - -) XOs yield.

Xylobiose (DP X2) has been found to be an important oligosaccharide in the food industry and was reported to have 30% sweetness of sucrose, while other XOs exhibited less sweetness. This has resulted in xylobiose as the main target in food-related products. Besides that, the use of XOs as a food ingredient can help to produce specific food to promote health and reduce the risk of side effect.

2.4. Enzymatic Hydrolysis of Pretreated Solids

Essentially, most of the cellulose component remained in the pretreated solids and only a small portion was solubilized in the pretreatment liquid [6,21]. Thus, instead of high XOs obtained from the pretreatment liquid samples, the conversion of sugars from solid samples was also studied. Table 3 summarizes chemical compositions and physical properties of untreated, subcritical water and subcritical water-CO_2 pretreatments of OPMF. It was observed that cellulose contents were increased towards higher pretreatment severities and recorded the highest cellulose content, 36.67% at CS_{PCO2} = 0.06 (190 °C, 60 min, 3 MPa). In contrast, hemicellulose content was decreased when higher pretreatment severities were applied. The lowest hemicellulose content obtained was 3.14% at CS_{PCO2} = 0.06, indicating successful removal of hemicellulose.

Sugar yields from xylose and glucose obtained from enzymatic hydrolysis of selected pretreated solid samples are presented in this section. Xylose and glucose yields were compared with untreated, subcritical H_2O and subcritical H_2O-CO_2 pretreatments. Xylose and glucose yields from untreated OPMF recorded the lowest compared to all pretreated samples. Approximately 4% increment of glucose yield was obtained when OPMF was hydrothermally pretreated at 150 °C, for 60 min. The enzymatic hydrolysis of the untreated sample yielded lower glucose concentration since *Acremonium cellulase* has less accessibility to cellulose when high hemicellulose present in the biomass. This finding was in agreement with the study by Hsu et al. [32], where higher sugar yield can be achieved at higher hemicellulose removal which provided more accessibility of cellulase to cellulose. Zakaria et al. [4] reported obvious hemicellulose dissolution of OPEFB and OPFF was obtained when using hot compressed water (HCW) treatment at condition temperature ranges from 170–190 °C for 20 min and 10 min, respectively that resulted in high conversion yield of cellulose to glucose.

Table 3. Effect of subcritical H_2O-CO_2 pretreated solids on cellulose crystallinity index, specific surface area and sugar yields.

Treatment Conditions	Untreated OPMF	Subcritical H_2O	Subcritical H_2O-CO_2		
Temperature (°C)	-	150	150	170	190
Time (min)	-	60	180	40	60
Pressure (MPa)	-	0	5	3	3
Log, *Ro*	-	3.25	3.73	3.66	4.43
CS_{PCO2}	-	−1.16	−0.34	−0.61	0.06
pH	-	4.41	4.18	4.27	4.43
Solid recovery (%)	-	85.75	73.73	70.22	61.88
Cellulose (%)	23.58	22.61	28.29	29.24	36.67
Hemicellulose (%)	22.34	17.94	12.14	12.31	3.14
CrI (%)	52.35	62.35	58.92	59.10	63.47
SSA (m^2 g^{-1})	2.33	8.17	17.11	8.18	20.22
Pore volume (cm^3 g^{-1})	0.01	0.04	0.08	0.04	0.01
		* Sugar yield (%)			
Glucose	15.60 ± 7.5	31.83 ± 3.9	68.72 ± 11.0	70.26 ± 4.4	84.65 ± 2.5
Xylose	5.65 ± 0.6	16.99 ± 3.2	28.05 ± 2.8	29.96 ± 0	5.43 ± 0.9

* Sugar yield obtained from the untreated sample of OPMF.

2.5. Cellulose Crystallinity Index

From Table 3, the Crl values of untreated OPMF were lower (52.35%) compared to subcritical H_2O treatment (62.35%). The CrI values for pretreated solids under subcritical H_2O-CO_2 pretreatment were increased as the pretreatment intensity increased. This can be explained by the removal of amorphous hemicellulose from the solid samples towards higher pretreatment condition. Besides the presence of hemicellulose, enzymatic attack on cellulose can also be influenced by cellulose crystallinity of the biomass [8]. The increase Crl of cellulose represented the disclosure amount of cellulose that susceptible to the enzymatic attack in the sample [5].

2.6. Specific Surface Area

Hsu et al. [32] found that SSA and PV of pretreated solid structure affected the enzymatic hydrolysis of biomass. It was obvious that subcritical H_2O-CO_2 improved xylose and glucose yields and this might due to higher SSA and PV obtained after removal of xylan from the cellulose-hemicellulose-lignin matrix. Therefore, inconsistent values of SSA and PV in the pretreated solids are probably due to errors caused by redeposition of lignin or pseudolignin onto the surface of pretreated solids [33]. From Table 3, glucose yield from pretreated solid samples increased in line with increasing CS_{PCO2}, indicating the higher efficiency of the enzymatic hydrolysis. Xylose yields were also increased in line with increasing CS_{PCO2} and started to decrease at CS_{PCO2} = 0.06 due to the lower xylan content in the pretreated solids. Morphological characteristics of pretreated solid samples were analyzed using SEM to observe the effect of subcritical H_2O-CO_2 pretreatment on the surface of pretreated solid samples. The glucose yields obtained from enzymatic hydrolysis of pretreated solids were compared with a previous work [19]. At the same pretreatment severities log, *Ro* = 3.66 and log, *Ro* = 3.73, an addition of initial pressure of CO_2 at 3 and 5 MPa have resulted in an increment of glucose yields by 12.9% and 26.5%, respectively (Table 4). This finding has proven the autohydrolysis process with CO_2 assisted improved accessibility of cellulase to cellulose, which is economically feasible at a commercial scale of production.

Table 4. Comparative analyses between subcritical H_2O and subcritical H_2O-CO_2 pretreatments of OPMF on sugar yields.

Pretreatment/ References	This Study	[19]	This study	[19]
Temperature (°C)	150	150	170	180
Time (min)	180	180	40	20
Log, *Ro*	3.73	3.73	3.66	3.66
Pressure (MPa)	5	-	3	-
Glucose yield (%)	68.7	50.0	70.3	61.0

2.7. SEM Analysis

Morphological characteristics of pretreated solid samples were analyzed using SEM to observe the effect of subcritical H_2O-CO_2 pretreatment on the surface of pretreated solid samples Figure 2. After hydrothermal treatment, physical changes can be noticed on the surface of the pretreated solids compared to untreated solid. The untreated OPMF (Figure 2a) showed a rigid and intact surface of biomass, which provides less accessibility of enzyme penetration into cellulose component. Meanwhile, for all pretreated solids (Figure 2b–e), rough surface and more porous fibers were observed. When comparing to subcritical H_2O pretreatment, impregnation of CO_2 was more likely have a rougher surface and the peeling-off of the outer layer of the cell wall compared to the sample without CO_2 effect (Figure 2b,c) and the surface changes at the more severe condition in CO_2 reaction (Figure 2d,e). The blending effect of H_2O and CO_2 help to disrupt and fractionated more fiber, increased SSA and PV and made cellulose accessible to enzymatic attack [26,34].

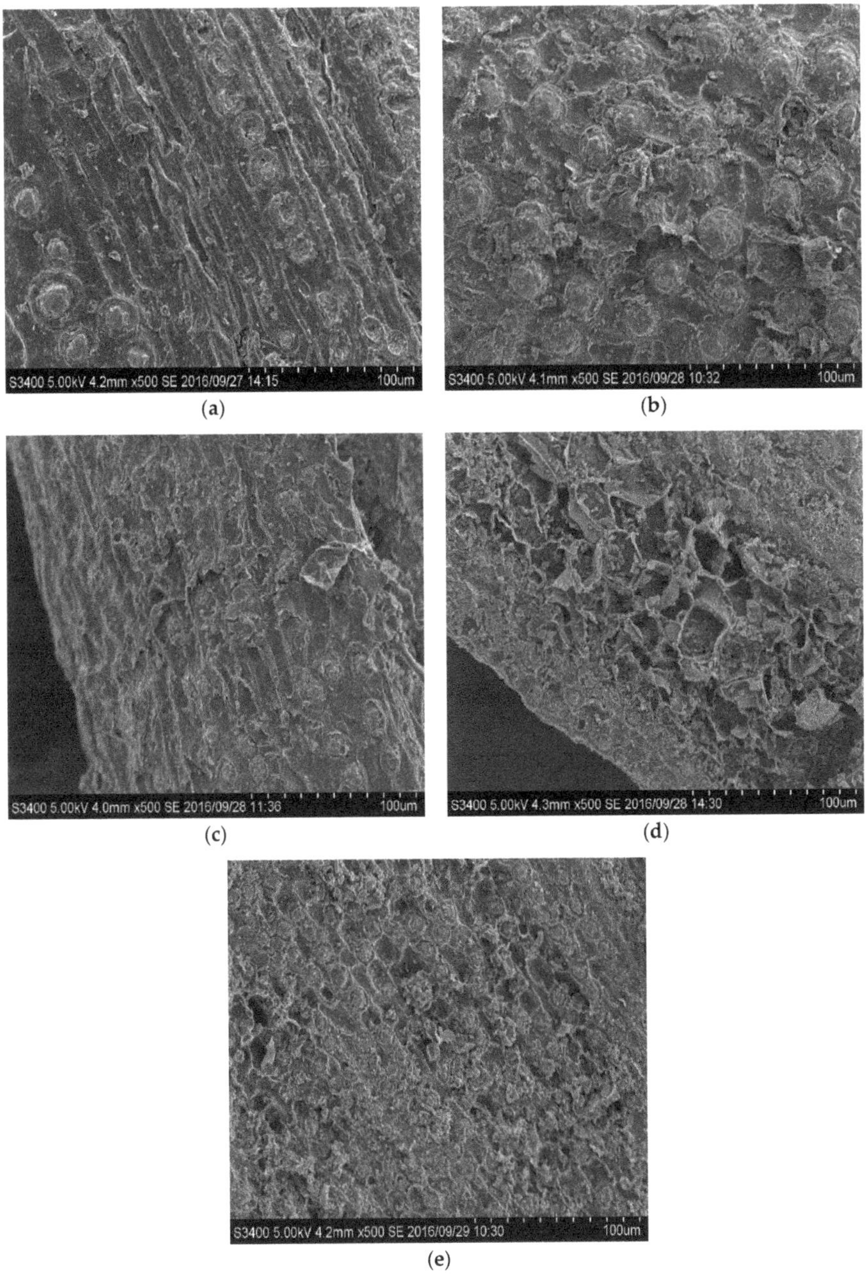

Figure 2. SEM micrographs of (**a**) untreated OPMF (**b**) subcritical H_2O treatment at 150 °C, 60 min (**c**) subcritical H_2O-CO_2 treatment at 150 °C, 60 min 5 MPa (**d**) subcritical H_2O-CO_2 treatment at 170 °C, 40 min, 3 MPa (**e**) subcritical H_2O-CO_2 treatment at 190 °C, 60 min, 3 MPa obtained with magnification 500×.

3. Materials and Methods

3.1. *Raw Material Preparation*

OPMF used in this study was collected from Seri Ulu Langat Palm Oil Mill, Dengkil, Selangor, Malaysia. The samples were sun-dried for two days and crushed kernels and shells were manually separated from OPMF fibers before compositional analysis and other experimental work to avoid error in data analysis. The samples were ground to 2 mm size using a Pulverisette 15 cutting mill (Fritsch, Idar-Oberstein, Germany) and dried overnight in *vacuo* at 40 °C. The samples were stored in a vacuum chamber at room temperature (24 °C) before further analysis.

3.2. *Chemical Compositional Analysis*

Chemical compositional analysis such as extractives, cellulose, hemicellulose, Klason lignin, ash and moisture content of untreated and pretreated OPMF were determined according to the method reported by Sluiter et al. [27].

3.3. *Subcritical H_2O and Subcritical H_2O-CO_2 Pretreatments*

Subcritical H_2O and subcritical H_2O-CO_2 pretreatments of OPMF were conducted in a 35 mL stainless steel tube reactor. In this study, 2 mm size OPMF was used with solid to liquid (S: L) ratio of 1:10. 3 g of oven dried OPMF samples and 30 mL of distilled water were filled inside the reactor. The reactor was tightened closely and purged with CO_2 at a pressure range from 0–5 MPa. The detection of gas leakage was performed to ensure there was no CO_2 leak from the reactor. The reactor was immersed in the sand bath, with controlled temperature (150–200 °C) using automatic temperature controller. The mixture was homogenized by agitation at 60 rpm at time range from 20–240 min. After completion of the heating process, the reactor was cooled down in the water reservoir. The pressure was released at the end of the pretreatment. The solid and liquid samples were separated using filter paper No. 2 with pore size 0.5 μm (Advantec, Tokyo, Japan) and pH of the pretreatment liquids was recorded using a digital pH meter (B-712 LAQUAtwin, Horiba, Kyoto, Japan). The pretreated solids were dried at 40 °C for 48 h in a vacuum drier. The pretreatment liquids were further filtered with 0.22 μm PTFE syringe filter (Whatman, Clifton, NJ, USA) and directly injected into the HPLC for sugars and by-products determination.

3.4. *Combined Severity Factor*

The intensity of hydrothermal process was expressed in terms of severity factor (log, *Ro*) that combined the reaction temperature and time factor according to Equation (1):

$$Ro = (t \exp\left[\frac{T-100}{14.75}\right]) \quad (1)$$

where t is time expressed in minutes, T is temperature expressed in °C and 14.75 is an empirical parameter related to temperature and activation energy [35]. Combined severity factor (CSF) include the pH effect on the pretreatment severity due to the presence of carbonic acid formed from the CO_2 pressure is according to Equation (2):

$$\mathrm{CSF} = \log Ro - \mathrm{pH} \quad (2)$$

because it was difficult to measure the pH during the reaction to show the influence of carbonic acid, the pH values were calculated from Equation (3):

$$\mathrm{pH} = 8.00 \times 10^{-6} \times (T^2 + 0.00209) \times (T - 0.126) \times \ln(P_{CO2}) + 3.92 \quad (3)$$

Thus, in this experiment, the combined severity factor was calculated as in Equation (4) below as recommended by van Walsum [9]:

$$CSF = \log(t \exp\left[\frac{(T-100)}{14.75}\right]) - 8.00 \times 10^{-6} \times T^2 + 0.00209 \times T - 0.126 \times \ln(P_{CO2}) + 3.92 \quad (4)$$

3.5. Determination of Monomeric and Total Monomeric Sugars from Pretreatment Liquids

Monomeric and total monomeric sugars such as glucose, xylose, galactose, mannose and arabinose in pretreatment liquid samples were determined by high-performance liquid chromatography (HPLC) according to the report by Inoue et al. [36]. For total monomeric sugar analysis, 5 mL of pretreatment liquid sample was hydrolyzed in diluted 4% (*v*/*v*) H_2SO_4 and autoclaved at 121 °C for 1 h. The sugar produced in the liquid sample was cooled and filtered using a Dionex OnGuard^{TM} 11A cartridge filter (Thermo Scientific, Waltham, MA, USA) to neutralize the pH before HPLC analysis [27]. The organic acids present in the liquid sample such as acetic acid, furfural, 5-HMF and formic acid were detected by HPLC as reported earlier [36].

3.6. Determination of Tannic Acid

Tannic acid concentration in the pretreatment liquids was determined using Folin-Ciolcalteu method by UV-VIS spectrophotometer (UV mini-1240, Shimadzu, Kyoto, Japan) following the method described by Makkar [37].

3.7. Determination Degree of Polymerization (DP) of XOs

The DP of XOs was determined by a Dionex ICS 3000 system equipped with an AS3000 auto sampler using high-performance anion exchange chromatography with a pulse amperometric detection (HPAED-PAD) system (Thermo Scientific). The types of XOs were identified by comparing the peak areas of standard xylobiose (X2), xylotriose (X3) (Wako, Osaka, Japan) xylotetraose (X4) (Biocon, Nagoya, Japan). A Carbopac PA1 column (4 × 250 mm, Dionex, Thermo Scientific) with PA1 guard column (4 × 50 mm, Dionex) was used at a flow rate of 1.0 mL/min and the column temperature was set at 35 °C. A pulsed amperometric detector with an Au electrode operating in the integrated amperometric mode (Dionex) was used for the detection of XOs which was separated with a gradient of 10–100 mM NaOH for 15 min, followed by 0–20 mM sodium acetate gradient in 100 mM NaOH for 25 min.

3.8. Enzymatic Hydrolysis

Enzymatic hydrolysis was performed by using enzyme cocktail constituting 40 FPU/mL *Acremonium* cellulase (Meiji Seika Co., Tokyo, Japan) and 10% Optimash BG (Genencor International, Rochester, CA, USA). The enzymatic assays were performed in 6% substrate loading. In a standard assay, 10 FPU/g substrate of *Acremonium* cellulase and 0.1% final concentration from Optimash BG stock were added to 0.09 g of the substrate in a final concentration of 50 mM sodium acetate buffer (pH 5.0). The mixture was added up to 1.5 mL total volume. The activities of enzyme cocktail in the reaction mixture as follows: FPase, 0.33 FPU/mL; xylanase, 32.5 U/mL; carboxymethyl cellulase, 7.4 U/mL; β-glucosidase, 1.8 U/mL; β-xylosidase, 0.03 U/mL. The enzymatic hydrolysis was performed at 50 °C for 72 h with shaking. The experiment was performed in triplicate and average results were presented. The sugar yield was calculated using Equation (5):

Sugar yield (%) = [weight of monomeric sugars after enzymatic hydrolysis/weight of total monomeric sugars from the untreated sample after hydrolysis using H_2SO_4] × 100 (5)

3.9. SEM, BET and CrI Analyses

The untreated and pretreated OPMF samples were sputtered with Pt-Pd for 100 s (Ion sputterer; Hitachi, Tokyo, Japan). The coated samples were examined by field emission scanning electron microscopy (S-3400N, Hitachi, Japan) at 1 kV. The solids were rinsed with ethanol then soaked with t-butyl alcohol and dried before SEM analysis [19].

The specific surface area of the sample was determined from the Brunauer-Emmett-Teller (BET) plot of nitrogen adsorption-desorption isotherms [38]. The total pore volume was determined at At $P/P_0 = 0.99$.

Wide angle X-ray diffraction (WAXD) patterns analysis of the untreated and hydrothermally treated OPMF samples were determined by RINT-TTR III X-ray diffractometer (Rigaku, Tokyo, Japan) as reported earlier [15].

The crystallinity index (CrI) was calculated using Equation (6) based on the method of Segal et al. [39]:

$$\textit{Crystallinity index}\ (\%) = [(I_{002} - I_{am})/I_{002}] \times 100 \tag{6}$$

I_{002}: The intensity at about $2\theta = 22.2°$
I_{am}: The intensity at $2\theta = 17.6°$

4. Conclusions

Subcritical H_2O-CO_2 pretreatment of OPMF was successfully performed under an optimal condition at $CSP_{CO2} = -0.19$ (180 °C, 60 min, 3 MPa). Approximately 8.16 g/L of XOs was produced, which equivalent to 36.59% of XOs yield from xylan. xylobiose, xylotriose and xylotetraose with DP X2–X4 were the XOs identified in the pretreatment liquids. Enzymatic hydrolysis of OPMF pretreated solids at higher pretreatment severities showed that high amount of glucose could be produced. To the best of our knowledge, this is the first report on subcritical H_2O-CO_2 pretreatment of OPMF for the production of XOs and its potential in industrial applications. Purification of XOs produced from OPMF using several methodologies and future application of XOs are in research in development progress.

Author Contributions: All the experimental work was conducted by N.A. The manuscript was prepared by N.A. and M.R.Z., M.Z.M.Y., S.F., H.I., H.A., and M.A.H. supervised this work and provided all experimental and analytical equipment. All authors have read and approved the final manuscript.

Acknowledgments: This work is partly supported by Science and Technology Research Partnership for Sustainable Development (SATREPS), organized by Japan Science and Technology Agency (JST) and Japan International Cooperation Agency (JICA), under the leadership of Prof. Yoshihito Shirai from Kyushu Institute of Technology, Japan. We are grateful to Seri Ulu Langat Palm Oil Mill for providing the raw materials. Norlailiza Ahmad was the recipient of Graduate Research Fellowship from Universiti Putra Malaysia.

Conflicts of Interest: The authors declare no conflict of interest.

References

1. Abdullah, N.; Sulaiman, F. The oil palm wastes in Malaysia. In *Biomass Now-Sustainable Growth and Use*; InTech: Rijeka, Croatia, 2013; pp. 75–93.
2. Iberahim, N.I.; Jahim, J.M.; Harun, S.; Nor, M.T.M.; Hassan, O. Sodium hydroxide pretreatment and enzymatic hydrolysis of oil palm mesocarp fiber. *Int. J. Chem. Eng. Appl. Sci.* **2013**, *4*, 101–105. [CrossRef]
3. Nordin, N.I.A.A.; Ariffin, H.; Andou, Y.; Hassan, M.A.; Shirai, Y.; Nishida, H.; Ibrahim, N.A. Modification of oil palm mesocarp fiber characteristics using superheated steam treatment. *Molecules* **2013**, *18*, 9132–9146. [CrossRef] [PubMed]
4. Zakaria, M.R.; Hirata, S.; Fujimoto, S.; Hassan, M.A. Combined pretreatment with hot compressed water and wet disk milling opened up oil palm biomass structure resulting in enhanced enzymatic digestibility. *Bioresour. Technol.* **2015**, *193*, 128–134. [CrossRef] [PubMed]

5. Zakaria, M.R.; Hirata, S.; Hassan, M.A. Combined pretreatment using alkaline hydrothermal and ball milling to enhance enzymatic hydrolysis of oil palm mesocarp fiber. *Bioresour. Technol.* **2014**, *169*, 236–243. [CrossRef] [PubMed]
6. Garrote, G.; Domí, H.; Parajó, J.C. Autohydrolysis of corncob: Study of non-isothermal operation for xylooligosaccharide production. *J. Food Eng.* **2002**, *52*, 211–218. [CrossRef]
7. Da Silva, S.P.M.; Morais, A.R.C.; Bogel-Łukasik, R. The CO_2-assisted autohydrolysis of wheat straw. *Green Chem.* **2014**, *16*, 238–246. [CrossRef]
8. Morais, A.R.; Mata, A.C.; Bogel-Lukasik, R. Integrated conversion of agroindustrial residue with high pressure CO_2 within the biorefinery concept. *Green Chem.* **2014**, *16*, 4312–4322. [CrossRef]
9. Van Walsum, G.P. Severity function describing the hydrolysis of xylan using carbonic acid. In *Twenty-Second Symposium on Biotechnology for Fuels and Chemicals*; Humana Press: New York, NY, USA, 2001; pp. 317–329.
10. Relvas, F.M.; Morais, A.R.C.; Bogel-Lukasik, R. Kinetic modeling of hemicellulose-derived biomass hydrolysis under high pressure CO_2–H_2O mixture technology. *J. Supercrit. Fluids* **2015**, *99*, 95–102. [CrossRef]
11. Fockink, D.H.; Morais, A.R.C.; Ramos, L.P.; Lukasik, R.M. Insight into the high-pressure CO_2 pre-treatment of sugarcane bagasse for a delivery of upgradable sugars. *Energy* **2018**, *151*, 536–544. [CrossRef]
12. Toscan, A.; Morais, A.R.C.; Paixão, S.M.; Alves, L.; Andreaus, J.; Camassola, M.; Dillon, A.J.P.; Lukasik, R.M. High-pressure carbon dioxide/water pre-treatment of sugarcane bagasse and elephant grass: Assessment of the effect of biomass composition on process efficiency. *Bioresour. Technol.* **2017**, *224*, 639–647. [CrossRef] [PubMed]
13. Bian, J.; Peng, P.; Peng, F.; Xiao, X.; Xu, F.; Sun, R.C. Microwave-assisted acid hydrolysis to produce xylooligosaccharides from sugarcane bagasse hemicelluloses. *Food Chem.* **2014**, *156*, 7–13. [CrossRef] [PubMed]
14. Cara, C.; Ruiz, E.; Carvalheiro, F.; Moura, P.; Ballesteros, I.; Castro, E.; Gírio, F. Production, purification and characterisation of oligosaccharides from olive tree pruning autohydrolysis. *Ind. Crops Prod.* **2012**, *40*, 225–231. [CrossRef]
15. Bragatto, J.; Segato, F.; Squina, F.M. Production of xylooligosaccharides (XOS) from delignified sugarcane bagasse by peroxide-HAc process using recombinant xylanase from *Bacillus subtilis*. *Ind. Crops Prod.* **2013**, *51*, 123–129. [CrossRef]
16. Ho, A.L.; Carvalheiro, F.; Duarte, L.C.; Roseiro, L.B.; Charalampopoulos, D.; Rastall, R.A. Production and purification of xylooligosaccharides from oil palm empty fruit bunch fibre by a non-isothermal process. *Bioresour. Technol.* **2014**, *152*, 526–529. [CrossRef] [PubMed]
17. Akpinar, O.; Erdogan, K.; Bostanci, S. Production of xylooligosaccharides by controlled acid hydrolysis of lignocellulosic materials. *Carbohydr. Res.* **2009**, *344*, 660–666. [CrossRef] [PubMed]
18. Sabiha-Hanim, S.; Noor, M.A.M.; Rosma, A. Effect of autohydrolysis and enzymatic treatment on oil palm (*Elaeisguineensis Jacq.*) frond fibres for xylose and xylooligosaccharides production. *Bioresour. Technol.* **2011**, *102*, 1234–1239. [CrossRef] [PubMed]
19. Zakaria, M.R.; Hirata, S.; Fujimoto, S.; Ibrahim, I.; Hassan, M.A. Soluble inhibitors generated during hydrothermal pretreatment of oil palm mesocarp fiber suppressed the catalytic activity of *Acremonium* cellulase. *Bioresour. Technol.* **2016**, *200*, 541–547. [CrossRef] [PubMed]
20. Zhu, Y.; Kim, T.H.; Lee, Y.Y.; Chen, R.; Elander, R.T. Enzymatic production of xylooligosaccharides from corn stover and corncobs treated with aqueous ammonia. In *Twenty-Seventh Symposium on Biotechnology for Fuels and Chemicals*; Humana Press: New York, NY, USA, 2006; pp. 586–598.
21. Nabarlatz, D.; Ebringerová, A.; Montané, D. Autohydrolysis of agricultural by-products for the production of xylo-oligosaccharides. *Carbohydr. Polym.* **2007**, *69*, 20–28. [CrossRef]
22. Otieno, D.O.; Ahring, B.K. A thermochemical pretreatment process to produce xylooligosaccharides (XOS), arabinooligosaccharides (AOS) and mannooligosaccharides (MOS) from lignocellulosic biomasses. *Bioresour. Technol.* **2012**, *112*, 285–292. [CrossRef] [PubMed]
23. Yang, B.; Wyman, C.E. Characterization of the degree of polymerization of xylooligomers produced by flow through hydrolysis of pure xylan and corn stover with water. *Bioresour. Technol.* **2008**, *99*, 5756–5762. [CrossRef] [PubMed]
24. Jönsson, L.J.; Martín, C. Pretreatment of lignocellulose: Formation of inhibitory by-products and strategies for minimizing their effects. *Bioresour. Technol.* **2016**, *199*, 103–112. [CrossRef] [PubMed]

25. Xiao, L.P.; Shi, Z.J.; Xu, F.; Sun, R.C. Hydrothermal treatment and enzymatic hydrolysis of *Tamarix ramosissima*: Evaluation of the process as a conversion method in a biorefinery concept. *Bioresour. Technol.* **2013**, *135*, 73–81. [CrossRef] [PubMed]
26. Sarkar, N.; Ghosh, S.K.; Bannerjee, S.; Aikat, K. Bioethanol production from agricultural wastes: An overview. *Renew. Energy* **2012**, *37*, 19–27. [CrossRef]
27. Sluiter, A.; Hames, B.; Ruiz, R.; Scarlata, C.; Sluiter, J.; Templeton, D. *Determination of Sugars, by-Products, and Degradation Products in Liquid Fraction Process Samples. NREL/TP-510-42623, Laboratory Analytical Procedure (LAPs)*; National Renewable Energy Laboratory: Golden, CO, USA, 2006.
28. Chen, M.H.; Bowman, M.J.; Dien, B.S.; Rausch, K.D.; Tumbleson, M.E.; Singh, V. Autohydrolysis of *Miscanthus x giganteus* for the production of xylooligosaccharides (XOS): Kinetics, characterization and recovery. *Bioresour. Technol.* **2014**, *155*, 359–365. [CrossRef] [PubMed]
29. Samanta, A.K.; Jayapal, N.; Kolte, A.P.; Senani, S.; Sridhar, M.; Suresh, K.P.; Sampath, K.T. Enzymatic production of xylooligosaccharides from alkali solubilized xylan of natural grass (*Sehima nervosum*). *Bioresour. Technol.* **2012**, *112*, 199–205. [CrossRef] [PubMed]
30. Reddy, S.S.; Krishnan, C. Production of high-pure xylooligosaccharides from sugarcane bagasse using crude β-xylosidase-free xylanase of *Bacillus subtilis* KCX006 and their bifidogenic function. *LWT-Food Sci. Technol.* **2016**, *65*, 237–245. [CrossRef]
31. Moure, A.; Gullón, P.; Domínguez, H.; Parajó, J.C. Advances in the manufacture, purification and applications of xylo-oligosaccharides as food additives and nutraceuticals. *Process Biochem.* **2006**, *41*, 1913–1923. [CrossRef]
32. Hsu, T.C.; Guo, G.L.; Chen, W.H.; Hwang, W.S. Effect of dilute acid pretreatment of rice straw on structural properties and enzymatic hydrolysis. *Bioresour. Technol.* **2010**, *101*, 4907–4913. [CrossRef] [PubMed]
33. Pu, Y.; Hu, F.; Huang, F.; Davison, B.H.; Ragauskas, A.J. Assessing the molecular structure basis for biomass recalcitrance during dilute acid and hydrothermal pretreatments. *Biotechnol. Biofuels* **2013**, *6*, 15. [CrossRef] [PubMed]
34. Gao, M.; Xu, F.; Li, S.; Ji, X.; Chen, S.; Zhang, D. Effect of SC-CO_2 pretreatment in increasing rice straw biomass conversion. *Biosyst. Eng.* **2010**, *106*, 470–475. [CrossRef]
35. Overend, R.P.; Chornet, E.; Gascoigne, J.A. Fractionation of lignocellulosics by steam-aqueous pretreatments. *Philos. Trans. R. Soc. Lond. A Math. Phys. Eng. Sci.* **1987**, *321*, 523–536. [CrossRef]
36. Inoue, H.; Yano, S.; Endo, T.; Sakaki, T.; Sawayama, S. Combining hot-compressed water and ball milling pretreatments to improve the efficiency of the enzymatic hydrolysis of eucalyptus. *Biotechnol. Biofuels* **2008**, *1*, 1. [CrossRef] [PubMed]
37. Makkar, H.P. *Quantification of Tannin in Tree and Shrub Foliage: A Laboratory Manual*; Kluwer Academic Publishers: Dordrecht, The Netherlands, 2003.
38. Ishiguro, M.; Endo, T. Addition of alkali to the hydrothermal- mechanochemical treatment of Eucalyptus enhances its enzymatic saccharification. *Bioresour. Technol.* **2014**, *153*, 322–326. [CrossRef] [PubMed]
39. Segal, L.; Creely, J.J.; Martin, A.E.; Conrad, C.M. An empirical method for estimating the degree of crystallinity of native cellulose using the X-ray diffractometer. *Text. Res.* **1959**, *29*, 786–794. [CrossRef]

Sample Availability: Samples of the raw OPMF is available from the authors.

molecules

MDPI

Article

Enhanced Enzymatic Hydrolysis and Structural Features of Corn Stover by NaOH and Ozone Combined Pretreatment

Wenhui Wang [1], Chunyan Zhang [1], Shisheng Tong [2], Zhongyi Cui [1] and Ping Liu [1,*]

1 College of Food Science and Nutritional Engineering, China Agricultural University, Beijing 100083, China; wwhui123321@163.com (W.W.); m18612203994@163.com (C.Z.); zhongyi.cui@dupont.com (Z.C.)

2 Bio-Pharmaceutical College, Beijing City University, Beijing 100094, China; shishengt@163.com

* Correspondence: LiuPing@cau.edu.cn; Tel.: +86-106-273-7131

Received: 19 March 2018; Accepted: 8 May 2018; Published: 29 May 2018

Abstract: A two-step pretreatment using NaOH and ozone was performed to improve the enzymatic hydrolysis, compositions and structural characteristics of corn stover. Comparison between the unpretreated and pretreated corn stover was also made to illustrate the mechanism of the combined pretreatment. A pretreatment with 2% (*w*/*w*) NaOH at 80 °C for 2 h followed by ozone treatment for 25 min with an initial pH 9 was found to be the optimal procedure and the maximum efficiency (91.73%) of cellulose enzymatic hydrolysis was achieved. Furthermore, microscopic observation of changes in the surface structure of the samples showed that holes were formed and lignin and hemicellulose were partially dissolved and removed. X-ray Diffraction (XRD), Fourier Transform Infrared Spectroscopy (FTIR) and Cross-Polarization Magic Angle Spinning Carbon-13 Nuclear Magnetic Resonance (CP/MAS ^{13}C-NMR) were also used to characterize the chemical structural changes after the combined pretreatment. The results were as follows: part of the cellulose I structure was destroyed and then reformed into cellulose III, the cellulose crystal indices were also changed; a wider space between the crystal layer was observed; disruption of hydrogen bonds in cellulose and disruption of ester bonds in hemicellulose; cleavage of bonds linkage in lignin-carbohydrate complexes; removal of methoxy in lignin and hemicellulose. As a result, all these changes effectively reduced recalcitrance of corn stover and promoted subsequent enzymatic hydrolysis of cellulose.

Keywords: corn stover; alkali; ozone; combined pretreatment; enzymatic hydrolysis; surface morphology; structural characteristics

1. Introduction

In an effort to reduce the energy crisis and the environmental pollution, preparation of recycled lignocellulosic biomass for the use of energy, materials and chemicals has become the focus of today's research. Due to the low degree of lignification, high carbohydrate content and easy absorption of carbohydrate, corn stover has high value in comparison with other lignocellulosic biomass [1]. Pretreatments with physical, chemical and biological methods, however, are necessary to change complex network structure among cellulose, hemicellulose and lignin in corn stover, ascertain pretreatment can release the closure and reduce the strong interchain between lignin and cellulose, compromise the crystalline structure of cellulose, enhance accessibility of enzyme and make corn stover fully utilized [2]. With existing pretreatment methods, sodium hydroxide can rupture the interchain between lignin and other carbohydrates significantly, saponify the inter-molecular ester bonds between hemicellulose and other components, make lignocellulosic swell to remove lignin effectively [3,4]. In addition, this pretreatment not only increases the porosity and internal specific surface area of fibrous materials to ensure effective contact of the enzyme with fibrous materials and degrade it but also changes the structure of lignocellulose and improves its digestibility of polysaccharide by increasing cellulose conversion rate [5]. Because of strong oxidation, low solubility and selective oxidation of ozone, ozone oxidation technology has some limitations in gas-liquid transfer such as the slow rate, high cost and low ozone utilization rate, which makes it difficult to be used alone [6,7]. Ben'koet et al. used ozone to pretreat aspen wood and found that the efficiency of enzymatic hydrolysis was determined by the absorption rate of ozone [8]. Panneerselvam et al. used different ozone concentrations of 40 mg/L, 50 mg/L, 58 mg/L to treat energy grass [9]. Pretreatment conditions and results showed that ozone treatment can remove lignin effectively without cellulose degradation. Bule et al. used ozone to pretreat wheat stover, the particle size of which was less than 60 mesh, for 2 h and the results showed that the lignin structure was modified significantly and the sugar recovery rate increased from 13.11% to 63.17% in comparison with untreated samples [10]. The previous experiment made corn stover treated in 2% NaOH solution at a temperature of 80 °C for 2 h. The specific surface area diameter of corn stover particles was reduced from 189.9 μm to 132.2 μm, and the specific surface area of stover decreased after ozone treatment at pH 5 for 50 min up to 93.11 μm, compared with the specific surface area diameter of the non-alkaline control group decreased by 51%, indicating that alkali combined with ozone made the stover particles smaller by removing lignin. This result was shown in Supplementary Materials.

In this work, a two-step pretreatment using NaOH and ozone was performed on corn stover to improve its enzymatic hydrolysis and changes in compositions and structural characteristics compared to unpretreated. Pretreated corn stover was also analyzed to illustrate the mechanism of the combined pretreatment.

2. Materials and Methods

2.1. Materials and Sample Preparation

The corn stover was obtained from the farm research fields at the Jilin Agricultural University (Changchun, Jilin, China). After the corn stover sample was cut into small pieces, it was oven-dried to bring down the moisture content, then milled and screened to particle size of less than 1 mm. The dry sample was kept at −20 °C for future use.

2.2. Methods

2.2.1. NaOH Treatment

Two gram of dry corn stover and 30 mL of 2% (*w*/*v*) NaOH were mixed completely in a 50 mL centrifugal tube reactor, which was then incubated in a water bath for 2 h–4 h at 40 °C, 60 °C and 80 °C. The pretreatment conditions including NaOH treatment temperatures and times, shown in Table 1. When the reaction was over, the tube reactor was cooled to room temperature and filtered via a 300-mesh sieve to separate the mixture into the solid residue and liquid hydrolysate. The solid residue was rinsed with deionized water or saturated carbon dioxide water until it reached neutral pH.

Table 1. Level of NaOH pretreatment factors.

Factor	NaOH Pretreatment Conditions		
NaOH Pretreatment Temperature	40 °C	60 °C	80 °C
NaOH Pretreatment Time	2 h	3 h	4 h

Note: In order to facilitate the description of the structure of typical pre-treated samples, a special nomenclature for NaOH pre-treatment conditions were as follows: A: 2% NaOH at 80 °C for 2 h; B: 2% NaOH at 80 °C for 3 h; C: 2% NaOH at 60 °C for 2 h; D: 2% NaOH at 80 °C for 4 h.

2.2.2. Ozone Treatment

Two gram of sample and 30 mL deionized water were placed in a 60 mL of beaker to prepare for ozone pretreatment. 2 mol/L of dilute sulphuric acid was also added to adjust the initial pH of the reaction liquid. Ozone was generated by an ozone generator (CF-10F, Beijing, China). During the reaction, ozone concentration maintained at 78 mg/L for different time with magnetic stirring (85-2, Shanghai, China) at a room temperature. Ozonation experimental conditions are shown in Table 2.

Table 2. Level of Ozone Pretreatment factors.

Factor	Ozone Pretreatment Conditions	
Ozone treatment Initial pH	5	9
Ozone treatment time	25 min	35 min

2.2.3. Combined Sodium Hydroxide and Ozone Pretreatment

The experiment is divided into two processes: NaOH Pretreatment and Ozone Pretreatment. The experimental design of the alkali treatment stage is a two-factor three-level, as shown in Table 1. Each group of experiments after alkali treatment was further treated with ozone. The experimental design of the ozone treatment stage is a two-factor two-level, as shown in Table 2. Combined pretreatment samples were prepared for enzymatic hydrolysis and other analysis. The pretreatment conditions, including NaOH and ozone treatment, were optimized for high delignification and high cellulose composition. For convenient description of structure characterization, the special nomenclature for combined pre-treatment conditions were as follows: A-5-25: 2% NaOH at 80 °C for 2 h and the ozone initial pH 5 for 25 min. A-9-25: 2% NaOH at 80 °C for 2 h and the ozone initial pH 9 for 25 min. A-9-35: 2% NaOH at 80 °C for 2 h and the ozone initial pH 9 for 35 min. B-9-25: 2% NaOH at 80 °C for 3 h and the ozone initial pH 9 for 25 min. C-9-35: 2% NaOH at 60 °C for 2 h and the ozone initial pH 9 for 35 min. D-9-35: 2% NaOH at 80 °C for 4 h and the ozone initial pH 9 for 35 min. Blank: untreated degreased stover.

2.2.4. Enzymatic Hydrolysis

Pretreated stover samples, of 0.2 g in 100 mL, were placed in an Erlenmeyer flask and added to 10 mL acetate buffer (0.1 mol/L, pH 4.8), which was prepared with sterile water and contained 40 μg/mL tetracycline, 30 μg/mL cycloheximide and 40 μL xylanase solution. The mixture was incubated in shaking bath (120 rpm) at 70 °C for 24 h. After the reaction, the sample was cooled down to room temperature, 40 μL cellulase and 30 μL β-glucosidase were added and it was then incubated at 50 °C for 72 h. Cycloheximide could inhibit the DNA translation of eukaryotes to stop cell growth or even cause death. The purpose of adding cycloheximide and tetracycline hydrochloride was to inhibit the growth of microorganism which influenced the pH value during the enzymatic process and affected enzyme activity. Enzymatic hydrolysate was filtered through 0.22 μm membrane and then analyzed by HPLC to determine the glucose content and calculated the cellulase hydrolysis rate.

2.2.5. Determination of the Composition Content of the Corn Stover Samples

In this paper, three components (cellulose, hemicellulose and lignin) in the stover before and after pretreatment were determined by two-step acid hydrolysis method (NREL, 2008b). The content of cellulose and hemicellulose, the lignin removal rate was determined by the following equation:

$$\text{Cellulose concent } (\%) = \frac{C_1 \times V \times 0.9}{m} \times 100, \quad (1)$$

$$\text{Hemicellulose content } (\%) = \frac{C_2 \times V \times 0.88}{m} \times 100, \quad (2)$$

In the equation, C_1 was the concentration of glucose measured by HPLC (mg/mL); C_2 was the concentration of xylose measured by HPLC (mg/mL); V was the total volume of the reaction system (87 mL); m was the dry weight of the sample (300 mg); 0.9 was the conversion of glucose to cellulose, 0.88 was the conversion of xylose to hemicellulose.

$$W_1\ (\%) = \frac{m_1 - m_2}{0.3} \times 100, \tag{3}$$

In the equation, W_1 was the acid-insoluble lignin in the stover; m_1 was the total weight of the sand core funnel and the residue; m_2 was the weight of the sand core funnel; 0.3 was the dry weight of the sample. The unit of measurement was g.

$$W_2\ (\%) = \frac{OD_{320} \times V \times n}{300} \times 100, \tag{4}$$

In the equation, W_2 was the acid-soluble lignin in the stover; V was the total volume of the reaction system (87 mL); 300 was the dry weight (units: mg) of the sample; OD_{320} was the absorbance at 320 nm, 30 L/g·cm; n was the dilution factor.

$$W_3\ (\%) = \frac{m_4}{m_3} \times 100, \tag{5}$$

In the equation, W_3 was the lignin removal rate, the total weight of acid-soluble lignin and the acid-insoluble lignin were the total lignin content of the stover; m_3 was the total lignin weight of the untreated stover; m_4 was the total lignin weight of the pretreatment stover.

2.3. *Structural Analysis*

2.3.1. Scanning Electron Microscope (SEM) Analysis

Measured stover samples were placed in an oven for 24 h at 50 °C to remove moisture, imaged with S-3400n scanning electron microscope with a voltage of 20 kV, current of 30 mA and distance of 11.3 mm. Electron microscopy was amplified at different rates to observe the surface morphology of the sample.

2.3.2. X-ray Diffraction (XRD) Analysis

The samples were examined by X-ray diffractometer with CuKa radiation (λ = 0.154 nm). CuKa radiation was eliminated with nickel. The operation voltage and current was 40 kV and 40 mA respectively. The measurement method was $\theta/2\theta$ linkages scanning. The range of 2θ was 5° to 70°. The step was 0.02° and the time interval was 0.2 s. The sample was pressed at 40 °C and subjected to a 2θ intensity curves. Using Origin and MDI jade 5.0 for data analysis.

2.3.3. Fourier Transform Infrared Spectroscopy (FTIR) Analysis

The samples were placed in an oven at 50 °C for 24 h to remove moisture. 10 mg of dry sample was mixed with 200 mg KBr, manually ground in an agate mortar and pressed at 20 MPa for 2 min in oil pressure. The tablets were placed on a sample rack for FTIR spectra spectroscopy and the spectra was recorded between 4000 and 400 cm^{-1}. The PerkinElmer Spectrum and Origin software were used for data analysis.

2.3.4. Cross-Polarization Magic Angle Spinning Carbon-13 Nuclear Magnetic Resonance (CP/MAS ^{13}C-NMR) Analysis

Solid-state cross-polarization magic angle spinning was performed on an Agilent 600 M spectrometer operating. The cellulose-rich solid residue sample was packed tightly into the 4 mm ZrO_2 rotor, 150.81 MHz, spun at 12 kHz at 40 °C. The contact time for cross-polarization was set to 1 ms and delayed for 3 s.

3. Results and Discussion

3.1. Enzymatic Hydrolysis and Composition of Pretreated Corn Stover

The stover was co-pretreated by NaOH and ozone and the three compontent content and cellulose enzymolysis were shown in Figure 1. The initial pH at 9 before ozone treatment, which was more conducive to cellulose enzymatic hydrolysis, than pH 5 and the ozone treatment time that conducive to cellulose enzymatic hydrolysis was 25 min > 35 min. When the stover was treated with 2% NaOH at 80 °C for 2 h and ozone treatment condition was the initial pH 5 for 25 min, the maximum enzymatic hydrolysis rate was 86.84%. When the stover was treated with 2% NaOH at 80 °C for 2 h and ozone treatment conditions were the initial pH 9 for 25 min, the maximum enzymatic hydrolysis rate was 91.73%. The effect of the three components in the pretreated stover on the enzymatic hydrolysis of cellulose was different due to the pretreatment conditions. The relative content of cellulose in the stover was 62.48%, the removal rate of lignin was 84.35% and the relative content of hemicellulose was 13.74% after the best pretreatment combination. The correlation between hemicellulose content and cellulose enzymatic hydrolysis was significant ($p = 0.037 < 0.05$). The correlation between hemicellulose content and cellulose enzymatic hydrolysis was significant ($p = 0.037 < 0.05$). The significant difference between the cellulose content and cellulose enzymatic hydrolysis rate was found to be $p = 0.000$ (<0.05) which meant their relevance was extremely significant. The significant difference between the lignin removal and the cellulose enzymolysis rate was found to be $p = 0.017$ (<0.05), indicating that the enzymatic hydrolysis of cellulose was significantly affected by lignin removal.

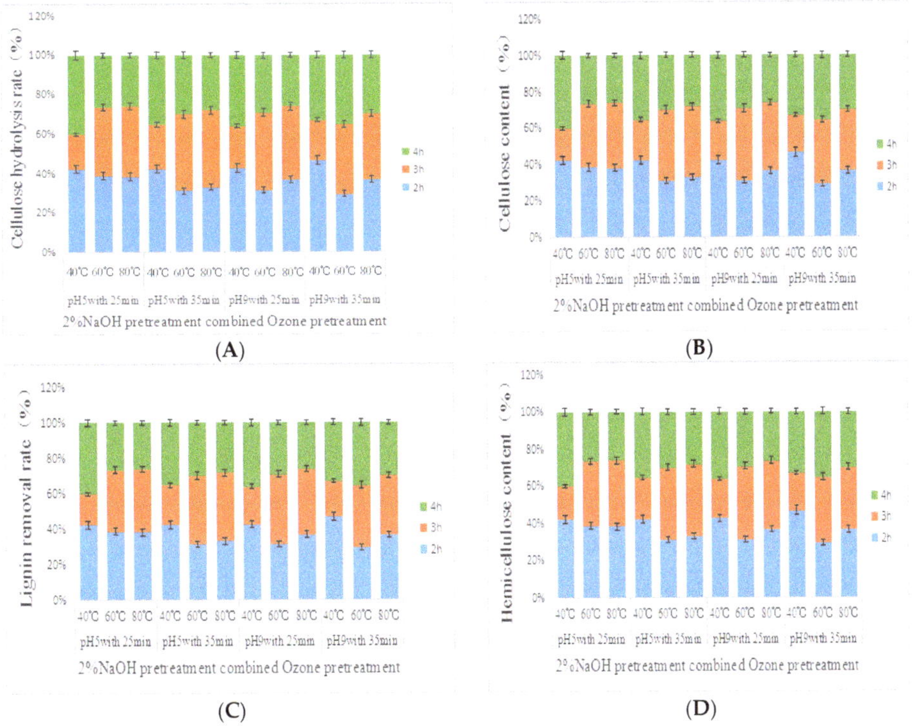

Figure 1. Results of enzymatic hydrolysis and content of cellulose, lignin and hemicellulose in corn stover after NaOH-ozone pretreatment. (**A**) Cellulose hydrolysis rate; (**B**) Cellulose content; (**C**) Lignin removal rate; (**D**) Hemicellulise content.

3.2. SEM Analysis

The surface structure of the stover before and after pretreatment is shown in Figure 2. It was found that the surface of untreated (blank) degreased stover was smooth, intact, dense. After the synergistic treatment, change in the surface of the stover was obvious. The density structure was damaged to different degrees, the surface of the stover was fluffy and full of holes, depressions and cracks that increased its specific surface area. In addition, a significant peeling phenomenon appeared on the surface, which indicated that the silica protrusions, waxes and bolts on the outer surface of corn stover were basically cleaned up after synergistic treatment. In A-9-25, we could see fluffy, neat and ordered fiber bundles along the fiber, which indicated that the synergistic treatment could effectively remove ingredients wrapped outside cellulose and break the complex network structure of lignocellulos.

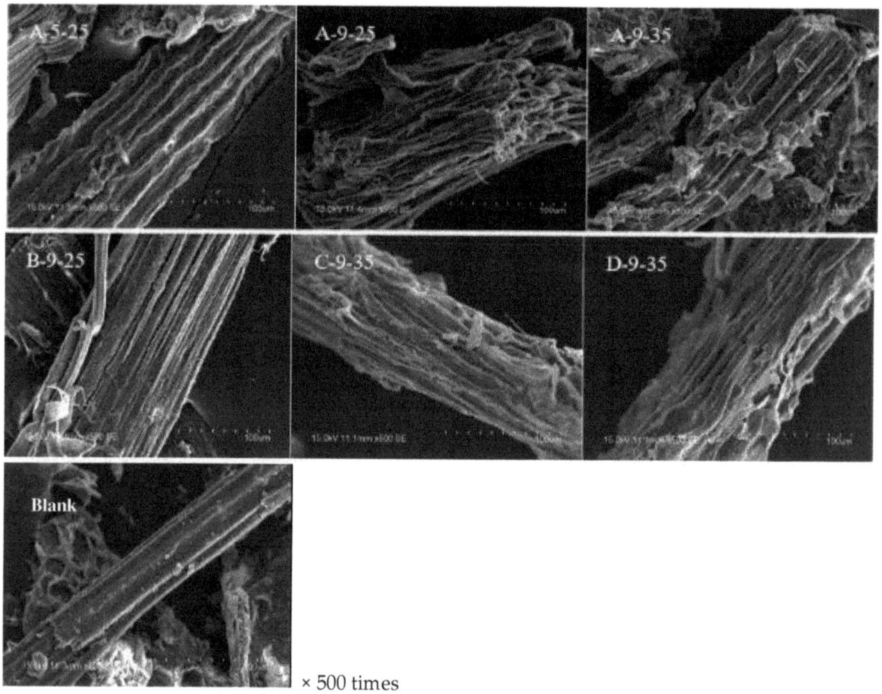

× 500 times

Figure 2. Scanning electron microscopy (SEM) images of samples before/after the combined pretreatment. A-5-25: 2% NaOH at 80 °C for 2 h and the ozone initial pH 5 for 25 min. A-9-25: 2% NaOH at 80 °C for 2 h and the ozone initial pH 9 for 25 min. A-9-35: 2% NaOH at 80 °C for 2 h and the ozone initial pH 9 for 35 min. B-9-25: 2% NaOH at 80 °C for 3 h and the ozone initial pH 9 for 25 min. C-9-35: 2% NaOH at 60 °C for 2 h and the ozone initial pH 9 for 35 min. D-9-35: 2% NaOH at 80 °C for 4 h and the ozone initial pH 9 for 35 min. Blank: untreated degreased stover.

In comparison with A-9-25 and B-9-25, we could see that the stover surface of A-9-25 had more pores, less fiber bundles filler and larger gap, the surface mechanical tissue outside stover was exposed, the cell wall was relaxed, the outer wall specific surface area increased, indicating that 2% NaOH was capable to expand the fiber structure than 4% NaOH. As a result, the enzymatic hydrolysis had better penetration into the cellulose and improved the accessibility of the enzyme [11]. Cellulose content showed that A-9-25 (62.48%) < B-9-25 (69.64%). This angle indicated that the factors affected the contact of cellulose and the enzyme, such as the swelling of the fiber material, the impact of the pores on the cellulose hydrolysis rate was greater than the increase of the fiber content in the stover. Studies have also shown that the enzymatic hydrolysis rate of cellulose and cellulose swelling degree had a linear relationship [12]. In comparison with A-9-25, B-9-25 and A-9-35, it could be roughly concluded that the effect of ozone treatment time on cellulose content and subsequent enzymatic hydrolysis was higher than that of NaOH treatment concentration.

In comparison with A-9-35, C-9-35 fiber bundle surface had a translucent thin layer of material and fluffy scaly structure, the degree of damage was less than A-9-35, indicating that 2% NaOH treatment at 80 °C for 2 h compared to 60 °C treatment 2 h on the stover surface structure damage was greater and it was consistent with the results of low lignin and lower lignin removal rate and lower cellulose enzyme hydrolysis rate (61.45%) in the C-9-35 stover. This may be attributed to the fact that the 80 °C solution allowed the NaOH solution to penetrate better into the cellulose crystallization zone, better weakened the intermolecular or intramolecular hydrogen bonding forces of the cellulose, resulting in better defatting of the degreased stover [13].

The degree of destruction of A-9-35 stover was greater than D-9-35. Combined cellulose content A-9-35 (65.76%) > D-9-35 (58.98%), hemicellulose content A (11.17%) > D-9-35 (9.15%) and lignin removal rate A-9-35 (81.65%) < D-9-35 (88.43%), we could see that pretreatment at a high temperature (80 °C) with a long time could remove more lignin and reduce cellulose and it was consistent with the result that the enzymatic hydrolysis rate of A-9-35 was about 18% higher than that of D-9-35.

In comparison with A-5-25 and A-9-25, surface structural damage degree was A-9-25 > A-5-25. It was consistent with the result of enzymatic hydrolysis A-9-25 > A-5-25, cellulose content A-9-25 > A-5-25, lignin removal rate was A-9-25 > A-5-25. It also showed that the pretreatment effect at the initial pH 9 for 25 min was better than that at pH 5 for 25 min.

3.3. XRD Analysis

Both the crystalline structure and crystal grain index of cellulose played an important role in the enzymolysis efficiency. In order to study the structural changes of stover cellulose after co-treatment, X-ray diffraction analysis of stover before and after pretreatment was showed in Figure 3.

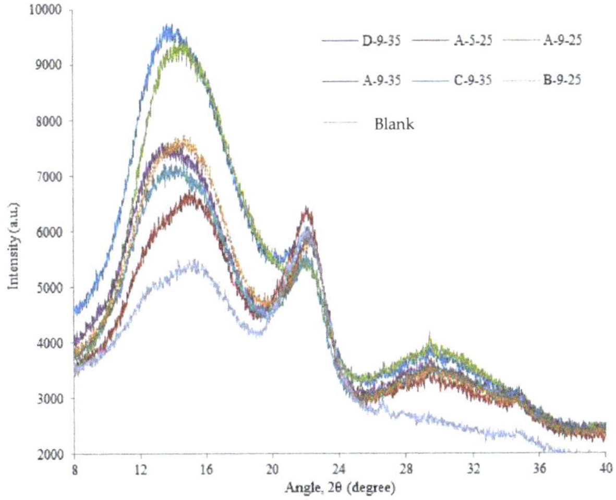

Figure 3. X-ray diffraction (XRD) patterns of the samples before/after combined pretreatment. A-5-25: 2% NaOH at 80 °C for 2 h and the ozone initial pH 5 for 25 min. A-9-25: 2% NaOH at 80 °C for 2 h and the ozone initial pH 9 for 25 min. A-9-35: 2% NaOH at 80 °C for 2 h and the ozone initial pH 9 for 35 min. B-9-25: 2% NaOH at 80 °C for 3 h and the ozone initial pH 9 for 25 min. C-9-35: 2% NaOH at 60 °C for 2 h and the ozone initial pH 9 for 35 min. D-9-35: 2% NaOH at 80 °C for 4 h and the ozone initial pH 9 for 35 min. Blank: untreated degreased stover.

The peaks at 2θ of 13–17° and 20–23° in Figure 3 exhibited more homogeneous polycrystalline of cellulose [14,15]. All tested samples had significant cellulose surface peaks at 2θ of 15.2° and 22.1°. According to the literature, cellulose I had two crystalline forms, named cellulose Iα and Iβ, different XRD spectra depended on the proportion of these two fiber morphology [16]. After pretreatment, the diffraction peak near 15.2° changed obviously. The diffusion peak of untreated stover shifted to the lower position and the peak shape became high and sharp, which indicating that the spacing of the cellulose microcrystals increased and the stacking density decreased. Specifically, the peak of untreated stover and A-5-25 treatment group were close to 15.2° (100, Iα) but the peak angle after the treatment of B-9-25 and A-9-25 shifted to 14.7° (100, Iα), the peak angle after pretreatment of C-9-35 shifted to 14.1° (10-1, Iα) and the peak angle of A-9-35 and D-9-35 were reduced to 13.8° (011, Iβ). By comparison, NaOH treatment at 80 °C, ozone initial pH 9 for 35 min could reduce the intergranular layer spacing but the ozone initial pH 5 did not have this effect. It also showed that synergistic pretreatment had a great effect on the change of stover crystal grain index. The peak intensity in the 100 crystal plane of A-9-25 was stronger than that of 011 crystals in A-9-35, which may be the key reason for the difference of enzymatic hydrolysis effects.

After pretreatment, the crystal diffraction peak amplitude of 020 near 22.1° at 2θ was small and the peak intensity reduced obviously, indicating that the pretreatment did not have a significant effect on the distance of the crystal layer of the crystal grain. The pretreatment group showed weak peaks near 26.7° (201, Iα), 27.8° (20-1, Iα) and 34.7° (004, Iβ), showing the characteristic structure of natural cellulose I. The stover sample after co-pretreatment of NaOH-ozone, 201 and 20-1 crystal faces disappeared. The new diffraction peak (022) formed at 2θ of 29.5° and proved the presence of cellulose II, which indicated that the stover sample after pretreatment was a mixed crystal structure of cellulose I and II [17]. The change of crystal structure and grain index promoted the hydrolysis of the cellulose and the enzymatic hydrolysis of cellulose occurred more easily in crystal face 100 and newly formed crystal face 022 [18].

D-9-35 had the highest peak intensity at 2θ of 13.8°, mainly due to its high lignin removal rate. Zhao et al. pretreated bagasse with peracetic acid and found that CrI increased because of the removal of lignin [18]. This was consistent with the result of our study that peak intensity of D-9-35 at 2θ = 22.1°, was higher than A-9-35, indicating that 2% NaOH treatment was more conducive than 4% NaOH to the subsequent increase in the rate of enzymatic hydrolysis. The peak intensity of A-9-25 at 2θ = 14.7° and 29.5° was significantly higher than that of B-9-25 and the peak intensity at 2θ = 22.1° was significantly lower than that of B-9-25, indicating that 2% NaOH was more conducive to the hydrolysis than 4% NaOH.

In summary, the peak intensity of A-9-25 near 14.7° and 29.5 at 2θ were much higher than that of other pretreatment groups, the peak intensity of 22.1° at 2θ was the lowest. These results indicated that the microcrystalline structure of stover treated after 2% NaOH at 80 °C for 2 h and the initial pH 9 of ozone for 25 min had shifted, which was conducive to enzymatic hydrolysis.

3.4. FTIR Analysis

The characteristic absorption peaks of cellulose, hemicellulose and lignin in infrared spectrum was shown in Table 3. The FTIR spectra of untreated and pretreated stover were measured in Figure 4. The carbonyl at 1737 cm^{-1} was esterified (polyxylose C=O conjugate) and came from the ester bond between the acetyl group attached to xylose and glucuronic acid, the peak was much stronger in spectra of untreated stover than that of the treated. This indicated that co-pretreatment could remove hemicellulose ester linkages [19].

Table 3. Assignments of characteristic absorption of samples.

Number	Wavelength/cm^{-1}	Absorption Band Attribution
1	898	Vibration of β-glycosidic bonds in cellulose and hemicellulose
2	1051	The bending of hydroxyl groups in lignin
3	1250	Ether bond between lignin and carbohydrates (β-*O*-4)
4	1370	Phenolic hydroxyl groups in lignin
5	1427	Methoxy in lignin (–OCH_3)
6	1454	Methoxy in lignin (–OCH_3)
7	1515	Extension of C=C on Lignin Aromatic Rings
8	1605	Lignin aromatic skeleton vibration
9	1654	Conjugated carbonyls in lignin
10	1704	Non-conjugated carbonyls in lignin degradation products
11	1732	Ether bond between lignin and carbohydrate (non-conjugated ketone and carboxyl group C=O stretch)

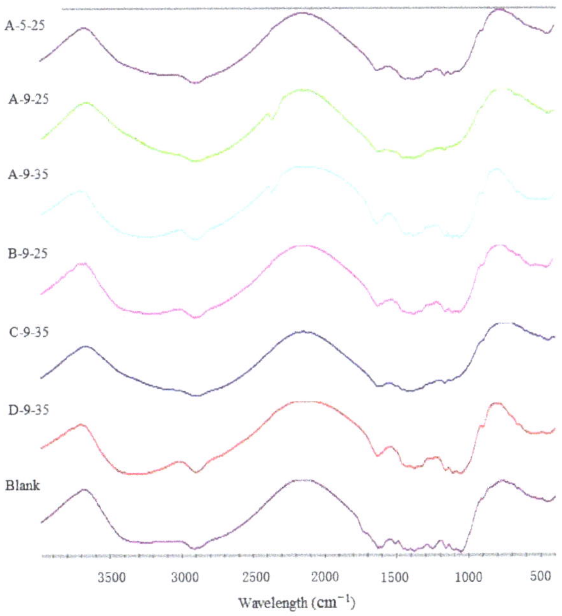

Figure 4. Fourier transform infrared (FTIR) spectra of the samples before/after combined pretreatment. A-5-25: 2% NaOH at 80 °C for 2 h and the ozone initial pH 5 for 25 min. A-9-25: 2% NaOH at 80 °C for 2 h and the ozone initial pH 9 for 25 min. A-9-35: 2% NaOH at 80 °C for 2 h and the ozone initial pH 9 for 35 min. B-9-25: 2% NaOH at 80 °C for 3 h and the ozone initial pH 9 for 25 min. C-9-35: 2% NaOH at 60 °C for 2 h and the ozone initial pH 9 for 35 min. D-9-35: 2% NaOH at 80 °C for 4 h and the ozone initial pH 9 for 35 min. Blank: untreated degreased stover.

Wavelength 1512 cm^{-1} belonged to the stretching of lignin aromatic ring –C=C– aromatic skeleton. The peak was characteristic in lignin indicating G > S [20]. The sharp band almost disappeared in spectra of treated stover but had stronger absorbance in spectra of the untreated sample. The decrease or disappearance of peak intensity could be attributed to the removal of aromatic ring lignin and the destruction of the lignocellular structure in the residue under the corresponding pretreatment conditions [21]. This was consistent with the chemical composition of the sample. It showed that ozone treatment could reduce the content of –C=C– in wheat stover [10].

Due to the breakage of the bond between the lignin-carbohydrate after pretreatment, the peak at 1250 cm^{-1} was evident in the control group and was weak in the other groups [16]. The peak at 1320 cm^{-1} was much stronger in the spectra of the control group than that of other groups, suggesting that the guaiacyllignin (G) structure of the lignin in the residue was destroyed after pretreatment. Compared to the peak intensities of A-9-25 and A-9-25, it showed that 25 min ozone treatment was more favorable for removing the G structure. According to the literature, the toughness of G structure was higher than that of S structure [22,23], so the destruction of G structure was more conducive to subsequent enzymatic hydrolysis.

3.5. ^{13}C-NMR Analysis

It could be seen from the figure that most of the signals in the ^{13}C-NMR spectrum of the samples after the pretreatment were similar. Compared with the non-pretreated samples, the peaks in the 20–35 ppm region were weakened in the A-5-25 group and almost disappeared in the other groups. The disappearance of the peak in the region or the decrease of the peak intensity showed that stover pretreated only trace amounts of residual lignin. It was also noted that the initial treatment with ozone at pH 9 from the initial ozone at pH 5 removed the lignin more. The result was consistent with the 65.40% removal rate of lignin treated in A-5-25 in chapter 3. Compared A-9-25 and A-9-35, $C_{1–6}$ signal strength corresponding to the former peak to the latter were significantly stronger and sharper peak shape, indicating that the pretreatment of ozone for 25 min compared to 35 min, the greater degree of damage to the stover, the cellulose could be better separated, the higher relative content of cellulose was more conducive to subsequent enzymatic hydrolysis.

Compared A-9-35 and D-9-35, no new peak appeared and no old peak disappeared, indicating that the type of carbon in the carbohydrate compound did not change when the treatment time of 2% NaOH at 80 °C increased from 2 h to 4 h. D-9-35 compared with the non-pretreatment group, the carbon signal peak was sharper, indicating that the treatment of 2% NaOH at 80 °C for 4 h with the ozone initial pH 9 for 35 min resulted in the high purity separation of the components in the stover.

The peaks in the 106–153 ppm region were significantly higher in A-9-25 than in B-9-25 and the peaks in B-9-25 almost disappeared, indicating that 2% NaOH at 80 °C for 2 h had less lignin removal and lower cellulose relative content than 4% NaOH. Compared with the above conclusions, the effect of NaOH concentration on composition of stover was greater than that of NaOH treatment time. Compared B-9-25 and untreated group, the peak intensity of 4% NaOH treatment was higher and sharper than that of untreated group, suggesting that 4% NaOH at 80 °C for 2 h with ozone initial pH 9 for 25 min made stover component separated in high purity. The spectra of A-9-35 and C-9-35 were similar and there was no change in peak number and intensity. For ^{13}C-NMR, it was impossible to determine the difference between NaOH treatment temperatures 80 °C and 60 °C.

Comparing the spectra in Figure 5, the peak was obvious at 113 ppm in untreated stover. In A-5-25 and A-9-25, the peak at 106–153 ppm area was obvious, the peak in A-5-25 was mainly at 124 ppm, the peak in A-9-25 shifted to 134.2 ppm. It showed that with the initial pH of co-pretreatment changed from 5 to 9, the aromatic ether bond fragmented and free phenolic hydroxyl group formed. But the peak of disappeared in B-9-25 and the peak was not present in all ozone treatments for 35 min, it presumed that the corresponding aromatic lignin in this area was sensitive to the ozone initial pH 9 for 35 min and 4% NaOH. The untreated stover and A-5-25 treatment group had significant levels of aliphatic hydroxy lignin in the 20–35 ppm area but disappeared after ozone treatment at pH 9, indicating that the corresponding aromatic substances in this region were more sensitive to the ozone initial pH 9.

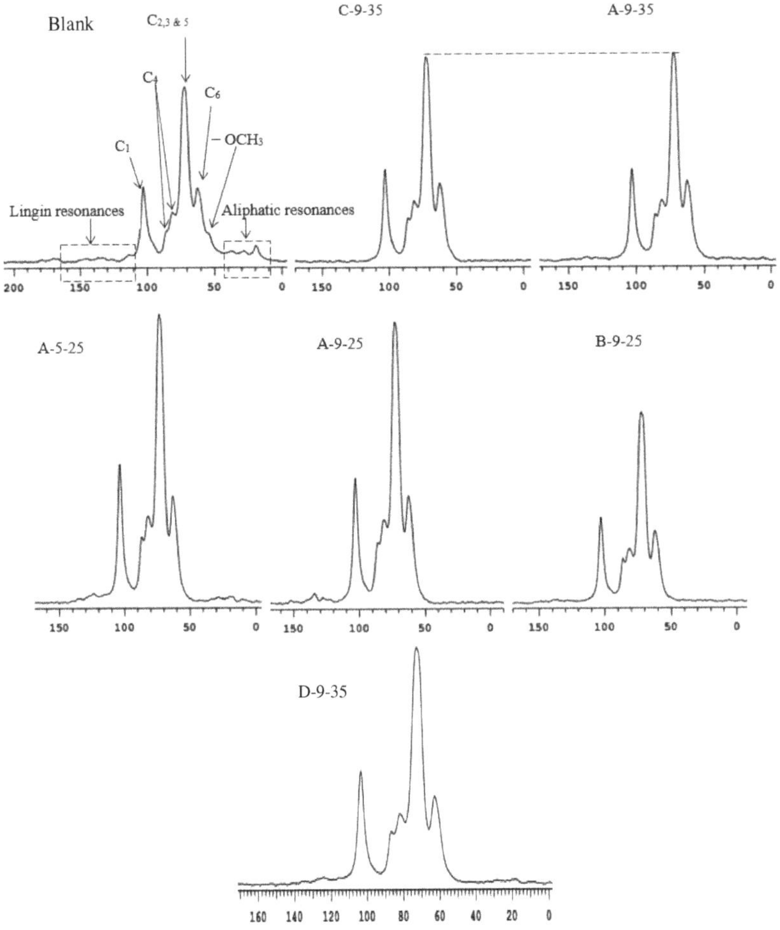

Figure 5. Cross-Polarization Magic Angle Spinning Carbon-13 Nuclear Magnetic Resonance (CP/MAS ^{13}C-NMR) spectra of the samples before/after combined pretreatment. A-5-25: 2% NaOH at 80 °C for 2 h and the ozone initial pH 5 for 25 min. A-9-25: 2% NaOH at 80 °C for 2 h and the ozone initial pH 9 for 25 min. A-9-35: 2% NaOH at 80 °C for 2 h and the ozone initial pH 9 for 35 min. B-9-25: 2% NaOH at 80 °C for 3 h and the ozone initial pH 9 for 25 min. C-9-35: 2% NaOH at 60 °C for 2 h and the ozone initial pH 9 for 35 min. D-9-35: 2% NaOH at 80 °C for 4 h and the ozone initial pH 9 for 35 min. Blank: untreated degreased stover.

4. Conclusions

The optimal pretreatment condition was found to be 2% (*w*/*w*) NaOH treated at 80 °C for 2 h followed by ozone treatment for 25 min with an initial pH 9 and the maximum efficiency (91.73%) of enzymatic hydrolysis of cellulose was achieved. The promoting effect of three components in corn stover on the cellulose enzymatic hydrolysis was different with pretreatment conditions. Under the optimum pretreatment condition, the relative content of cellulose in the treated stover was 62.48%, the removal rate of lignin was 84.35% and the relative content of hemicellulose was 13.74%.

The results of SEM observation showed that the intensive structure of stover fiber changed to different degrees after synergistic treatment, many pores appeared on the surface and the fiber bundle was exposed. All of these increased the substrate accessibility of enzyme. The XRD characterization of the cellulose crystalline state showed that the synergistic pretreatment could change the crystal structure and crystallographic index, expand the interlayer spacing and so that the crystalline state of cellulose was more conducive to enzymatic hydrolysis. The FTIR characterization of chemical bond properties of stover before and after pretreatment showed that the pretreatment could effectively break the hemicellulose bond, the linkage bond between lignin and other carbohydrate and the intra-/inter-molecular hydrogen bond between the cellulose and other carbohydrate. The CP/MAS ^{13}C-NMR determination of different positions of carbon in stover showed that pretreatment was beneficial to the removal of acetyl groups in hemicellulose and -OCH_3 in lignin. All changes were conducive to the promotion of cellulose enzymatic hydrolysis. Finally, it emphasized that in this paper, corn stover was a representative of lignocellulose and the pretreatment method used in this article was suitable for other lignocellulosic materials [24].

Supplementary Materials: This study optimized conditions of sodium hydroxide synergistic ozone pretreatment. Preconditioning conditions of this experiment were determined based on a large number of previous experiments in our laboratory. The most representative is the analysis of surface area and porosity. The following is available in supplementary materials: http://www.mdpi.com/1420-3049/23/6/1300/s1. Figure S1 showed the specific surface area changes with the increase of alkali treatment temperature. Figure S2: the specific surface area of corn stover changed with the NaOH treatment time increases. Figure S3 was about the particle size changes in the different pH value of ozone treatment in NaOH combined with ozone treatment. Figure S4: the effect of ozone treatment time on the specific surface area of corn stover was carried out. The mechanism of ozonation is revealed clearly by the analysis of specific surface area.

Author Contributions: W.W. responsible for conducting experiments, data collection and analysis, manuscript writing and revision; C.Z. responsible for the conduct of pretreatment experiments; S.T. responsible for experimental guidance and data analysis guidance; Z.C. responsible for conducting pre-experiment and literature search; P.L. responsible for the design of experimental ideas and manuscript writing instructions. All authors have read and approved the final manuscript.

Acknowledgments: The authors are deeply grateful for the support provided by Aidong Sun at the Beijing Forestry University (Beijing, China), as he offered us the experimental materials.

Conflicts of Interest: The authors declare no conflict of interest.

Nomenclature

AT	single NaOH pretreatment
OT	single ozone pretreatment
AT-OT	the combine pretreatment with sodium hydroxide and ozone
G	guaiacyllignin
S	syringyllignin

References

1. Chen, X.; Ning, X.Q.; Zhang, B.B.; Wang, Y.L.; Zeng, G.M. Effect of steam explosion and ionic liquid pretreatment technology on the enzymatic hydrolysis of corn stalk. *Asian J. Chem.* **2012**, *24*, 1015–1018.
2. Li, M.F.; Chen, C.Z.; Sun, R.C. Effect of pretreatment severity on the enzymatic hydrolysis of bamboo in hydrothermal deconstruction. *Cellulose* **2014**, *21*, 4105–4117. [CrossRef]
3. Ramos, L.P.; Nazhad, M.M.; Saddler, J.N. Effect of enzymatic hydrolysis on the morphology and fine structure of pretreated cellulosic residues. *Enzyme Microb. Technol.* **1993**, *15*, 821–831. [CrossRef]
4. Zhang, Y.-H.P.; Lynd, L.R. Toward an aggregated understanding of enzymatic hydrolysis of cellulose: Noncomplexed cellulase systems. *Biotechnol. Bioeng.* **2004**, *88*, 797–824. [CrossRef] [PubMed]
5. Pang, Y.; Li, X.; Luo, Q. Effect of temperature and chemical pretreatment on anaerobic biogasification of corn stalk. *Chin. J. Bioprocess Eng.* **2005**, *3*, 37–41.
6. Ikeura, H.; Hamasaki, S.; Tamaki, M. Effects of ozone microbubble treatment on removal of residual pesticides and quality of persimmon leaves. *Food Chem.* **2013**, *138*, 366–371. [CrossRef] [PubMed]
7. Zhang, P.Y.; Gang, Y.U.; Sun, H.T.; Jiang, Z.P. Preliminary study on the degradation of organic compound by integrated ozone/activated carbon. *China Environ.* **2000**, *20*, 159–162.
8. Ben'ko, E.M.; Manisova, O.R.; Lunin, V.V. Effect of ozonation on the reactivity of lignocellulose substrates in enzymatic hydrolysis to sugars. *Russ. J. Phys. Chem.* **2013**, *87*, 1108–1113. [CrossRef]

9. Panneerselvam, A.; Sharma-Shivappa, R.R.; Kolar, P.; Clare, D.A.; Ranney, T. Hydrolysis of ozone pretreated energy grasses for optimal fermentable sugar production. *Bioresour. Technol.* **2013**, *148*, 97–104. [CrossRef] [PubMed]
10. Bule, M.V.; Gao, A.H.; Hiscox, B.; Chen, S. Structural modification of lignin and characterization of pretreated wheat stover by ozonation. *J. Agric. Food Chem.* **2013**, *61*, 3916–3925. [CrossRef] [PubMed]
11. Li, H.Y.; Huang, K.L.; Jin, M.; Wei, Q.Q.; Li, Z.G. Effects of alkaline ozone pretreatment on rice stover. *Trans. Chin. Soc. Agric. Eng.* **2010**, *26*, 264–268.
12. Ogiwara, Y.; Arai, K. Swelling degree of cellulose materials and hydrolysis rate with cellulase. *Text. Res. J.* **1968**, *38*, 885–891. [CrossRef]
13. Li, Z.; Wang, Y.; Cheng, L.; Guo, W.; Wu, G. Effect of Nano-$CaCO_3$ on the structure and properties of holocellulose-fiber/polypropylene biomass composites. *J. Wood Chem. Technol.* **2017**, *37*, 62–74. [CrossRef]
14. Zheng, Q.F.; Wang, Z.M.; Chen, B.G.; Liu, G.F.; Zhao, J. Analysis of XRD Spectral Structure and Carbonization of the Biochar Preparation. *Spectrosc. Spectr. Anal.* **2016**. [CrossRef]
15. Nanda, S.; Azargohar, R.; Kozinski, J.A.; Dalai, A.K. Characteristic studies on the pyrolysis products from hydrolyzed Canadian lignocellulosic feedstocks. *Bioenergy Res.* **2014**, *7*, 174–191. [CrossRef]
16. Chao, Z.; Shao, Q.; Ma, Z.; Li, B.; Zhao, X. Physical and chemical characterizations of corn stalk resulting from hydrogen peroxide presoaking prior to ammonia fiber expansion pretreatment. *Ind. Crops Prod.* **2016**, *83*, 86–93.
17. Li, X.; Ye, J.; Chen, J.; Yu, J.; Ding, M. Dissolution of wheat stover with aqueous NaOH/Urea solution. *Fibers Polym.* **2015**, *16*, 2368–2374. [CrossRef]
18. Zhao, X.B.; Wang, L.; Liu, D.H. Technical note peracetic acid pretreatment of sugarcane bagasse for enzymatic hydrolysis: A continued work. *J. Chem. Technol. Biotechnol.* **2008**, *83*, 950–956. [CrossRef]
19. Steinbeiss, S.; Schmidt, C.M.; Heide, K.; Gleixner, G. δ 13 C values of pyrolysis products from cellulose and lignin represent the isotope content of their precursors. *J. Anal. Appl. Pyrolysis* **2006**, *75*, 19–26. [CrossRef]
20. Faix, O. Classification of Lignins from Different Botanical Origins by FT-IR Spectroscopy. *Holzforsch.-Int. J. Biol. Chem. Phys. Technol. Wood* **1991**, *45*, 21–28. [CrossRef]
21. Zhong, R.; Morrison, W.H.; Himmelsbach, D.S.; Poole, F.L.; Ye, Z.-H. Essential role of caffeoyl coenzyme A O-methyltransferase in lignin biosynthesis in woody poplar plants. *Plant Physiol.* **2000**, *124*, 563–577. [CrossRef] [PubMed]
22. Kim, T.H.; Kim, J.S.; Sunwoo, C.; Lee, Y.Y. Pretreatment of corn stover by aqueous ammonia. *Bioresour. Technol.* **2003**, *90*, 39–47. [CrossRef]
23. Studer, M.H.; Demartini, J.D.; Davis, M.F.; Sykes, R.E.; Davison, B.; Keller, M.; Tuskan, G.A.; Wyman, C.E. Lignin content in natural Populus variants affects sugar release. *Proc. Natl. Acad. Sci. USA* **2011**, *108*, 6300–6305. [CrossRef] [PubMed]
24. Aznm, A.M. Pretreatment of cane bagasse with alkaline hydrogen peroxide for enzymatic hydrolysis of cellulose and ethanol ferment-atiom. *J. Environ. Sci.* **2009**, *100*, 1608–1613.

Sample Availability: Samples of the compounds corn stover are available from the authors.

molecules

MDPI

Review

Efficient Anaerobic Digestion of Microalgae Biomass: Proteins as a Key Macromolecule

Jose Antonio Magdalena [1], Mercedes Ballesteros [1,2] and Cristina González-Fernandez [1,*]

[1] Biotechnological Processes Unit, IMDEA Energy, 28040 Madrid, Spain; joseantonio.magdalena@imdea.org (J.A.M.); mercedes.ballesteros@imdea.org (M.B.)
[2] Biofuels Unit, CIEMAT, 28040 Madrid, Spain
* Correspondence: cristina.gonzalez@imdea.org

Academic Editor: Ivet Ferrer

Received: 12 March 2018; Accepted: 3 May 2018; Published: 6 May 2018

check for updates

Abstract: Biogas generation is the least complex technology to transform microalgae biomass into bioenergy. Since hydrolysis has been pointed out as the rate limiting stage of anaerobic digestion, the main challenge for an efficient biogas production is the optimization of cell wall disruption/hydrolysis. Among all tested pretreatments, enzymatic treatments were demonstrated not only very effective in disruption/hydrolysis but they also revealed the impact of microalgae macromolecular composition in the anaerobic process. Although carbohydrates have been traditionally recognized as the polymers responsible for the low microalgae digestibility, protease addition resulted in the highest organic matter solubilization and the highest methane production. However, protein solubilization during the pretreatment can result in anaerobic digestion inhibition due to the release of large amounts of ammonium nitrogen. The possible solutions to overcome these negative effects include the reduction of protein biomass levels by culturing the microalgae in low nitrogen media and the use of ammonia tolerant anaerobic inocula. Overall, this review is intended to evidence the relevance of microalgae proteins in different stages of anaerobic digestion, namely hydrolysis and methanogenesis.

Keywords: microalgae; anaerobic digestion; proteins; biogas; inhibition

1. Introduction

Environmental issues and energy self-sufficiency concerns related to fossil fuels have led to research on new approaches to improve renewable energy production to substitute them. Anaerobic digestion is one of those technologies devoted to the production of biofuels, which involves the degradation of organic matter through the action of different microorganisms. Anaerobic digestion exhibits many advantages such as its efficiency for organic matter removal, its applicability at any scale and the wide variety of substrates that can be used as feedstock. Likewise, the multiproduct generation attained during digestion is also a major benefit of this technology. Those end-products, including biogas and digestate, are easy to separate and can be a source of energy and fertilizers, respectively [1].

Among the different substrates that can be employed, microalgae are being recently studied since this biomass can be grown in residual effluents, do not need arable land to be cultivated while contributing to CO_2 mitigation and wastewater bioremediation [2]. Previous studies have demonstrated the technoeconomic and environmental benefits of microalgae biomass for bioenergy purposes when considered as by-product in other technologies [3–8]. In the same manner, out of the bioenergy producing technologies where microalgae can be used as feedstocks, anaerobic digestion is probably the most economically feasible since it does not require highly concentrated biomass [9] and anaerobes can use proteins, carbohydrates and lipids for methane production purposes [10]. Microalgae biomass has a wide range of compositions, depending on growth conditions and species [11,12]. In general terms, biochemical profile of chlorophytes range 30–60% of proteins, 20–40% of carbohydrates, and 4–57% of lipids [13,14]. Each macromolecule has different achievable methane yields [10]. Thus, in principle, different microalgae compositions produce different methane yields [12]. At the same time, microalgae composition varies depending not only among strains but also on the growth conditions (nutrients availability and operational conditions) [15,16]. In addition to the different macromolecular composition that microalgae might exhibit, this biomass also differs in structural features. Most of the microalgae able to thrive in wastewater effluents have a chemically complex and structurally robust cell wall composed of low biodegradable substances that hinder the anaerobic digestion [17,18]. Some of these compounds are sporopollenin, algaenan, and polymeric carbohydrates that offer a barrier towards anaerobes [19,20]. During anaerobic digestion, cell walls are degraded by extracellular enzymes of hydrolytic bacteria. Nevertheless, this process might be too slow and thus, a limited hydrolysis rate renders the anaerobic digestion as a lengthy and inefficient bioprocess. Pretreatments are used in order to facilitate the accessibility of these extracellular enzymes whereby improving hydrolysis stage. Different microalgae pretreatments have been studied such as thermal, chemical, mechanical or biological. Methane yields improvements achieved with those different pretreatments can be found elsewhere [21–24]. Out of the different pretreatments, biological approach is the most environmentally friendly [25]. Opposite to other pretreatments, the additional benefits of biological pretreatments are the absence of inhibiting by-products [26] and the high selectivity of the reactions [27]. This approach might not only be used for biomass hydrolysis but also to provide crucial information related to the macromolecule that reduces the anaerobic biodegradability of microalgae biomass. In this manner, this review summarizes the main results attained during the last years of research devoted to microalgae pretreatments in the biogas production context. Moreover, this period of research highlighted the importance of proteins on different stages of the digestion. This review attempts to provide comprehensive evidences of the key role of microalgae proteins.

2. Pretreatment of Microalgae Biomass to Improve Biogas Production

Since low biodegradability is a common issue in anaerobic digestion of different substrates (such as activated sludge, lignocellulose and photosynthetic microorganisms), a wide range of pretreatments are available to enhance the hydrolysis step [28]. Cell wall rupture or hydrolysis is needed to make available microalgae organic matter to anaerobic microorganisms [29]. Pretreatments are classified in four main groups, namely thermal, mechanical (ultrasound and microwave), chemical (acidic, alkaline, solvents and ozonation) and thermo-chemical (acid or alkali reagents addition combined with high temperatures) and biological (enzymes and microorganisms). Those pretreatments have been intensively studied during the last decade to improve biogas production of microalgae biomass (Table 1). Most of them have been assessed in Biochemical Methane Potential (BMP) assays (batch digestion mode).

Table 1. Studied pretreatments to improve biogas production using microalgae as substrates.

High Demanding Energy Pretreatments	Operation Mode	Biomass	Conditions	Methane Yield Increase	References
Thermal	Batch	*Scenedesmus* sp.	75 °C for 10 h 95 °C for 10 h	58% 69%	[29–31]
	Batch	*Scenedesmus* sp.	80 °C for 15 min	60%	[32]
	Batch	*Chlorella* sp.	70 °C for 30 min 90 °C for 30 min	37%48%	[33]
	Batch	*Stigeoclonium* sp. *Monoraphidium* sp and *Nitzschia*	130 °C for 15–30 min	28%	[31]
	Semi-continuous	*Chlorella* sp.	120 °C40 min	1.5-fold	[34]
Mechanical	Batch	*Scenedesmus* sp.	128.9 KJ/g TS for 30 min	87%	[32]
	Batch	*Monoraphidium* sp. and *Stigeoclonium* sp.	26.7 KJ/g TS for 30 min	85%	[31]
	Batch	*Mixture of microalgae biomass*	10; 27; 40; 57 KJ/g TS	6-24%	[35]
Chemical	Batch	*Chlorella* sp. and *Scenedesmus* sp.	CaO (4 and 10% w/w) at 25, 55 and 72 °C	25%	[36]
	Batch	*Chlorella* sp.	4 M H_2SO_4 at 120 °C for 20–40 min	72.5%	[37]
Low Demanding Energy Pretreatments		**Biomass**	**Solubilization**	**Methane Yield**	**References**
Proteases	Batch	*C. reinhardtii* *C. vulgaris*	86-96% for both biomasses	51% in *Chlorella* biomass 7% C. *reindhartii*	[38]
	Batch	*Scenedesmus* sp.	30%	1.53-fold	[39]
	Semi-continuous	*C.vulgaris*	47%	2.6-fold	[39]
	Semi-continuous	*C. vulgaris*	54%	5 and 6.3-fold (OLR= 1.5 g/L d and OLR= 3 g/L d)	[40]
Carbohydrases	Batch	*C. vulgaris* and *Scenedesmus* sp.	84% 36%	1.2-fold	[41]

2.1. High Energy Demanding Pretreatments

Thermal, thermo-chemical and mechanical pretreatments are considered as high energy demanding processes and, in order to evaluate its efficiency, the final energy balance of the pretreatment process has to be addressed. Given that thermal energy is available in biogas production facilities, the most used pretreatment is thermal application. Thermal pretreatments involve biomass heat up in a wide range of temperatures (50–270 °C) and reaction time (from minutes to hours). With regard to thermal application, the effect on the biomass depends on the microalgae strain and applied temperature [30]. Passos et al. [31] and Passos and Ferrer [42] applied thermal pretreatment to *Scenedesmus* sp. biomass at 75 °C and 95 °C for 10 h resulting in methane yield enhancement of 58% and 69%, respectively . Similar values were attained by González-Fernández et al. [43] when treating *Scenedesmus* at 80 °C for only 15 min, highlighting the impact of temperature rather than the heating time as the most relevant parameter in thermal pretreatment. Similar temperatures were tested in *Chlorella* biomass (70 and 90 °C) for 30 min resulting in an enhanced methane yield of 37% and 48% compared to the raw biomass (322 mL CH_4/g VS_{in}) [32]. These results evidenced that thermal pretreatments are strain specific and thus, at the same temperature applied, different methane yields enhancement can be attained among the different biomass used. Higher temperatures (130 °C for 15–30 min) were also tested, resulting in 28% methane yield increase when compared to a raw biomass composed by a mixture of green algae (*Stigeoclonium* sp. and *Monoraphidium* sp.) and diatoms (*Nitzschia*) (105 mL CH_4/g VS_{in}) [31]. Due to the potential formation of Maillard compounds at higher temperatures, moderate temperatures in the range of 80–120 °C are most widely tested. Moreover, thermal pretreatments have been tested not only in batch mode, but also in semicontinuous mode. Méndez et al. [33] reported a methane yield enhancement of 1.5-fold compared to raw *Chlorella* biomass (84 mL CH_4/g COD_{in}) when using 120 °C for 40 min for feeding a Completely Stirred Tank Reactor (CSTR). Although no common inhibitors were identified, the results obtained in the CSTR were considerably lower (50% less) than the ones obtained in batch mode digestion. This experimentation corroborated the need to test each pretreatment in different feeding modes. Although thermal pretreatments normally present positive results in terms of methane yield, the values attained are very diverse depending on different variables such as the pretreated biomass, temperature, pretreatment time employed and operation mode during the digestion. Moreover, as mentioned above, these methods involved some drawbacks such as the formation of recalcitrant compounds that could potentially decrease the performance of the process [34,35].

Mechanical pretreatments are commonly employed to disrupt different kind of organic substrates in industrial processes [44,45]. Ultrasound treatment has been applied to disrupt microalgae cell wall in different bioprocess devoted to biofuel production, such as ethanol production from *Chlorella* biomass [46] and biodiesel generation from *Spirulina* biomass [47]. In the case of anaerobic digestion, ultrasound pretreatment has also shown positive results in terms of methane yield enhancement. González-Fernández et al. [43] applied 128.9 kJ/g TS at 85 °C and 30 min to enhance methane yield of *Scenedesmus* biomass from 81 mL CH_4/g COD_{in} to 153 mL CH_4/g COD_{in} (87% enhancement). Nevertheless, those authors also pointed out the fact that ultrasound application is having associated an increase in temperature which also acts as a pretreatment. As a matter of fact, when it comes to the pretreatment of *Scenedesmus* sp., the benefits of ultrasound application were rather questionable compared to the enhancement in methane yield attained only with the application of temperature. Ultrasound pretreatment (26.7 KJ/g TS for 30 min) was also applied to *Monoraphidium* sp. and *Stigeoclonium* sp. biomass and their methane yields were enhanced from 105 mL CH_4/g COD_{in} to 196 mL CH_4/g COD_{in} [42]. When testing different energy inputs (10; 27; 40; 57 KJ/g TS), applied to different mixtures of microalgae biomass (mixture A: 40% *Chlamydomonas*, 20% *Scenedesmus* and 40% *Nannocloropsis*; mixture B: 58% *Acutodesmus obliquus*, 36% *Oocystis* sp., 1% *Phormidium* and 5% *Nitzschia* sp; Mixture C: *Microspora* $\approx$ 100%), an increase in methane yield ranging from 6 to 24% at 10 MJ/kg TS was determined, while higher energy inputs did not report any significant increase [34]. Despite all those positive results in terms of methane yields enhancement, the main limitation of ultrasound pretreatment is the high energy input required when compared to thermal, chemical or biological methods [21].

Chemical methods are often combined with heat pretreatment. Thermochemical pretreatments have been less employed than thermal and mechanical pretreatments due to its potential toxicity for the anaerobes. Cell wall disruption with alkali and acid pretreatments has been tested with positive results for the production of ethanol, butanol and biomethane when using microalgae biomass as a feedstock [48,49]. Studies related to microalgae biomass solubilization using thermo-alkaline methods include for instance the use of reagents such as NaOH or CaO. Different doses of CaO (4 and 10% *w*/*w*) and different temperatures (25, 55 and 72 °C) resulted in maximum proteins and carbohydrates solubilization of 32.4% and 31.4%, respectively, and methane yield enhancement of 25% compared to the raw biomass (260 mL CH_4/g VS_{in}) at the highest temperature and lime dose tested (72 °C and 10% *w*/*w*) [50]. When using NaOH (0.5, 2 and 5% *v*/*v*) in *Chlorella* and *Scenedesmus* biomass, the conducted experiments revealed that despite of the biomass solubilisation, the methane yield enhancement was really low (10%, [36]). Thermo-acid pretreatments have been less employed than thermo-alkali. For instance, *Chlorella* biomass was heated at 120 °C either for 20 min and 40 min. Sulphuric acid addition combined with 120 °C for 40 min enhanced carbohydrates solubilization by 7-fold, although no solubilization of the protein fraction was reported. In terms of methane production, this thermo-acid pretreatment improved the methane yield from the untreated biomass from 139 mL CH_4 g/COD_{in} to 230 mL CH_4 g/COD_{in} [51]. Since anaerobic digestion is taking place at around pH 7, one of the main limitations of chemical pretreatments is the need to readjust the pH previously to the digestion. In this manner, chemical costs limit the use of these pretreatments. Moreover, some of the chemicals need to be removed previously to the anaerobic digestion as they can be toxic for anaerobes [27].

In conclusion, high energy demanding pretreatment methods report high values in terms of methane yield. However, they are energetically unbalanced. This means that the energy required to carry out the pretreatment is higher than the one obtained in form of biogas. This is why research has been directed towards the use of low energy demanding pretreatments

2.2. Low Energy Demanding Pretreatments

Compared to other pretreatments, the biological approach presents some advantages such as lower energy demand and high specificity [37]. These pretreatments include the use of suitable enzymes or microorganisms to hydrolyze microalgae biomass. Information about the cell wall composition is scarce, but necessary in order to select the most suitable enzyme for the pretreatment. For that reason, a wide range of biocatalysts have been tested. In principle, given the similarities between higher plants and microalgae, the most studied catalysts are carbohydrases. Among them, cellulases, hemicellulases, amylases and pectinases are the most tested ones [37,52]. Some other enzymatic cocktails employed for microalgae biomass hydrolysis include lysozyme (catalyzing the hydrolysis of 1,4-beta-linkages between *N*-acetylmuramic acid and *N*-acetyl-D-glucosamine residues in peptidoglycan [53]), proteases (hydrolyzing peptide bonds [39]) and laccases [25]. Overall, the best results have been evidenced by adding commercial proteases cocktails. For instance, carbohydrases and proteases were compared hydrolyzing *Chlamydomonas reinhardtii* and *Chlorella vulgaris* [38]. Enzyme doses applied for carbohydrases and proteases were 0.3 mL/g DW and 0.2 mL/g DW, respectively. The enzymatic pretreatment lasted for 5 h and results obtained after carbohydrases addition were 86% and 96% carbohydrate solubilization for *C. vulgaris* and *C. reindhartii* while in the case of protease addition both biomass resulted in 96% protein solubilization. However, the authors pointed out that despite of the high carbohydrate solubilization, only a 14% enhancement methane yield was observed in *Chlorella* biomass, whereas no improvement was observed in *Chlamydomonas*. In the case of protease pretreated biomass, methane yield was enhanced by 51% in the *C. vulgaris* and 7% for *C. reindhartii*. The reason for the low methane yield enhancement recorded for *C. reindhartii* was due to the inherent high anaerobic biodegradability of this strain (75%, 263 mL CH_4 g/COD_{in}). Methane yield is limited by the inherent methane yield that the biomass can attain. However, the kinetics might be enhanced by the use of pretreatments. More specifically, methane yield might be enhanced by protease pretreatment in the range of 1.07 to 6.3 fold depending on the targeted microalgae biomass within 10–15 days of digestion [38,40].

An alternative to improve economically the enzymatic pretreatment and avoid the addition of high cost cocktails is the addition of hydrolytic secretomes released by other microorganisms. For instance, 0.7 g/L of cellulase-secreting bacteria was added to *Chlorella vulgaris* for 48 h resulting in an increase of 18% organic matter solubilization and 2-fold methane yield compared to the raw biomass [54]. Non-specific extracellular enzymes of *Anthracophyllum discolor* were employed to disrupt the cell wall of *Botryococcus braunii*, resulting in an improvement of 60% methane yield, when enzymatic concentration of 1000 U/mL was applied [55]. Likewise, cellulolytic marine bacteria were applied to *Botryococcus braunii* and *Nannochloropsis gaditana* biomass 1:1 ratio DW resulting in a methane enhancement of 140% and 150%, respectively compared to the raw biomass [56].

As it is observed in Table 1, almost all tested pretreatments improved methane production yields although a direct linkage between solubilization and methane enhancement still requires in-depth research in continuous systems to determine the energy balance and costs of the overall process [57]. Even though this pretreatment is economically unfeasible yet, enzymatic pretreatments, targeting at specific molecules, could provide important information in order to identify which is the microalgae macromolecule hampering biogas production when using this biomass [23].

3. Biological Approach to Enhance Biogas Production: Enzymatic Pretreatment

Opposite to other pretreatments, biological reactions show high selectivity and absence of inhibitory compounds. Biocatalysts do not only disrupt the cell wall, but they also hydrolyze the macromolecules during biological pretreatment. As it was indicated above, these methods are energetically competitive since they require soft temperatures and smooth shaking. Different parameters must be taken into account such as pH, temperature, enzyme dose, and exposure time [21]. Given the different macromolecular composition, structural features and cell wall composition among microalgae strains, a wide range of biocatalysts can be used. Despite of the high economic cost of the enzymatic cocktails, the use of biocatalysts can provide crucial information to identify the macromolecule hampering anaerobic digestion of microalgae biomass. Moreover, the costs could be reduced either by in situ enzymes production [54,58] or by particular enzymes secreted by bacteria and fungi via sludge bioaugmentation [23,59,60].

3.1. Carbohydrases

Carbohydrases are in charge of hydrolysing carbohydrates polymers present within the cell wall and inside the cells into simple sugars. Since it is believed that carbohydrates are the responsible of cell wall toughness, cellulaseshave been tested in microalgae biomass to enhance the hydrolysis. Cellulases from *Trichoderma reseei* were mixed with metal oxides to treat *Chlorella* biomass resulting in glucose yield of 91% of the theoretical maximum [61]. Furthermore, enzymatic cocktails aimed at degrading the compartmentalized cell material such as amylases and amyloglucosidases have been tested to promote the efficiency of the hydrolysis step. As a matter of fact, a combination of amylases and cellulases was tested to degrade the cell wall and the cell material with acid hydrolysis in *Chlorella sorokiniana, Nannochloropsis gaditana,* and *Scenedesmus.* This treatment produced a sugar release of 128 mg/g DW, 129 mg/g DW and 60 mg/g DW, respectively against control values for the different biomass (70 mg/g DW, 20 mg/g DW and 25 mg/g DW) [62]. Carbohydrases have also been tested to facilitate lipid extraction by using exoglucanase, endoglucanase, xylanase and laccase produced by different biomass-degrading bacteria, improving lipid extraction up to 40% [63]. All those studies are mainly focused on carbohydrates solubilisation while, only recently, the biomass subjected to carbohydrases has been investigated for biogas production purposes. Ometto et al. [9] tested different enzymatic cocktails on three different biomass, namely *Scenedesmus obliquus, Chlorella sorokiniana* and *Arthrospira maxima* [5]. Out of the tested enzymatic cocktails, mixtures of cellulase plus pectinase and esterase plus protease were the most effective catalysts for organic matter hydrolysis of all three biomass. In the same manner, commercial cocktails hydrolyzing the carbohydrate fraction such as Viscozyme, Celluclast and Pectinase (from Novozymes, Bagsværd, Denmark) have been employed in *C. vulgaris* and *Scenedesmus*. The use of Viscozyme provided carbohydrate fraction solubilization of 84% and 36% for *C. vulgaris* and *Scenedesmus* respectively, while the methane yield enhancement was 1.2-fold for both of them, despite of the different biomass composition and strain [41]. This experimentation suggested that the carbohydrate fraction cannot be understood as a whole to elucidate the relation between solubilization efficiency and the methane yield achievable. Instead of this, an in-depth research must be done concerning the carbohydrates composition of microalgae cell wall.

3.2. Lipases

When compared to other macromolecular constituents, lipids could be very useful substrates for anaerobic digestion due to its high potential methane yield. More specifically, theoretical methane yield for lipids is 1.014 L CH_4/g VS compared to 0.496 and 0.415 L CH_4/g VS for proteins and carbohydrates, respectively [10]. However, instability of the system can easily occur due to the formation of long chain fatty acids when lipids are hydrolyzed [64]. As a matter of fact, studies are mainly focused on the optimal concentration of lipids that makes possible to carry out anaerobic digestion without inhibition. In this way, it has been highlighted that lipid fraction should not be over 30% to avoid process inhibition [65]. To overcome such an inhibition, different strategies have been developed. For instance, Palatsi et al. [66] tested different recovery strategies to reduce the negative effect of long chain fatty acids by using different feeding patterns and adsorbents addition. Despite of the high lipid potential to enhance methane yield, microalgae biomass grown in wastewater does not present high lipid content [67,68]. At this point, it should be stressed that microalgae grown in residual effluents is the only feasible way to produce biofuel using this feedstock. In this manner, really limited information on lipases treatment of microalgae biomass for biogas production can be found in literature. For instance, an enzymatic mixture containing protease, α-amylase, xylanase, lipase and cellulase employed for *Rhizoclonium* biomass (filamentous green algae) hydrolysis resulted in 40% yield enhancement [69]. In this case, the mixture of enzymes made difficult the identification of the enzymatic activity responsible for such an enhancement. Ometto et al. [9] also tested esterases in different lipid rich microalgae biomass. Moreover, this investigation reported the use of esterases alone and the mixture of esterases and proteases. No biogas production was attempted for the biomass pretreated with esterases alone and thus, no conclusion could be withdrawn. Nevertheless, their work revealed that this later enzymatic mixture supported much higher organic matter solubilization than the values attained for esterases application alone, highlighting the importance of microalgae proteins.

3.3. Proteases

Microalgae biomass is normally prevailing in protein content. As a matter of fact, this polymer might represent approximately 40–60% of the microalgae dry weight [24,70]. Protein fraction might be degraded by proteases since they hydrolyze peptides into amino acids. The use of proteases is receiving particular interest in last years, especially in combination with other pretreatments or other commercial enzymatic cocktails [71,72]. Some examples on the use of proteases in different microalgae biomass were evaluated in terms of organic matter solubilization and methane yields [38–40]. In the context of anaerobic digestion, methane yields of *C. vulgaris* and *Scenedesmus* sp. were enhanced by 2.6-fold and 1.53-fold, respectively, when pretreated with protease [39]. It is important to note that those results were attained with proteins rich biomass. More specifically, *Chlorella vulgaris* exhibited 64% protein and 22% carbohydrate content. When dealing with carbohydrate rich *C. vulgaris* biomass (39.6%), protease hydrolysis efficiency (54%) displayed higher organic matter values than carbohydrolase hydrolysis (approx. 26%). The different effect of both enzymatic cocktails was also observed in the methane yields attained by both pretreated biomass. In that case, methane yield achieved with the biomass pretreated with proteases was 137 mL CH_4 g/COD_{in} while 65 mL CH_4 g/COD_{in} was obtained for the biomass pretreated with carbohydrases [40]. This fact showed that even working with carbohydrate rich *C. vulgaris*, the proteolytic cocktail supported high organic matter hydrolysis and methane yields.

Comparison of different studies regarding enzymatic pretreatments suggested that proteins are the molecules that hindered the access of anaerobic bacteria to microalgae organic matter in the anaerobic digestion process to a greater extent than carbohydrates or lipids. Therefore, the protein fraction has been carefully analyzed during the anaerobic digestion process of microalgae biomass in the subsequent section

4. Biomass Proteins in Anaerobic Digestion of Microalgae

Anaerobic digestion is divided in four different stages including hydrolysis, acidogenesis, acetogenesis and methanogenesis (Figure 1). When protein rich microalgae are subjected to anaerobic digestion, the bioprocess can be affected at different stages.

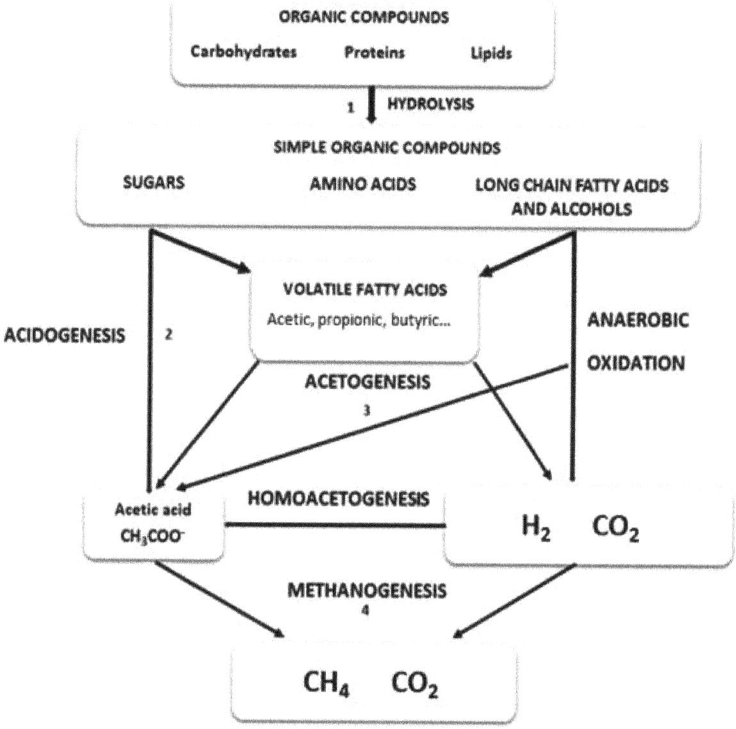

Figure 1. Reactive scheme for the anaerobic digestion of polymeric microalgal biomass.

Anaerobic degradation of proteins and lipids has not been investigated in depth compared to that of carbohydrates. Proteins are hydrolyzed to aminoacids by extracellular enzymes. Anaerobic and facultatively anaerobic bacteria, mainly *Clostridium*, are responsible of aminoacids fermentation. Clostridia obtain energy by coupled oxidation-reduction reaction between aminoacids via the so-called Stickland reaction. This reaction entails the oxidation (dehydrogetation) of one aminoacid and the reduction of a second aminoacids (hydrogenation) (Figure 2).

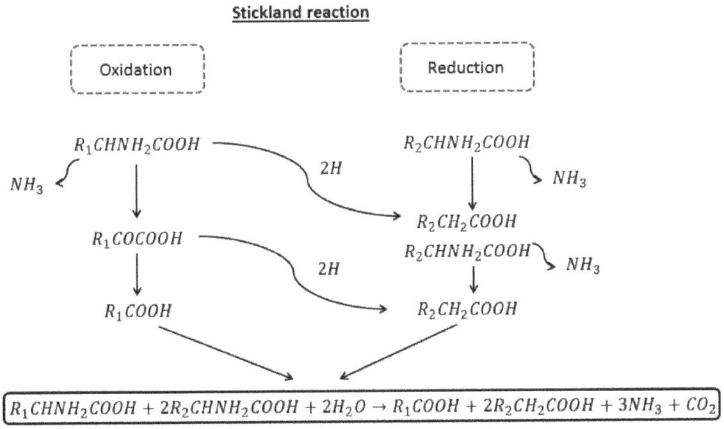

Figure 2. Stickland reactions scheme.

Aminoacids can act as electron acceptors or donors. In the first case, the aminoacid form a carboxylic acid with one carbon shorter than the original acid (e.g alanine to acetate) while when acting and electron acceptor, it retains the carbon to form a carboxylic acid with the same chain length as the original aminoacid (e.g., glycine to acetate). The aminoacid is de-ammonified by anaerobic oxidation, yielding volatile fatty acids and hydrogen, as shown in Table 2 [73].

Table 2. Aminoacid products based on Stickland reaction (modified from [73]).

Amino Acid	Formula	HAc	HProp	HBu	HVa	IN	IC	Other	H_2	ATP
Arginine	$C_6H_{14}O_2N_4$	0.5	0.5	0	0.5	4	1	0	−1	1
Histidine	$C_6H_9O_2N_3$	1	0	0.5	0	3	1	1	0	2
Lysine	$C_6H_{14}O_2N_2$	1	0	1	0	2	0	0	0	1
Tyrosine	$C_9H_{11}O_3N$	1	0	0	0	1	1	0.882	1	1
Tryptophan	$C_{11}H_{12}O_3N$	0	0	0	0	1	1	1.471	2	1
Phenylalanine	$C_9H_{11}O_2N$	0	0	0	0	1	1	1.176	2	1
Cysteine	$C_3H_6O_2NS$	1	0	0	0	1	1	0	1	1
Methionine	$C_5H_{11}O_2NS$	0	1	0	0	1	1	0	1	1
Threonine	$C_4H_9O_3N$	1	0	0.5	0	1	0	0	−1	1
Serine	$C_3H_7O_3N$	1	0	0	0	1	1	0	1	1
Leucine/Isoleucine	$C_6H_{13}O_2N$	0	0	0	1	1	1	0	2	1
Valine	$C_5H_{11}O_2N$	0	0	1	0	1	1	0	2	1
Glutamine	$C_5H_9O_4N$	1	0	0.5	0	1	1	0	0	2
Aspartate	$C_4H_7O_4N$	1	0	0	0	1	2	0	2	2
Glycine	$C_2H_5O_2N$	1	0	0	0	1	0	0	−1	0
Alanine	$C_3H_7O_2N$	1	0	0	0	1	1	0	2	1
Proline	$C_5H_9O_2N$	0.5	0.5	0	0.5	1	0	0	−1	0

4.1. The Relevance of Microalgae Proteins in the Hydrolysis Stage of Anaerobic Digestion

The first biological process involved in anaerobic digestion is hydrolysis, which is the limiting step and its effectiveness is crucial for the overall process [9,74]. Focusing on proteins, they are hydrolyzed into amino acids by extracellular enzymes secreted by different bacteria such as *Clostridium*, *Vibrio*, *Peptococcus*, *Bacillus*, *Proteus*, or *Bacteroides* [23]. As reviewed above, research devoted to microalgae digestion conducted over last years showed higher methane production in protease pretreated biomass compared to raw biomass and biomass treated with carbohydrases [40]. Methane production of protease pretreated *C. vulgaris* was enhanced by 51% compared to the raw biomass, showing the benefits of having proteins in the soluble phase. Similarly, methane yield enhancement (37%) of cyanobacteria was also attributed to the proteolytic activity developed upon biomass storage [74].

Even though protease addition has revealed the importance of microalgae proteins in microalgae digestion, it is clear that the use of commercial cocktails would not make biogas production profitable. In this manner, the use of commercial proteases helped in the identification of the macromolecule opposing more resistance to an optimal anaerobic digestion but cheaper alternatives should be investigated for avoiding the addition of commercial enzymes. Two main strategies can be applied for such a purpose. The first one entails the use of in-situ released enzymes by fungi or bacteria. Through the so-called bioaugmentation, microorganisms can be added to the anaerobic sludge used as degradation consortium. In this manner, once identified the microorganisms producing the enzymatic cocktail required for the targeted microalgae biomass, it can be added to the anaerobic sludge. Obviously, the appropriate microbial species should be carefully selected to be effective, not only for microalgae hydrolysis, but also to be viable and present good activity within the anaerobic microbiome. The potential of bioaugmentation, including the main benefits and limitations, has been recently reviewed [75]. This approach has been applied in more conventional substrates while literature available on bioaugmentation strategies devoted to microalgae anaerobic digestion is scarce. Nevertheless, this strategy was successfully applied to improve methane production of *C. vulgaris* biomass [60]. Those researchers showed an enhanced methane yield (18–38%) after adding *Clostridium thermocellum* at various inoculum ratios to degrade the carbohydrate fraction of microalgae biomass. Likewise, the same bacteria, *C. thermocellum*, was reported to enhance methane yield (18–38%) when degrading *Haematococcus pluvialis*. Therefore, this acidogenic phase bacteria is nowadays considered as a promising biotechnological tool to improve anaerobic digestion of microalgae through bioaugmentation.

The second alternative to increase the hydrolytic activity of anaerobic sludge is the use of metals. The addition of trace metals as micronutrients have been proven to stimulate methane production. The dosing needs to be well balanced to support the desired microbial activity or growth rate above which the trace metals become inhibitory or toxic. These metals are essential in the anaerobic reactions, since most of them are part of the active site of enzymes. The effect on different trace metal on anaerobic digestion can be found elsewhere [76]. Even though the use of trace elements is beneficial in most cases, the response of the system is uncertain due to the complexity of the anaerobic digestion process. It is recommended for substrates which initially have low trace element content. For instance, Kim et al. [77] evaluated the effect of trace elements at different range temperatures highlighting the benefits of using Fe, Co. or Ni for the hydrolysis step due to the increase of COD solubilization and organic acids production.

4.2. The Relevance of Microalgae Proteins in the Methanogenesis Stage of Anaerobic Digestion

Out of the subsequent stages involved in anaerobic digestion, hydrogen and acetic acid are converted to methane gas and carbon dioxide during methanogenesis. This last stage is performed by archaea. When compared to anaerobic bacteria involved in anaerobic digestion, archaea are more sensitive to toxic compounds and also exhibit lower growth rates. Acidifiers present ten to twenty-fold higher growth rates and five-fold conversion rates than methanogens [1,69]. With regard to their sensibility toward toxic compounds, methanogens exhibit low tolerance against ammonium nitrogen. Depending on digester pH and operation temperature, the ammonium/ammonia equilibrium might shift. This latter component has been claimed to be highly toxic for methanogens. Ammonia diffuses freely through the permeable membrane of methanogens cells causing changes in intracellular pH and resulting in potassium deficiency and/or proton imbalance [78]. Moreover, ammonium can also inhibit enzymes that are involved in methane production [79]. Yenigün and Demirel [80] reported inhibition of the methanogenesis stage at total ammonia nitrogen (TAN) and ammonia concentrations of 1700–1800 mg/L and 150 mg/L, respectively. As a result, the high concentration of TAN (NH_3 and NH_4^+) can lead to volatile fatty acids accumulation. This last process involves acidification of the anaerobic broth, which in turns inhibits methanogens activity. Therefore, the main drawback of protein rich biomass, such as microalgae, during digestion is the high amount of nitrogen released in form of ammonium that can inhibit methane formation. In fact, this inhibition has been already evidenced by Mahdy et al. [38] during the digestion of protein rich *Chlorella vulgaris*. Those authors attributed the stepwise methane production decrease to the high nitrogen mineralization (77%) taking place during the digestion of protease pretreated microalgae biomass. In this manner, microalgae proteins are not only limiting the hydrolysis stage of the anaerobic digestion but they might also be detrimental in methanogenesis stage. Similar to the developed strategies to overcome the negative effect of microalgae proteins in hydrolysis, some solutions have been proposed to overcome the issues that proteins might cause in methanogenesis during those last years of research.

To avoid inhibition by ammonium, different strategies can be implemented. One of them entails the use of nitrogen poor media for microalgae cultivation. Due to the low nitrogen availability in the medium, proteins accumulation is restricted while lipids and carbohydrates fractions become more abundant in the grown biomass [81,82]. Biogas production was modified using this method in different studies [80,83]. This strategy can be easily applied by using urban wastewater as culture media, which normally contains considerable lower nitrogen concentrations than synthetic salt media (≈60 vs. 300–600 mg N/L). The benefit of this strategy has been evidenced recently using *Spirulina* biomass for biogas production [12]. Similar results were obtained with *C. vulgaris*, where a higher accumulation of carbohydrates (40%) was observed when microalgae was grown in urban wastewater while only 22% was obtained in biomass grown in synthetic medium. Concomitantly with the increase in carbohydrates, protein biomass content was reduced (from 64 to 33%) and thus, methane production was enhanced [40].

A second approach to avoid ammonium inhibition is through sludge bioaugmentation. This approach consists in introducing or enriching specific anaerobic microorganisms with special features. Thus, anaerobic microorganisms that are tolerant to high NH_4^+ concentrations should be used within the anaerobic sludge to accomplish this goal. Although it is generally believed that total ammonia levels above 3 g/L have toxic effect on the methanogens, the resistance of methanogens can be increased by exposing the microorganisms to high nitrogen concentrations [83]. The use ammonia tolerant inocula has been recently demonstrated as an efficient option for digestion of *C. vulgaris* and cattle manure [84]. In this study, the effectiveness of adapted methanogens resulted in a 33% methane yield increase. This approach allowed operating the digester at 3.7–4.2 g NH_4^+-N/L. Tian et al. [85] operated an acclimation experiment in continuous anaerobic reactors fed with substrate rich in the protein fraction such as microalgae and cattle slurry manure. Results showed a stable biomethanization process despite of the high ammonium concentration (10 g NH_4^+-N/L). Authors stressed the changes on the anaerobic population taking place as the responsible feature to handle high ammonium concentration. Even though this biological strategy is very promising, it is necessary to do further research due to the challenges that might arise such as the different behavior that the bioaugmented inocula under different operational conditions imposed in the reactors. Attention must be directed to microorganism's population since they might fail to thrive or be washed out from the reactors.

5. Conclusions

Anaerobic digestion of microalgae has been presented as a promising alternative for generation of bioenergy. The implementation of this process requires pretreatment of the rigid algae cell wall in order to make available the organic matter to anaerobes. Enzymatic pretreatment with proteases showed the best performance in terms of organic matter solubilization and methane production. This feature already highlighted the importance of proteins in the hydrolysis stage of anaerobic digestion. Solving this problem with protease addition could result in methanogens inhibition mediated by high ammonium concentrations reached during nitrogen mineralization. Two solutions are proposed to overcome potential inhibition, namely the reduction of nitrogen levels of microalgae biomass using a low nitrogen concentration culture media and the use of ammonium tolerant anaerobic inocula. This fact showed that protein embedded in microalgae cell wall might be responsible for their inherent low biodegradability. Microalgae proteins might be crucial not only in the hydrolytic phase but also during methanogenesis.

Author Contributions: J.A.M. and C.G.-F. were responsible for the manuscript preparation. M.B. was responsible for revising the manuscript. The final publication was prepared with contribution from all authors.

Acknowledgments: The authors wish to thank the Spanish Ministry of Economy and Competitiveness for the financial support provided through the grants ENE2013-45416-R and RYC-2014-16823 and the Community of Madrid for the support offered in the framework of the project INSPIRA-1 (S2013/ABI-2783).

Conflicts of Interest: The authors declare no conflict of interest.

References

1. Henze, M.; van Loosdrecht, M.C.M.; Ekama, G.A.; Brdjanovic, D. *Biological Wastewater Treatment*; IWA Publishing: London, UK, 2008.
2. Abdel-Raouf, N.; Al-Homaidan, A.A.; Ibraheem, I.B.M. Microalgae and wastewater treatment. *Saudi J. Biol. Sci.* **2012**, *19*, 257–275. [CrossRef] [PubMed]
3. Xin, C.; Addy, M.M.; Zhao, J.; Cheng, Y.; Cheng, S.; Mu, D.; Liu, Y.; Ding, R.; Chen, P.; Ruan, R. Comprehensive techno-economic analysis of wastewater-based algal biofuel production: A case study. *Bioresour. Technol.* **2016**, *211*, 584–593. [CrossRef] [PubMed]
4. Passos, F.; Gutiérrez, R.; Uggetti, E.; Garfí, M.; García, J.; Ferrer, I. Towards energy neutral microalgae-based wastewater treatment plants. *Algal Res.* **2017**, *28*, 235–243. [CrossRef]
5. Milledge, J.J.; Heaven, S. Energy Balance of Biogas Production from microalgae: Development of an energy and mass balance model. *Curr. Biotechnol.* **2015**, *4*, 554–567. [CrossRef]
6. Davis, R.; Aden, A.; Pienkos, P.T. Techno-economic analysis of autotrophic microalgae for fuel production. *Appl. Energy* **2011**, *88*, 3524–3531. [CrossRef]
7. Arashiro, L.T.; Montero, N.; Ferrer, I.; Acién, F.G.; Gómez, C.; Garfí, M. Life cycle assessment of high rate algal ponds for wastewater treatment and resource recovery. *Sci. Total Environ.* **2018**, *622*, 1118–1130. [CrossRef]
8. Muñoz, R.; Gonzalez-Fernandez, C. *Microalgae-Based Biofuels and Bioproducts: From Feedstock Cultivation to End-Products*; Woodhead Publishing Series in Energy; Elsevier Science: New York, NY, USA, 2017.
9. Ometto, F.; Quiroga, G.; Psenicka, P.; Whitton, R.; Jefferson, B.; Villa, R. Impacts of microalgae pre-treatments for improved anaerobic digestion: Thermal treatment, thermal hydrolysis, ultrasound and enzymatic hydrolysis. *Water Res.* **2014**, *65*, 350–361. [CrossRef] [PubMed]
10. Angelidaki, I.; Sanders, W. Assessment of the anaerobic biodegradability of macropollutants. *Re/Views Environ. Sci. Bio/Technol.* **2004**, *3*, 117–129. [CrossRef]
11. Sialve, B.; Bernet, N.; Bernard, O. Anaerobic digestion of microalgae as a necessary step to make microalgal biodiesel sustainable. *Biotechnol. Adv.* **2009**, *27*, 409–416. [CrossRef] [PubMed]
12. Markou, G.; Angelidaki, I.; Georgakakis, D. Carbohydrate-enriched cyanobacterial biomass as feedstock for bio-methane production through anaerobic digestion. *Fuel* **2013**, *111*, 872–879. [CrossRef]
13. Yu, W.-L.; Ansari, W.; Schoepp, N.G.; Hannon, M.J.; Mayfield, S.P.; Burkart, M.D. Modifications of the metabolic pathways of lipid and triacylglycerol production in microalgae. *Microb. Cell Fact.* **2011**, *10*, 91. [CrossRef] [PubMed]
14. Mata, T.M.; Martins, A.A.; Caetano, N.S. Microalgae for biodiesel production and other applications: A review. *Renew. Sustain. Energy Rev.* **2010**, *14*, 217–232. [CrossRef]
15. Fábregas, J.; Maseda, A.; Domínguez, A.; Otero, A. The cell composition of *Nannochloropsis* sp. changes under different irradiances in semicontinuous culture. *World J. Microbiol. Biotechnol.* **2004**, *20*, 31–35. [CrossRef]
16. Pancha, I.; Chokshi, K.; George, B.; Ghosh, T.; Paliwal, C.; Maurya, R.; Mishra, S. Nitrogen stress triggered biochemical and morphological changes in the microalgae *Scenedesmus* sp. CCNM 1077. *Bioresour. Technol.* **2014**, *156*, 146–154. [CrossRef] [PubMed]
17. Golueke, C.G.; Oswald, W.J.; Gotaas, H.B. Anaerobic digestion of algae. *Appl. Microbiol.* **1957**, *5*, 47–55. [PubMed]
18. Demuez, M.; Mahdy, A.; Tomás-Pejó, E.; González-Fernández, C.; Ballesteros, M. Enzymatic cell disruption of microalgae biomass in biorefinery processes. *Biotechnol. Bioeng.* **2015**, *112*, 1955–1966. [CrossRef] [PubMed]

19. De Leeuw, J.W.; Versteegh, G.J.M.; van Bergen, P.F. Biomacromolecules of algae and plants and their fossil analogues. In *Plants and Climate Change*; Rozema, J., Aerts, R., Cornelissen, H., Eds.; Springer: Dordrecht, The Netherlands, 2006; pp. 209–233.
20. Kodner, R.B.; Summons, R.E.; Knoll, A.H. Phylogenetic investigation of the aliphatic, non-hydrolyzable biopolymer algaenan, with a focus on green algae. *Org. Geochem.* **2009**, *40*, 854–862. [CrossRef]
21. Passos, F.; Uggetti, E.; Carrère, H.; Ferrer, I. Pretreatment of microalgae to improve biogas production: A review. *Bioresour. Technol.* **2014**, *172*, 403–412. [CrossRef] [PubMed]
22. Klassen, V.; Blifernez-Klassen, O.; Wobbe, L.; Schluter, A.; Kruse, O.; Mussgnug, J.H. Efficiency and biotechnological aspects of biogas production from microalgal substrates. *J. Biotechnol.* **2016**, *234*, 7–26. [CrossRef] [PubMed]
23. Gonzalez-Fernandez, C.; Sialve, B.; Molinuevo-Salces, B. Anaerobic digestion of microalgal biomass: Challenges, opportunities and research needs. *Bioresour. Technol.* **2015**, *198*, 896–906. [CrossRef] [PubMed]
24. González-Fernández, C.; Sialve, B.; Bernet, N.; Steyer, J.-P. Impact of microalgae characteristics on their conversion to biofuel. Part II: Focus on biomethane production. *Biofuels Bioprod. Biorefin.* **2012**, *6*, 205–218. [CrossRef]
25. Hom-Diaz, A.; Passos, F.; Ferrer, I.; Vicent, T.; Blánquez, P. Enzymatic pretreatment of microalgae using fungal broth from *Trametes versicolor* and commercial laccase for improved biogas production. *Algal Res.* **2016**, *19*, 184–188. [CrossRef]
26. Cordova, O.; Passos, F.; Chamy, R. Physical pretreatment methods for improving microalgae anaerobic biodegradability. *Appl. Biochem. Biotechnol.* **2017**. [CrossRef] [PubMed]
27. Pandey, A.; Negi, S.; Binod, P.; Larroche, C. *Pretreatment of Biomass: Processes and Technologies*; Elsevier Science: New York, NY, USA, 2014.
28. Zhen, G.; Lu, X.; Kato, H.; Zhao, Y.; Li, Y.Y. Overview of pretreatment strategies for enhancing sewage sludge disintegration and subsequent anaerobic digestion: Current advances, full-scale application and future perspectives. *Renew. Sustain. Energy Rev.* **2017**, *69*, 559–577. [CrossRef]
29. Passos, F.; Ferrer, I. Microalgae conversion to biogas: Thermal pretreatment contribution on net energy production. *Environ. Sci. Technol.* **2014**, *48*, 7171–7178. [CrossRef] [PubMed]
30. González-Fernández, C.; Méndez, L.; Ballesteros, M.; Tomás-Pejó, E. Hydrothermal Processing of Microalgae. In *Hydrothermal Processing in Biorefineries: Production of Bioethanol and High Added-Value Compounds of Second and Third Generation Biomass*; Ruiz, H.A., Hedegaard Thomsen, M., Trajano, H.L., Eds.; Springer: Cham, The Netherlands, 2017.
31. Passos, F.; Carretero, J.; Ferrer, I. Comparing pretreatment methods for improving microalgae anaerobic digestion: Thermal, hydrothermal, microwave and ultrasound. *Chem. Eng. J.* **2015**, *279*, 667–672. [CrossRef]
32. Wang, M.; Lee, E.; Dilbeck, M.P.; Liebelt, M.; Zhang, Q.; Ergas, S.J. Thermal pretreatment of microalgae for biomethane production: Experimental studies, kinetics and energy analysis. *J. Chem. Technol. Biotechnol.* **2017**, *92*, 399–407. [CrossRef]
33. Mendez, L.; Mahdy, A.; Ballesteros, M.; González-Fernández, C. Biomethane production using fresh and thermally pretreated *Chlorella vulgaris* biomass: A comparison of batch and semi-continuous feeding mode. *Ecol. Eng.* **2015**, *84*, 273–277. [CrossRef]
34. Alzate, M.E.; Muñoz, R.; Rogalla, F.; Fdz-Polanco, F.; Pérez-Elvira, S.I. Biochemical methane potential of microalgae: Influence of substrate to inoculum ratio, biomass concentration and pretreatment. *Bioresour. Technol.* **2012**, *123*, 488–494. [CrossRef] [PubMed]
35. Mendez, L.; Mahdy, A.; Demuez, M.; Ballesteros, M.; González-Fernández, C. Effect of high pressure thermal pretreatment on *Chlorella vulgaris* biomass: Organic matter solubilisation and biochemical methane potential. *Fuel* **2014**, *117*, 674–679. [CrossRef]
36. Mahdy, A.; Mendez, L.; Ballesteros, M.; González-Fernández, C. Autohydrolysis and alkaline pretreatment effect on *Chlorella vulgaris* and *Scenedesmus* sp. methane production. *Energy* **2014**, *78*, 48–52. [CrossRef]
37. Carrillo-Reyes, J.; Barragán-Trinidad, M.; Buitrón, G. Biological pretreatments of microalgal biomass for gaseous biofuel production and the potential use of rumen microorganisms: A review. *Algal Res.* **2016**, *18*, 341–351. [CrossRef]
38. Mahdy, A.; Mendez, L.; Ballesteros, M.; González-Fernández, C. Enhanced methane production of *Chlorella vulgaris* and *Chlamydomonas reinhardtii* by hydrolytic enzymes addition. *Energy Convers. Manag.* **2014**, *85*, 551–557. [CrossRef]

39. Mahdy, A.; Mendez, L.; Ballesteros, M.; González-Fernández, C. Protease pretreated *Chlorella vulgaris* biomass bioconversion to methane via semi-continuous anaerobic digestion. *Fuel* **2015**, *158*, 35–41. [CrossRef]
40. Mahdy, A.; Ballesteros, M.; González-Fernández, C. Enzymatic pretreatment of *Chlorella vulgaris* for biogas production: Influence of urban wastewater as a sole nutrient source on macromolecular profile and biocatalyst efficiency. *Bioresour. Technol.* **2016**, *199*, 319–325. [CrossRef] [PubMed]
41. Mahdy, A.; Mendez, L.; Tomás-Pejó, E.; del Mar Morales, M.; Ballesteros, M.; González-Fernández, C. Influence of enzymatic hydrolysis on the biochemical methane potential of *Chlorella vulgaris* and *Scenedesmus* sp. *J. Chem. Technol. Biotechnol.* **2016**, *91*, 1299–1305. [CrossRef]
42. Passos, F.; Ferrer, I. Influence of hydrothermal pretreatment on microalgal biomass anaerobic digestion and bioenergy production. *Water Res.* **2015**, *68*, 364–373. [CrossRef] [PubMed]
43. González-Fernández, C.; Sialve, B.; Bernet, N.; Steyer, J.P. Comparison of ultrasound and thermal pretreatment of *Scenedesmus* biomass on methane production. *Bioresour. Technol.* **2012**, *110*, 610–616. [CrossRef] [PubMed]
44. Carrere, H.; Antonopoulou, G.; Affes, R.; Passos, F.; Battimelli, A.; Lyberatos, G.; Ferrer, I. Review of feedstock pretreatment strategies for improved anaerobic digestion: From lab-scale research to full-scale application. *Bioresour. Technol.* **2016**, *199*, 386–397. [CrossRef] [PubMed]
45. Rodriguez, C.; Alaswad, A.; Mooney, J.; Prescott, T.; Olabi, A.G. Pre-treatment techniques used for anaerobic digestion of algae. *Fuel Process. Technol.* **2015**, *138*, 765–779. [CrossRef]
46. Hirano, A.; Ueda, R.; Hirayama, S.; Ogushi, Y. CO_2 fixation and ethanol production with microalgal photosynthesis and intracellular anaerobic fermentation. *Energy* **1997**, *22*, 137–142. [CrossRef]
47. Martínez, N.; Callejas, N.; Morais, E.G.; Vieira Costa, J.A.; Jachmanián, I.; Vieitez, I. Obtaining biodiesel from microalgae oil using ultrasound-assisted in-situ alkaline transesterification. *Fuel* **2017**, *202*, 512–519. [CrossRef]
48. Wang, Y.; Guo, W.; Cheng, C.-L.; Ho, S.-H.; Chang, J.-S.; Ren, N. Enhancing bio-butanol production from biomass of *Chlorella vulgaris* JSC-6 with sequential alkali pretreatment and acid hydrolysis. *Bioresour. Technol.* **2016**, *200*, 557–564. [CrossRef] [PubMed]
49. Efremenko, E.N.; Nikolskaya, A.B.; Lyagin, I.V.; Senko, O.V.; Makhlis, T.A.; Stepanov, N.A.; Maslova, O.V.; Mamedova, F.; Varfolomeev, S.D. Production of biofuels from pretreated microalgae biomass by anaerobic fermentation with immobilized *Clostridium acetobutylicum* cells. *Bioresour. Technol.* **2012**, *114*, 342–348. [CrossRef] [PubMed]
50. Solé-Bundó, M.; Carrère, H.; Garfí, M.; Ferrer, I. Enhancement of microalgae anaerobic digestion by thermo-alkaline pretreatment with lime (CaO). *Algal Res.* **2017**, *24*, 199–206. [CrossRef]
51. Mendez, L.; Mahdy, A.; Timmers, R.A.; Ballesteros, M.; González-Fernández, C. Enhancing methane production of *Chlorella vulgaris* via thermochemical pretreatments. *Bioresour. Technol.* **2013**, *149*, 136–141. [CrossRef] [PubMed]
52. Chng, L.M.; Lee, K.T.; Chan, D.J.C. Synergistic effect of pretreatment and fermentation process on carbohydrate-rich *Scenedesmus dimorphus* for bioethanol production. *Energy Convers. Manag.* **2017**, *141*, 410–419. [CrossRef]
53. Khan, M.I.; Lee, M.G.; Shin, J.H.; Kim, J.D. Pretreatment optimization of the biomass of *Microcystis aeruginosa* for efficient bioethanol production. *AMB Express* **2017**, *7*, 19. [CrossRef] [PubMed]
54. Kavitha, S.; Subbulakshmi, P.; Rajesh Banu, J.; Gobi, M.; Tae Yeom, I. Enhancement of biogas production from microalgal biomass through cellulolytic bacterial pretreatment. *Bioresour. Technol.* **2017**, *233*, 34–43. [CrossRef] [PubMed]
55. Ciudad, G.; Rubilar, O.; Azócar, L.; Toro, C.; Cea, M.; Torres, Á.; Ribera, A.; Navia, R. Performance of an enzymatic extract in *Botrycoccus braunii* cell wall disruption. *J. Biosci. Bioeng.* **2014**, *117*, 75–80. [CrossRef] [PubMed]
56. Muñoz, C.; Hidalgo, C.; Zapata, M.; Jeison, D.; Riquelme, C.; Rivas, M. Use of cellulolytic marine bacteria for enzymatic pretreatment in microalgal biogas production. *Appl. Environ. Microbiol.* **2014**, *80*, 4199–4206. [CrossRef] [PubMed]
57. Passos, F.; Hom-Diaz, A.; Blanquez, P.; Vicent, T.; Ferrer, I. Improving biogas production from microalgae by enzymatic pretreatment. *Bioresour. Technol.* **2016**, *199*, 347–351. [CrossRef] [PubMed]
58. He, S.; Fan, X.; Katukuri, N.R.; Yuan, X.; Wang, F.; Guo, R.-B. Enhanced methane production from microalgal biomass by anaerobic bio-pretreatment. *Bioresour. Technol.* **2016**, *204*, 145–151. [CrossRef] [PubMed]

59. Arenas, E.G.; Rodriguez Palacio, M.C.; Juantorena, A.U.; Fernando, S.E.L.; Sebastian, P.J. Microalgae as a potential source for biodiesel production: Techniques, methods, and other challenges. *Int. J. Energy Res.* **2017**, *41*, 761–789. [CrossRef]
60. Aydin, S. Enhancement of microbial diversity and methane yield by bacterial bioaugmentation through the anaerobic digestion of *Haematococcus pluvialis*. *Appl. Microbiol. Biotechnol.* **2016**, *100*, 5631–5637. [CrossRef] [PubMed]
61. Velmurugan, R.; Incharoensakdi, A. MgO-Fe_3O_4 linked cellulase enzyme complex improves the hydrolysis of cellulose from *Chlorella* sp. CYB2. *Biochem. Eng. J.* **2017**, *122*, 22–30. [CrossRef]
62. Hernández, D.; Riaño, B.; Coca, M.; García-González, M.C. Saccharification of carbohydrates in microalgal biomass by physical, chemical and enzymatic pre-treatments as a previous step for bioethanol production. *Chem. Eng. J.* **2015**, *262*, 939–945. [CrossRef]
63. Guo, H.; Chen, H.; Fan, L.; Linklater, A.; Zheng, B.; Jiang, D.; Qin, W. Enzymes produced by biomass-degrading bacteria can efficiently hydrolyze algal cell walls and facilitate lipid extraction. *Renew. Energy* **2017**, *109*, 195–201. [CrossRef]
64. Ward, A.J.; Lewis, D.M.; Green, F.B. Anaerobic digestion of algae biomass: A review. *Algal Res.* **2014**, *5*, 204–214. [CrossRef]
65. Cirne, D.G.; Paloumet, X.; Björnsson, L.; Alves, M.M.; Mattiasson, B. Anaerobic digestion of lipid-rich waste—Effects of lipid concentration. *Renew. Energy* **2007**, *32*, 965–975. [CrossRef]
66. Palatsi, J.; Laureni, M.; Andrés, M.V.; Flotats, X.; Nielsen, H.B.; Angelidaki, I. Strategies for recovering inhibition caused by long chain fatty acids on anaerobic thermophilic biogas reactors. *Bioresour. Technol.* **2009**, *100*, 4588–4596. [CrossRef] [PubMed]
67. Passos, F.; Solé, M.; García, J.; Ferrer, I. Biogas production from microalgae grown in wastewater: Effect of microwave pretreatment. *Appl. Energy* **2013**, *108*, 168–175. [CrossRef]
68. Pittman, J.K.; Dean, A.P.; Osundeko, O. The potential of sustainable algal biofuel production using wastewater resources. *Bioresour. Technol.* **2011**, *102*, 17–25. [CrossRef] [PubMed]
69. Ehimen, E.A.; Holm-Nielsen, J.-B.; Poulsen, M.; Boelsmand, J.E. Influence of different pre-treatment routes on the anaerobic digestion of a filamentous algae. *Renew. Energy* **2013**, *50*, 476–480. [CrossRef]
70. Juneja, A.; Ceballos, R.M.; Murthy, G.S. Effects of environmental factors and nutrient availability on the biochemical composition of algae for biofuels production: A review. *Energies* **2013**, *6*, 4607–4638. [CrossRef]
71. Choi, S.P.; Nguyen, M.T.; Sim, S.J. Enzymatic pretreatment of *Chlamydomonas reinhardtii* biomass for ethanol production. *Bioresour. Technol.* **2010**, *101*, 5330–5336. [CrossRef] [PubMed]
72. Gerken, H.G.; Donohoe, B.; Knoshaug, E.P. Enzymatic cell wall degradation of *Chlorella vulgaris* and other microalgae for biofuels production. *Planta* **2013**, *237*, 239–253. [CrossRef] [PubMed]
73. Batstone, D.J.; Keller, J.; Angelidaki, I.; Kalyuzhnyi, S.V.; Pavlostathis, S.G.; Rozzi, A.; Sanders, W.T.M.; Siegrist, H.; Vavilin, V.A. The IWA anaerobic digestion model no 1 (ADM1). *Water Sci. Technol.* **2002**, *45*, 65–73. [PubMed]
74. Miao, H.; Lu, M.; Zhao, M.; Huang, Z.; Ren, H.; Yan, Q.; Ruan, W. Enhancement of Taihu blue algae anaerobic digestion efficiency by natural storage. *Bioresour. Technol.* **2013**, *149*, 359–366. [CrossRef] [PubMed]
75. Nzila, A. Mini review: Update on bioaugmentation in anaerobic processes for biogas production. *Anaerobe* **2017**, *46*, 3–12. [CrossRef] [PubMed]
76. Fermoso, F.G.; Bartacek, J.; Jansen, S.; Lens, P.N.L. Metal supplementation to UASB bioreactors: From cell-metal interactions to full-scale application. *Sci. Total Environ.* **2009**, *407*, 3652–3667. [CrossRef] [PubMed]
77. Kim, M.; Gomec, C.Y.; Ahn, Y.; Speece, R.E. Hydrolysis and acidogenesis of particulate organic material in mesophilic and thermophilic anaerobic digestion. *Environ. Technol.* **2003**, *24*, 1183–1190. [CrossRef] [PubMed]
78. Chen, Y.; Cheng, J.J.; Creamer, K.S. Inhibition of anaerobic digestion process: A review. *Bioresour. Technol.* **2008**, *99*, 4044–4064. [CrossRef] [PubMed]
79. Cabanelas, I.T.D.; Arbib, Z.; Chinalia, F.A.; Souza, C.O.; Perales, J.A.; Almeida, P.F.; Druzian, J.I.; Nascimento, I.A. From waste to energy: Microalgae production in wastewater and glycerol. *Appl. Energy* **2013**, *109*, 283–290. [CrossRef]
80. Yenigün, O.; Demirel, B. Ammonia Inhibition in Anaerobic Digestion: A Review. *Process Biochem.* **2013**, *48*, 901–911. [CrossRef]

81. Illman, A.M.; Scragg, A.H.; Shales, S.W. Increase in *Chlorella* strains calorific values when grown in low nitrogen medium. *Enzyme Microb. Technol.* **2000**, *27*, 631–635. [CrossRef]
82. An, M.; Wang, Y.; Liu, F.; Qi, X.; Zheng, Z.; Ye, N.; Sun, C.; Miao, J. Biomass, nutrient uptake and fatty acid composition of *Chlamydomonas* sp. ICE-L in response to different nitrogen sources. *Acta Oceanol. Sin.* **2017**, *36*, 105–110. [CrossRef]
83. Nakakubo, R.; Møller, H.B.; Nielsen, A.M.; Matsuda, J. Ammonia inhibition of methanogenesis and identification of process indicators during anaerobic digestion. *Environ. Eng. Sci.* **2008**, *25*, 1487–1496. [CrossRef]
84. Mahdy, A.; Fotidis, I.A.; Mancini, E.; Ballesteros, M.; Gonzalez-Fernandez, C.; Angelidaki, I. Ammonia tolerant inocula provide a good base for anaerobic digestion of microalgae in third generation biogas process. *Bioresour. Technol.* **2017**, *225*, 272–278. [CrossRef] [PubMed]
85. Tian, H.; Fotidis, I.A.; Mancini, E.; Treu, L.; Mahdy, A.; Ballesteros, M.; González-Fernández, C.; Angelidaki, I. Acclimation to extremely high ammonia levels in continuous biomethanation process and the associated microbial community dynamics. *Bioresour. Technol.* **2018**, *247*, 616–623. [CrossRef] [PubMed]

molecules

MDPI

Article

Effective Saccharification of Corn Stover Using Low-Liquid Aqueous Ammonia Pretreatment and Enzymatic Hydrolysis

Nguyen Phuong Vi Truong [1] and Tae Hyun Kim [2,*]

[1] Faculty of Natural Sciences, Thu Dau Mot University, Thu Dau Mot City 820000, Binh Duong Province, Vietnam; vitnp@tdmu.edu.vn

[2] Department of Materials Science and Chemical Engineering, Hanyang University, Ansan, Gyeonggi-do 15588, Korea

* Correspondence: hitaehyun@hanyang.ac.kr; Tel.: +82-31-400-5222

Received: 27 March 2018; Accepted: 28 April 2018; Published: 1 May 2018

Abstract: Low-liquid aqueous ammonia (LLAA) pretreatment using aqueous ammonia was investigated to enhance enzymatic saccharification of corn stover. In this method, ground corn stover was simply contacted with aqueous ammonia mist (ammoniation step), followed by pretreatment at elevated temperature (90–150 °C) for an extended period (24–120 h) at different solid/liquid (S/L) ratios (0.29, 0.47 or 0.67), termed a pretreatment step. After that, excess (unreacted) ammonia was removed by evaporation, and the pretreated material was immediately saccharified by an enzyme without a washing step. The effects of key reaction parameters on both glucan digestibility and XMG digestibility were evaluated by analysis of variance (ANOVA). Under the best pretreatment conditions [S/L = 0.47, 0.16 (g NH_3)/(g biomass), 90 °C, 24 h], LLAA pretreatment enhanced enzymatic digestibility from 23.1% for glucan and 11.3% for XMG (xylan + galactan + mannan) of untreated corn stover to 91.8% for glucan and 72.6% for XMG in pretreated solid.

Keywords: aqueous ammonia; alkaline pretreatment; enzymatic digestibility; lignocellulosic biomass; cellulosic sugar

1. Introduction

Limited supplies of fossil resources, climate change due to carbon dioxide accumulation in the atmosphere, and increased demand for fuels and chemicals have triggered an increase in utilization of diverse renewable feedstock. To implement the production of a wide range of fuels, chemicals, and materials from renewable sources, most green research in recent years has focused on the development of renewable fuels and bio-based chemicals as a substitute for conventional fossil fuels (gasoline and diesel) and petroleum-based chemicals. In particular, cellulosic fuel ethanol, a second-generation biofuel, has the potential to solve several problems, including limited feedstock availability and food competition with fuel, that are currently associated with first-generation biofuels such as fuel ethanol from corn starch or sugarcane [1]. Cellulosic ethanol can be produced from inexpensive and abundant lignocellulosic materials such as woody biomass and herbaceous biomass [2]. Therefore, it is currently believed that cellulosic ethanol can meet a larger proportion of global transportation fuel demand in the near future. Production of ethanol from lignocellulosic biomass is still challenging because of the recalcitrant nature of the latter; for example, lignin is an inhibitor of enzymatic and microbial reactions and has high crystallinity and complex chemical composition [3,4]. Unlike sugar and starch, carbohydrates of lignocellulosic biomass consist of five different sugar units (glucose, xylose, arabinose, galactose, and mannose). To utilize lignocellulosic biomass effectively, production of fuels and chemicals from all sugars is necessary [5].

Currently, most of fuel ethanol is being produced from corn starch or sugarcane in many countries, such as China, Brazil, and the United States. Corn stover includes husk, leaves, and stalk that are left in the field after grain harvest and is a co-product of corn grain production. Therefore, manufacture of fuel ethanol from corn stover may be a reasonable approach to commercialization of the first cellulosic ethanol process at present [1].

Because of the aforementioned difficulties with utilization of lignocellulosic biomass, pretreatment is necessary to disrupt the recalcitrant structure of the plant cell walls, thus enabling easy access to production of fermentable sugar, which is then fermented to produce ethanol [6]. Therefore, study in recent years has been focused on the development of effective pretreatment method intended to make the lignocellulosic sugars available for ethanol conversion. Nonetheless, it is known that most of pretreatment methods involving various acids and alkalis at high temperature typically generate inhibitory products such as phenolic compounds, furfural, 5-hydroxymethylfurfural, and aldehydes. Therefore, some alkaline pretreatments under mild reaction conditions are considered viable pretreatment methods for different types of lignocellulosic biomass such as wood biomass and herbaceous biomass with high lignin content [7]. In a large-scale biomass conversion process involving a pretreatment unit, the chemical and water inputs can be a critical factor for the development of a commercially viable biochemical method. Nevertheless, a washing step is typically required in both acid and alkali pretreatment methods for the removal of the remaining chemical reagents from the chemically treated biomass, and the recovery and reuse of water and chemicals significantly affect the total energy cost of the biomass conversion process.

To reduce the water and chemical inputs into biomass processing, our laboratory previously reported that a pretreatment method using anhydrous ammonia (low-moisture anhydrous ammonia; LMAA) effectively improves the enzyme saccharification yield of agricultural biomass [8–10]. Although the LMAA method has been developed to eliminate the washing step, one of the drawbacks of anhydrous (gaseous) ammonia is that it must be stored and handled under high pressure, which requires specially designed and well-maintained high-pressure equipment and systems during biomass processing.

In our present study, low-liquid aqueous ammonia (LLAA) pretreatment was proposed to solve such problems associated with a process using gaseous ammonia. This pretreatment method consists of ammoniation, pretreatment, and evaporation steps; i.e., corn stover is well contacted with aqueous ammonia mist using nozzle spray and tumbler mixer (Figure 1a) (ammoniation step), followed by pretreatment step at an elevated temperature (up to 150 °C) for an extended period (up to 120 h) using a tight-sealed batch reactor (Figure 1b). After that, excess (unreacted) ammonia is removed by evaporation, and the resulting material can be immediately saccharified by a commercial cellulase without a washing step. LLAA pretreatment can be expected to lower the operating cost because it requires low input of liquid (reagents and water). Furthermore, aqueous ammonia is easy to handle, making this method a more industrially adoptable process for an upcoming biomass-processing facility.

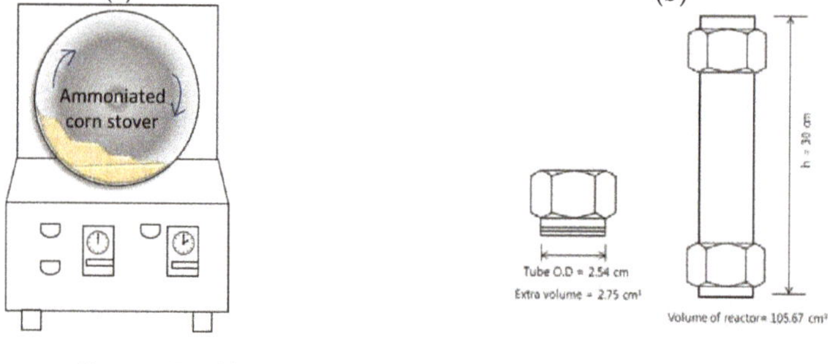

Figure 1. Tumbler mixer (**a**) and batch-type pretreatment reactor (**b**).

2. Results and Discussion

2.1. Effects of Reaction Temperature and Time on the Chemical Composition of Pretreated Corn Stover

The initial composition of the untreated corn stover is summarized in Table 1.

Table 1. Chemical composition of untreated corn stover.

Sample ID	Corn Stover [wt %]
Extractive	
Glucose	0.1 ± 0.2
Sucrose	1.1 ± 0.0
Fructose	1.2 ± 0.0
Soluble lignin	1.3 ± 0.0
Other extractives	16.5 ± 0.0
Extractives-free solid	
Glucan	33.0 ± 0.6
Xylan	17.9 ± 0.1
Arabinan	3.2 ± 0.5
Mannan	0.2 ± 0.0
Galactan	1.9 ± 0.2
AIL [1]	14.5 ± 0.1
ASL [2]	2.1 ± 0.0
Ash	0.9 ± 0.1
Protein	6.0 ± 0.0
Total	100

Note: All weight percentages were calculated on the basis of oven-dried biomass weight; [1] AIL: acid-insoluble lignin; [2] ASL: acid-soluble lignin SD: standard deviation (n = 3).

The effects of reaction temperature and time were evaluated, and Figure 2 presents the changes in chemical composition at various pretreatment temperatures with extended pretreatment periods. Three pretreatment temperatures (90, 120 and 150 °C) were applied during extended pretreatment periods (24–120 h) while we kept other conditions constant [0.16 (g NH_3)/(g biomass), S/L = 0.47]. As shown in Figure 2a,b, pretreatment at lower temperatures (90 and 120 °C) did not result in significant changes in carbohydrates (glucan and XMG) and lignin (acid-insoluble lignin, AIL and acid-soluble lignin, ASL) even with a prolonged reaction period (up to 120 h). On the other hand, there was a marginal change in both XMG and lignin contents at 150 °C (Figure 2c), in particular, after 72–96 h of pretreatment. Pretreatment at a high temperature (150 °C) for 120 h increased both AIL and ASL contents to 15.8% and 4.1%, respectively, which represented 1.3% and 2.0% increases as compared to untreated corn stover (Figure 2c). On the contrary, XMG content decreased from 20.0% of untreated corn stover to 18.4% after pretreatment at 150 °C for 120 h. Glucan content was maintained well at all three temperatures of pretreatment.

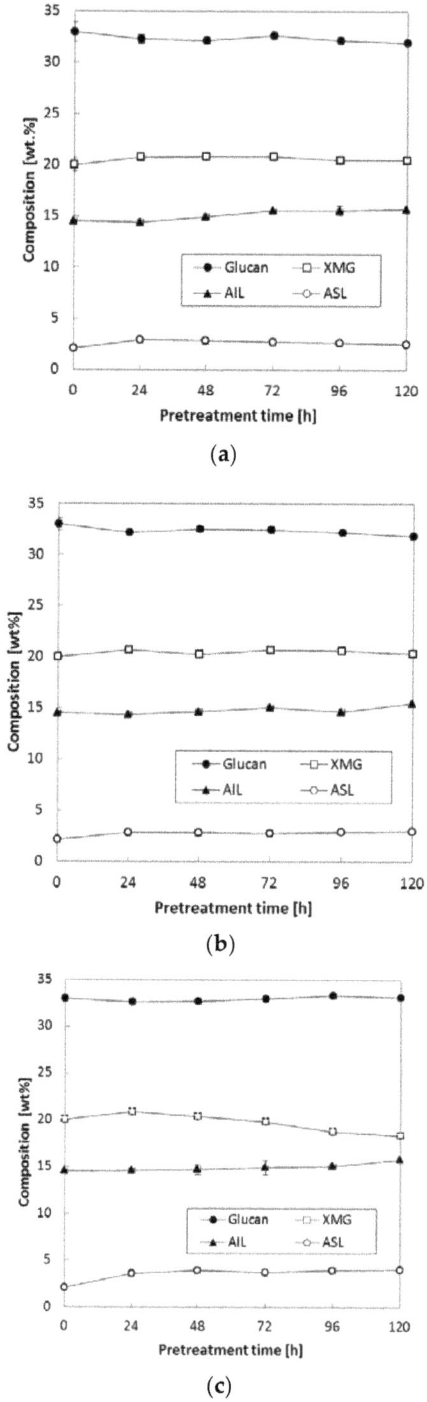

(**a**)

(**b**)

(**c**)

Figure 2. Effects of pretreatment temperature and time on the composition of pretreated corn stover. Pretreatment: 0.16 (g NH_3)/(g biomass), S/L ratio = 0.47, 24–120 h. (**a**) 90 °C, (**b**) 120 °C, (**c**) 150 °C. The data in the figure show mean values.

2.2. *The Effect of the S/L Ratio on Chemical Composition of Pretreated Corn Stover*

In the above test, various temperatures (90–150 °C) were tested while we kept NH_3 loading at 0.16 (g NH_3)/(g biomass) and S/L ratio at 0.47. Because it was found that chemical composition was more affected at 150 °C than other temperature (90 °C and 120 °C), another set of experiments to study the compositional changes during pretreatment was conducted at low NH_3 loading [0.08 (g NH_3)/(g biomass)] at various S/L ratios. When the S/L ratio was varied between 0.29 and 0.47, both XMG and lignin were slightly affected; i.e., as the reaction time increased, XMG content gradually decreased from 20.0% of untreated corn stover to 18.7–19.0% of pretreated corn stover, while both AIL and ASL increased accordingly; in particular, AIL increased from 14.5% of untreated corn stover to 15.8–16.6% of pretreated corn stover (Figure 3a,b). Nevertheless, it was found that the increase in the S/L ratio did not result in a considerable change in glucan content under all the tested conditions. Most significant changes in XMG and AIL occurred in case of the pretreatment at the highest S/L ratio (S/L = 0.67) and reaction time >72 h (Figure 3c). XMG content decreased from 20.0% of untreated corn stover to 2.9% of 120-h pretreated corn stover, whereas AIL increased from 14.5 to 28.0%. ASL content was slightly increased by ammonia pretreatment (from 2.1% of untreated corn stover to 4.0% of pretreated corn stover) as reaction time was extended to 120 h. The reason for the lignin upregulation during pretreatment under the harsh conditions (Figure 3c) was not clear at this stage. This observation was consistent with our previous report about the pretreatment of herbaceous biomass using gaseous ammonia; i.e., pretreated corn stover at 130–150 °C showed a considerable change in the composition of treated solids [8]. Nevertheless, it could be hypothesized according to the literature that the pretreatment reaction in the presence of water and the chemical depolymerize the linkages in the lignin–carbohydrate complex; this action results in removal of lignin along with other fiber fragments from cellulose and hemicellulose and, if they are not removed promptly, causes its subsequent repolymerization [11]. XMG is the main component of hemicellulose in herbaceous plants [12] and can easily be degraded during chemical pretreatment at a high temperature with a long reaction period [13,14]. This repolymerized lignin contains residual xylan and other degradation products becoming acid-insoluble complexes that are not hydrolyzed by sulfuric acid during chemical composition analysis following standard laboratory analytical procedure (LAP) of the National Renewable Energy Laboratory (NREL; Golden, CO, USA), thus resulting in increased measured lignin amounts [15–18]. In addition, another study indicates that the degraded hemicellulose/cellulose forms pseudo-lignin [19], which can affect lignin analysis.

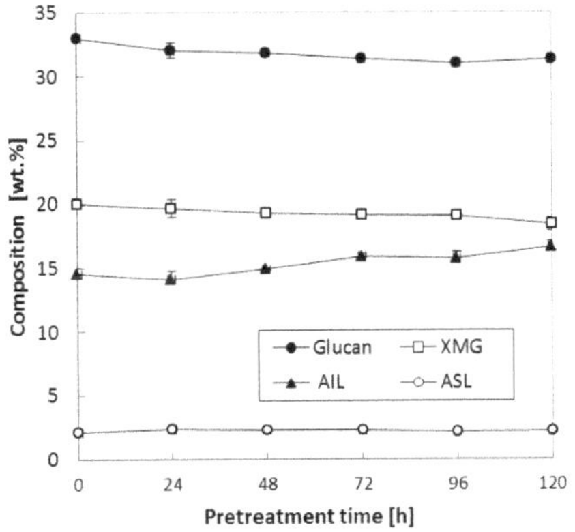

(a)

Figure 3. *Cont.*

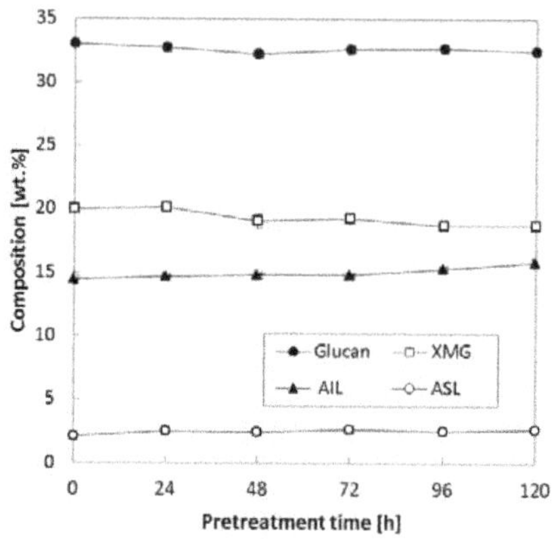

(b)

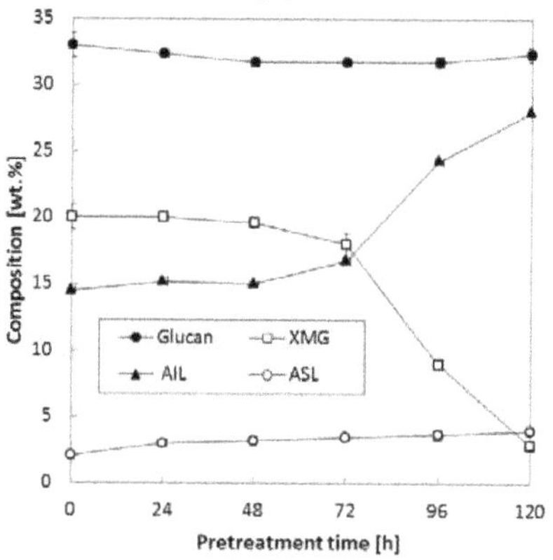

(c)

Figure 3. Effects of the S/L ratio on composition of corn stover. Pretreatment: 150 °C, 24–120 h, 0.08 (g NH_3)/(g biomass) (**a**) S/L ratio = 0.29, (**b**) S/L ratio = 0.47, (**c**) S/L ratio = 0.67. The data in the figure show mean values.

2.3. The Effect of NH_3 Loading on Enzymatic Digestibility of Pretreated Solids

In the above test (Section 2.2), high S/L ratio (0.67) resulted in significant decomposition of sugar, in particular, XMG during pretreatment, which was not desirable feature for an effective pretreatment for high sugar conversion yield [20,21]. To evaluate the effect of NH_3 loading on enzymatic digestibility, three different NH_3 loads [0.08, 0.16, or 0.24 (g NH_3)/(g biomass)] were applied while other conditions were kept constant (S/L = 0.47, reaction temperature 90 °C, and reaction time 24 h), and Table 2 summarizes the chemical composition data and enzymatic digestibility (at 72 h of the hydrolysis reaction) of the pretreated corn stover. An interesting trend was observed with increased NH_3 loading: glucan digestibility of the pretreated solid sample increased from 71.6 to 91.8% with NH_3 loading up to 0.16 (g NH_3)/(g biomass), then decreased to 84.7% at 0.24 (g NH_3)/(g biomass) loading. The XMG digestibility showed a similar trend: it increased from 66.7 to 72.6% when NH_3 loading was increased from 0.08 to 0.16 (g NH_3)/(g biomass) and decreased again above that NH_3 loading [66.5% at 0.24 (g NH_3)/(g biomass)]. Although it was unclear in the present step, it was assumed that a change in chemical composition may play a role in enzymatic saccharification.

Table 2. Effects of ammonia loading on composition and enzymatic digestibility.

NH_3 Loading	Composition			Enzymatic Digestibility (at 72 h)	
	Glucan	XMG	Lignin	Glucan	XMG
(g NH_3)/(g Biomass)	(wt %)	(wt %)	(wt %)	(%)	(%)
Untreated	33.0 ± 0.8	20.0 ± 0.4	16.6 ± 0.9	23.0 ± 2.1	11.3 ± 1.2
0.08	32.7 ± 0.4	20.1 ± 0.7	17.2 ± 0.9	71.6 ± 0.7	66.7 ± 0.8
0.16	32.3 ± 0.6	20.7 ± 1.0	17.2 ± 0.5	91.8 ± 0.5	72.6 ± 0.6
0.24	33.7 ± 0.5	19.7 ± 0.6	18.4 ± 0.6	84.7 ± 0.5	66.8 ± 1.0

Note: Pretreatment: 0.08–0.24 (g NH_3)/(g biomass), S/L = 0.47, 24 h, 90 °C; Enzymatic hydrolysis: 15 (FPU (filter paper unit) CTec2)/(g glucan) loading, 50 °C, 150 rpm, 72 h.

To further evaluate the effect of various S/L ratios on enzymatic saccharification, two different S/L ratios (0.29 and 0.47) were applied. Ammonia loading of 0.16 (g NH_3)/(g biomass) was used because it resulted in the highest digestibility (91.8% for glucan and 72.6% for xylan in Table 2). In this set of tests, three temperatures (90, 120 and 150 °C) with increased pretreatment time (~120 h) were applied to each S/L ratio (0.29 and 0.47). Figure 4 indicates that pretreatment at 150 °C for an extended treatment period (>72 h) resulted in lower glucan digestibility (71–85% at S/L = 0.29, 65–72% at S/L = 0.47) in comparison with the samples treated for 24–48 h (88–90% at S/L = 0.29, 82–84% at S/L = 0.47). It was assumed that higher lignin content (AIL) of pretreated corn stover at the high temperature (150 °C) contributed to the reduced enzymatic digestibility, in agreement with results from another study [22]. Owing to the improved enzymatic digestibility (Figure 4), 90 °C and 24 h were selected as the best pretreatment conditions for a further experiment (described in the following section); these conditions were assumed to be desirable because milder reaction conditions (90 °C and 24 h) are preferred for a reduction in the operating cost in a large-scale biomass conversion process.

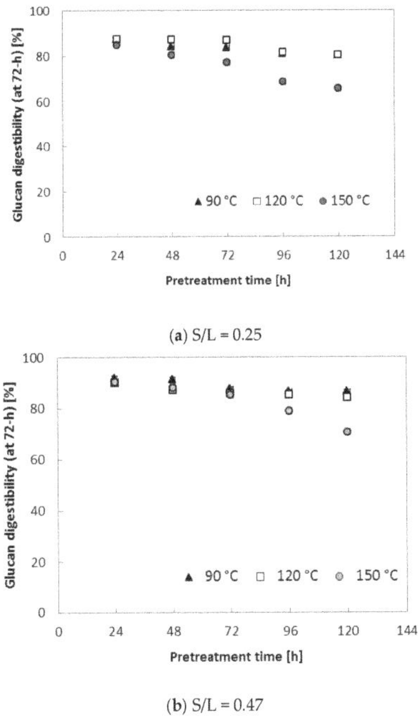

Figure 4. Glucan digestibility at elevated pretreatment temperature. Pretreatment: 0.16 (g NH_3)/(g biomass), 90–150 °C, 24–120 h, (**a**) S/L ratio = 0.29, (**b**) S/L ratio = 0.47. Enzymatic hydrolysis conditions: 15 (FPU CTec2)/(g glucan) loading, 50 °C, 150 rpm, hydrolysis time: 72 h. The data in the figure show mean values (standard deviation < 1.5).

2.4. Analysis of Variance (ANOVA)

To assess possible correlations of the effects between various reaction parameters and enzymatic digestibility, the single and combined effects of various factors on both glucan digestibility and XMG digestibility were evaluated by ANOVA, and the performance data are shown in Table 3. Among various reaction conditions, only the combined coefficient of "Temp × Time" had a p value less than 0.05 (p = 0.0233 for glucan digestibility and p = 0.0370 for XMG digestibility), implying that this coefficient significantly affects both glucan and XMG digestibility levels simultaneously, while other coefficients did not have a significant effect on enzymatic digestibility or influenced on either glucan or XMG digestibility. Therefore, the pretreatment temperature–time may be considered primary factors that can enhance the pretreatment effectiveness. In addition, the reaction temperature (Temp) seemed to have a significant effect on glucan digestibility (p = 0.0182) and showed a clear-cut tendency (close to significance) to affect XMG digestibility (p = 0.0511). On the other hand, the coefficient of time (reaction time), NH_3 (ammonia loading), and S/L and combined coefficient of "Time × S/L" and "S/L × NH_3" had lower influence on both glucan and XMG digestibility ($p > 0.05$). The combined coefficient of "Temp × S/L" had a p value less than 0.05, indicating that this coefficient significantly affects the glucan digestibility, whereas the combined coefficient of "Temp × NH_3" and "Time × NH_3" had a p value less than 0.05, suggesting that there is a significant effect on XMG digestibility.

Table 3. Effects of various parameters and their combinations on glucan and XMG digestibility (ANOVA).

Source	Enzymatic Digestibility [%]			
	Glucan		XMG	
	F Value	*p* Value	*F* Value	*p* Value
Temp	7.0205	0.0182	4.5518	0.0511
Time	1.1788	0.2947	0.1776	0.6799
NH_3	1.2794	0.2758	0.9883	0.3370
S/L	1.0458	0.3227	0.1709	0.6855
Temp × Time	6.4939	0.0233	5.3117	0.0370
Temp × S/L	22.7188	0.0002	0.1994	0.6620
Temp × NH_3	0.0025	0.9611	29.5164	<0.0001
Time × S/L	0.0466	0.8319	0.9617	0.3434
Time × NH_3	0.7513	0.3997	18.9960	0.0007
S/L × NH_3	3.1078	0.0983	1.03112	0.3271

Note: Pretreatment: 0.08, 0.16 and 0.24 (g NH_3)/(g biomass), S/L = 0.29, 0.47, and 0.67, 24–120 h, 90–150 °C. Enzymatic hydrolysis: 15 (FPU CTec2)/(g glucan) loading, 50 °C, 150 rpm, 72 h. The probability level of 0.05 (p = 0.05) was used to test the significance.

As discussed previously, the alkaline treatment such as the use of an ammonia solution can remove lignin and thereby increase the digestibility of biomass [18,23,24]. It was assumed that increasing the ammonia loading caused the breakdown of ester bonds in hemicellulose and lignin polymers at the elevated temperature; this situation consequently can improve the enzymatic hydrolysis of hemicellulose (XMG).

2.5. Residual Ammonia

Although ammonia can be evaporated and removed due to its high volatility, some of the impregnated ammonia cannot be easily removed and was assumed to affect the saccharification of fibers during enzymatic hydrolysis. The effect of residual ammonia content on enzymatic digestibility was evaluated, but it was assumed that residual ammonia content does not solely affect enzymatic digestibility because the level of residual ammonia content can be strongly influenced by other reaction parameters such as ammonia loading, pretreatment temperature, pretreatment time, the S/L ratio, and the combined effects of these parameters.

An evaluation assay of the effect of residual ammonia content on glucan digestibility was conducted for each reaction parameter. The effect of reaction severity on residual ammonia content was evaluated under various reaction conditions and the R^2 values as the predicted probability are summarized in Table 4. Because four different reaction parameters were compared, we categorized each different reaction condition into three different severity levels such as low, medium, and high severities. The higher severity means severe treatment conditions (see the note in Table 4). The R^2 values in Table 4 indicate that samples treated at S/L ratios corresponding to low and high severity resulted in a relatively strong correlation between residual ammonia content and glucan digestibility (R^2 = 0.3950 and 0.5607, respectively). In addition, samples treated with ammonia loading of medium severity showed R^2 = 0.4113, which indicated some correlation between residual ammonia content and glucan digestibility. Overall, the coefficients (R^2) of the trend lines were 0.027–0.5607; therefore, the model equations of the trend lines were not significant.

Table 4. Effects of residual ammonia content on glucan digestibility under various reaction conditions.

Reaction Conditions	R^2 Value for Glucan Digestibility		
	Low Severity	Medium Severity	High Severity
Time	0.0307	0.0298	0.0027
Temperature	0.0084	0.1352	0.0184
NH_3 loading	0.0037	0.4113	0.2717
S/L	0.3950	0.1374	0.5607

Note: Low severity: Time = 24 h, temp. = 90 °C, NH_3 loading = 0.08 (g NH_3)/(g biomass), S/L = 0.29. Medium severity: time = 48–96 h, temp. = 120 °C, NH_3 loading = 0.16 (g NH_3)/(g biomass); S/L = 0.47. High severity: time = 120 h, temp. = 150 °C, NH_3 loading = 0.24 (g NH_3)/(g biomass), S/L = 0.67.

Because it was found that the S/L ratio had the strongest effect (Figures 3 and 4), we next evaluated the effect of increasing S/L. Besides, residual ammonia was significantly affected ($p < 0.05$; data not shown) when S/L was changed. Figure 5 presents the relation of enzymatic digestibility (at 72 h of hydrolysis) of glucan with residual ammonia content at different S/L ratios. The R^2 values of three different S/L levels indicated that there was no clear trend between glucan digestibility and residual ammonia content even though the treated samples with high concentrations of residual ammonia seemed to have slightly lower digestibility than did the samples with low residual ammonia content. It should also be noted that the residual ammonia in the pretreated biomass can serve as an essential nitrogen source for microbial cell growth during fermentation if it is at an appropriate concentration.

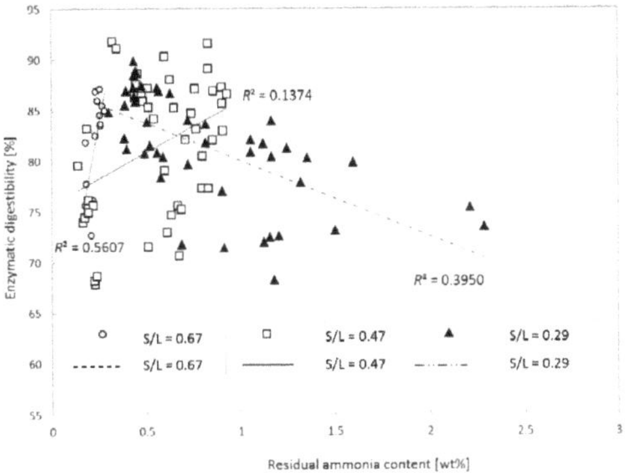

Figure 5. Effects of residual ammonia content on glucan digestibility of pretreated corn stover. Pretreatment: 0.08, 0.16 or 0.24 (g NH_3)/(g biomass), S/L ratio = 0.29, 0.47 or 0.67, 24–120 h, 90–150 °C. Enzymatic hydrolysis: 15 (FPU CTec2)/(g glucan) loading, 50 °C, 150 rpm, 72 h.

2.6. Mass Balance

Figure 6 summarizes the overall mass balance for the process of conversion of 100 g of corn stover to fermentable sugar by pretreatment under the best conditions [S/L = 0.47, 0.16 (g NH_3)/(g biomass), 90 °C, 24 h]. One of the features of LLAA method is that it does not solubilize any component during pretreatment and only modifies lignin and hemicellulose. Therefore, the input and output of the whole sugar conversion process are almost the same. The residual ammonia and ammonia recovery in this calculation were 1.7 wt % residual ammonia [0.16 (g NH_3)/(g biomass)] and 98.3% (15.7 g) after pretreatment and evaporation, respectively. Next, the pretreated solids were saccharified by means of 15 (FPU (filter paper unit) CTec2)/(g glucan) at 50 °C, 150 rpm, 72 h, and 1.0% (*w*/*v*) glucan loading. The highest glucan and XMG digestibility at an enzyme load of 15 FPU/(g glucan) was 91.8% and 72.6%, respectively. According to the mass balance in Figure 6, 33.7 g of glucose and 16.5 g of xmg were produced from 100 g of corn stover. The residue after enzymatic saccharification mostly consisted of lignin, ash, and unconverted polysaccharides.

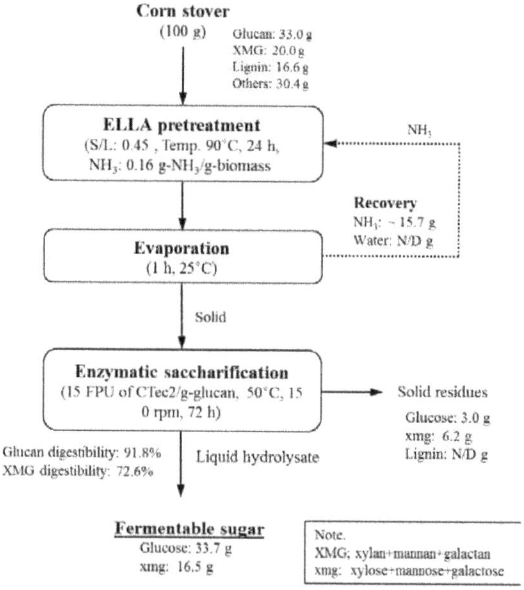

Figure 6. Schematic diagram and mass balance during conversion of corn stover to sugars.

2.7. Comparison of Various Ammonia Pretreatments

Alkaline pretreatment is considered an effective way to break down the structure of lignin and therefore to enhance the enzymatic hydrolysis of lignocellulosic biomass [25]. Table 5 shows a comparison of the features and reaction conditions of various alkaline pretreatment methods (in particular, methods involving ammonia). Pretreatment methods shown in Table 5 include low-liquid ammonia recycle percolation (LLARP), soaking in aqueous ammonia (SAA), LMAA, and LLAA [8,10,23,26,27].

Table 5. A comparison of various ammonia pretreatment methods.

		ARP/LLARP	SAA	LMAA	LLAA
Catalysts		Aqueous NH_3	Aqueous NH_3	Gaseous NH_3	Aqueous NH_3
Reaction type		Flow-through	Batch	Semi-batch	Batch
Chemical loading		0.5 (g NH_3)/(g biomass)	0.9 (g NH_3)/(g biomass)	0.1 (g NH_3)/(g biomass)	0.16 (g NH_3)/(g biomass)
Water consumption [1]		2.8 (g H_2O)/(g biomass)	5.1 (g H_2O)/(g biomass)	<1.0 (g H_2O)/(g biomass)	<1.0 (g H_2O)/(g biomass)
Temperature		170 °C	60 °C	90 °C	90 °C
Time		10 min	12 h	48 h	24 h
Pressure		2.5 MPa	-	-	-
Washing		Yes	Yes	No	No
Enzymatic digestibility [2]	Glucan	92.5% (ARP) [3] 90.1% (LLARP)	85.3%	84.1%	91.8%
	XMG	78.0% (LLARP)	75.3%	73.6%	72.6%
Reference		[26,27]	[23]	[8,10]	This study

Note: [1] Water consumption does not include water for washing after pretreatment. ARP: ammonia recycle percolation, LLARP: low-liquid ammonia recycle percolation, SAA: soaking in aqueous ammonia, LMAA: low-moisture anhydrous ammonia, LLAA: low-liquid aqueous ammonia; [2] enzyme loading; 15 FPU/g-glucan, enzymatic digestibility after 72 h of hydrolysis; [3] enzyme loading; 10 FPU/g-glucan.

Among the methods listed in Table 5, LMAA requires the least amount of chemical loading [0.1 (g NH_3)/(g biomass)], and LLAA is the next best method [0.16 (g NH_3)/(g biomass)] and shows the same water consumption [<1.0 (g H_2O)/(g biomass)]. The sugar production process using these two pretreatment methods can be considered more economical than those based on other pretreatment methods [0.5–0.9 (g catalyst)/(g biomass) and 2.8–10 (g H_2O)/(g biomass)]. Furthermore, in contrast to other methods (LLARP and SAA) in Table 5, the most desirable feature of LLAA and LMAA is that the washing step after pretreatment is not necessary; this feature can reduce the water consumption and thus reduce total energy cost in the biomass conversion process. In terms of severity of pretreatment conditions, LLAA, LMAA, and SAA processes involve mild reaction conditions. Although LLARP requires a short reaction period (~10 min), it should be carried out at high temperature (170 °C), while the other three ammonia pretreatment methods (LLAA, LMAA, and SAA) require more time (12–24 h) at a moderate temperature (60–90 °C). On the other hand, the longer pretreatment time and large water input in the SAA method are required even though it involves a mild reaction temperature; these characteristics are not considered desirable for an economically viable process [28].

3. Materials and Methods

3.1. Materials

3.1.1. Feedstock

Corn was grown and harvested in China in September 2015, and corn stover was then collected and provided by CJ Cheiljedang Co. (Seoul, Korea). The received corn stover was air-dried at ambient temperature (~25 °C), ground up, passed through a sieve with a mesh size of 10–35 mesh (US Standard, 0.5–2.0 mm of nominal sieve opening) sieves, and then stored in sealed plastic containers at ambient temperature. The initial composition of the biomass was determined by a standard LAP of the NREL (Table 1) [29]. It should be noted that glucan, xylan, and lignin are the main components among the various ones shown in Table 1; therefore, an evaluation of pretreatment effects was focused on those three components in this study.

Ammonium hydroxide (28.0–30.0%; lot number A29260I1) and sulfuric acid (ACS grade, 95–98%, lot number SZBF0140V) were purchased from Daejung Chemical & Metals Co., Ltd. (Shehung-si, Gyeonggi-do, Korea) and Sigma-Aldrich (St. Louis, MO, USA), respectively. Avicel® PH-101 (catalog number 900-3-6, lot number BCBJ029V, Sigma-Aldrich) was acquired and served as a control sample in the enzymatic-digestibility test.

3.1.2. Enzymes

Cellic® CTec2 (batch number: VCP10006, Novozymes Inc., Bagsvaerd, Denmark) was used for enzymatic saccharification of untreated and pretreated corn stover. The average activity of the enzyme, as determined by the LAP of the NREL was 88.91 FPU/mL [30].

3.2. *Pretreatment*

3.2.1. The First Step: Ammoniation

To apply ammonia loading at different target concentrations [0.08, 0.16 or 0.24 (g NH_3)/(g biomass)], an ammonium hydroxide (NH_4OH) solution at various solid/liquid (S/L) ratios (0.29, 0.47 or 0.67) was added in the form of mist using nozzle spray and tumbler mixer. The S/L ratio was calculated as follows:

$$S/L = \frac{\text{Total solids (g)}}{\text{Total solids (g)} + \text{water \& moisture (g)}}.$$

The initial moisture content of corn stover was approximately 8.5% and was loaded for ammoniation. After spraying of ammonium hydroxide mist, corn stover (100 g, dry basis) was homogenized at 30 rpm for 1 h in the tumbler mixer shown in Figure 1a.

3.2.2. The Second Step: Pretreatment

Ammoniated corn stover treated with aqueous ammonia (10 g, dry basis) was packed in a smaller sealed batch reactor (30.0-cm length, 2.54-cm internal diameter [ID], and 0.21-cm tube wall thickness [internal volume: 105.7 mL]; Figure 1b). Openings of the sealed batch reactor were tightened carefully enough to prevent ammonia leaking. The reactor was placed in the forced convection oven (model no. OF-22GW, Jeio Tech Co., Ltd., Daejeon, Korea) and then heated from ambient temperature to the target temperatures (90–150 °C) in 1 h and maintained at the desired temperature for 24–120 h.

3.2.3. The Third Step: Evaporation

After completion of the pretreatment process, the reactors were cooled down to ambient temperature. The reactors were then opened, and the treated sample was transferred into a tray. The collected sample was placed in the fume hood to remove excess ammonia by evaporation for 1 h at 25 °C. One portion of the sample was used for analysis of residual ammonia content, and the other portion was used for composition analysis.

3.3. *Analytical Methods*

Soxhlet extraction was applied to determine the water- and ethanol-soluble extractives of untreated corn stover. A two-step Soxhlet extraction was conducted; the first step of extraction with de-ionized (DI) water for 8 h was followed by the second step of extraction with ethanol (190 proof) for 24 h.

The chemical composition of untreated and pretreated corn stover was analyzed for carbohydrates, AIL, ASL (on a UV spectrophotometer at 320 nm), and ash (a gravimetric method involving a muffle furnace at 575 °C) following the NREL LAP [29]. Carbohydrate contents were determined by means of a high-performance liquid chromatography (HPLC) system (Shimadzu LC-10A, Shimadzu Inc., Kyoto, Japan) equipped with Bio-Rad Aminex HPX-87P (catalog number 1250098; Bio-Rad Inc., Hercules, CA, USA) and an 87H column (catalog number 1260140; Bio-Rad Inc., Hercules, CA, USA) and a refractive index detector (model RID-10A, Shimadzu Inc., Kyoto, Japan). Analytical conditions for HPLC were as follows: mobile phase of water (0.6 mL/min) at column temperature of 85 °C and 0.005 M H_2SO_4 (0.6 mL/min) at 65 °C for the HPX-87P column and HPX-87H column, respectively.

3.4. Enzymatic Digestibility

This property of pretreated and untreated corn stover was evaluated in duplicate in rubber-capped 250-mL Erlenmeyer flasks containing 100 mL of a liquid and 1.0 g of a glucan loading (3.0 g of pretreated solid loading, dry basis) according to the NREL-LAP [30]. The recovered solid samples obtained after the evaporation were used directly in the enzymatic digestibility tests without drying. Reaction conditions for the digestibility test were 50 °C, pH 4.8, and 150 rpm at 15 FPU/(g glucan) enzyme load in 0.05 M citrate buffer. Each sample in 100-mL working volume was saccharified in a shaking incubator (model number VS 8480SFN, Vision Scientific Co., Ltd., Daejeon, Korea). Total glucose content after 72 h of hydrolysis was used to calculate the enzymatic digestibility. Avicel® PH-101 was also put through the same digestibility test conditions and served as a control sample. The glucan and XMG digestibility values were calculated as follows:

$$\text{Glucan digestibility} = \frac{\text{Total released glucose (g)} \times 0.9}{\text{Initial glucan loading (g)}} \times 100,$$

where 0.9 is the factor for conversion of glucose to equivalents of glucan.

$$\text{XMG digestibility} = \frac{\text{Total released XMG (g)} \times 0.88}{\text{Initial XMG loading (g)}} \times 100,$$

where 0.88 is the factor for conversion of xylose to equivalents of XMG.

3.5. Residual Ammonia Analysis

One gram of untreated and pretreated samples was placed in a glass bottle with 80 mL of a 1.0% borate buffer solution. These glass bottles were placed in a convection oven at a stable temperature (80 °C) and incubated there for 24 h. After that, the glass bottles with residual ammonia in the liquid were removed from the oven. Liquid and solids were separated by filtration through filter paper (Fisher catalog number F2044-090, size: 90 mm Ø, pack: 100 units from CHmlab Group, Barcelona, Spain). Then, the filtrate was diluted to 100-mL working volume. The liquid, which contained ammonia, was reacted with a 10 N sodium hydroxide (NaOH) solution. Residual ammonia content in the liquid was determined by means of an ammonia analyzer (model Accumet®, XL250, Dual Channel pH/mV/Ion, Thermo Fisher Scientific Inc., Tampa, FL, USA) and an ion-selective electrode (ISE, Fisher catalog number 13-620-509).

3.6. ANOVA

The statistical analysis of the data was performed using SAS® software (version 9.4, SAS Institute Inc., Cary, NC, USA).

4. Conclusions

LLAA pretreatment can reduce energy use because it requires lesser inputs of ammonia and water as compared to other pretreatment technologies, and can enable economically viable processes. In addition, the LLAA pretreatment has advantages over previously developed ammonia pretreatment methods, e.g., it uses aqueous ammonia without washing. Therefore, this approach can be regarded as a more economically feasible technology for scaling up. Moreover, LLAA shows promise because of the effectiveness of this pretreatment at enhancing enzymatic digestibility of corn stover. The highest glucan and XMG digestibility levels were 91.8% and 72.6%, respectively, at 15 FPU/(g glucan) enzyme loading.

Author Contributions: N.P.V.T., the first author, performed all the experiments and analyzed the data. T.H.K., the corresponding author, designed the overall study and experiments, interpreted the results, and finalized the manuscript. All authors have read and approved the final manuscript.

Acknowledgments: This work was supported by the R&D program of Korea Institute of Energy Technology Evaluation and Planning (KETEP) grant funded by the Ministry of Trade, Industry & Energy (MOTIE), Republic of Korea (No. 20153010091990).

Conflicts of Interest: The authors declare no conflicts of interest.

References

1. Naik, S.N.; Goud, V.V.; Rout, P.K.; Dalai, A.K. Production of first and second generation biofuels: A comprehensive review. *Renew. Sustain. Energy Rev.* **2009**, *14*, 578–597. [CrossRef]
2. Badger, P.C. Ethanol from cellulose: A general review. In *Trends in New Crops and New Uses*; Jnick, J., Whipkey, A., Eds.; ASHS Press: Alexandria, VA, USA, 2002; pp. 17–21.
3. Tien, M.; Kirk, T.K. Lignin-degrading enzyme from *Phanerochaete chrysosporium*: Purification, characterization, and catalytic properties of a unique H_2O_2-requiring oxygenase. *Proc. Natl. Acad. Sci. USA* **1984**, *81*, 2280–2284. [CrossRef] [PubMed]
4. Chakar, F.S.; Ragauskas, A.J. Review of current and future softwood kraft lignin process chemistry. *Ind. Crops Prod.* **2004**, *20*, 131–141. [CrossRef]
5. Marshall, L.; Sugg, Z. *Corn Stover for Ethanol Production: Potential and Pitfalls. WRI Policy Note no. 4*; World Resources Institute: Washington DC, USA, 2009; pp. 1–10.
6. Charlier, L.K.; Mazeau, K. Molecular modeling of the structural and dynamical properties of secondary plant cell walls: Influence of lignin chemistry. *J. Phys. Chem.* **2012**, *116*, 4163–4174. [CrossRef] [PubMed]
7. Zhao, Y.; Wang, Y.; Zhu, J.Y.; Ragauskas, A.; Deng, Y. Enhanced enzymatic hydrolysis of spruce alkaline pretreatment at low temperature. *Biotechnol. Bioeng.* **2008**, *15*, 1320–1328. [CrossRef] [PubMed]
8. Cayetano, R.D.; Kim, T.H. Effect of low moisture anhydrous ammonia (LMAA) pretreatment at controlled ammoniation temperatures on enzymatic hydrolysis of corn stover. *Appl. Biochem. Biotechnol.* **2017**, *181*, 1257–1269. [CrossRef] [PubMed]
9. Yoo, C.G.; Lee, C.W.; Kim, T.H. Effect of low-moisture anhydrous ammonia (LMAA) pretreatment on biomass quality and enzymatic hydrolysis for long-term storage. *Appl. Biochem. Biotechnol.* **2014**, *174*, 2639–2651. [CrossRef] [PubMed]
10. Yoo, C.G.; Nghiem, N.P.; Hicks, K.B.; Kim, T.H. Pretreatment of corn stover using low-moisture anhydrous ammonia (LMAA) process. *Bioresour. Technol.* **2011**, *102*, 10028–10034. [CrossRef] [PubMed]
11. Li, J.; Henriksson, G.; Gellerstedt, G. Lignin depolymerization/repolymerization and its critical role for delignification of aspen wood by steam explosion. *Bioresour. Technol.* **2007**, *98*, 3061–3068. [CrossRef] [PubMed]
12. McMillan, J.D. Pretreatment of lignocellulosic biomass. In *Enzymatic Conversion of Biomass for Fuels Production*; Himmel, M.E., Baker, J.O., Overend, R.P., Eds.; ACS: Washington, DC, USA, 1993; pp. 292–323.
13. Moore, K.J.; Hatfield, R.D. Carbohydrates and forage quality. In *Forage Quality, Evaluation, and Utilization*; Fahey, G.C., Collins, M.C., Mertens, D.R., Moser, L.E., Eds.; ASA-CSSA-SSSA: Madison, WI, USA, 1994; pp. 229–280.
14. Satimanont, S.; Luengnaruemitchai, A.; Wongkasemjit, S. Effect of temperature and time on dilute acid pretreatment of corn cobs. *Int. J. Chem. Mol. Nucl. Mater. Metallurg. Eng.* **2012**, *6*, 216–320.
15. Young, R.A.; Akhtar, M. *Environmentally Friendly Technologies for the Pulp and Paper Industry*; Wiley: New York, NY, USA, 1997.
16. Garrote, G.; Dominguez, H.; Parajo, J.C. Hydrothermal processing of lignocellulosic materials. *Eur. J. Wood Wood Prod.* **1999**, *57*, 191–202. [CrossRef]
17. Tingyue, G. *Green Biomass Pretreatment for Biofuels Production*; Springer: Dordrecht, The Netherlands, 2013.
18. Kim, T.H.; Lee, Y.Y. Fractionation of corn stover by hot-water and aqueous ammonia treatment. *Bioresour. Technol.* **2006**, *97*, 224–232. [CrossRef] [PubMed]
19. Hu, F.; Jung, S.; Ragauskas, A. Pseudo-lignin formation and its impact on enzymatic hydrolysis. *Bioresour. Technol.* **2012**, *117*, 7–12. [CrossRef] [PubMed]

20. Drapcho, C.M.; Nghiem, N.P.; Walker, T.H. *Biofuels Engineering Process Technology*; McGraw Hill: New York, NY, USA, 2008; pp. 133–134.
21. Kim, T.H. Pretreatment of Lignocellulosic Biomass. In *Bioprocessing Technologies in Biorefinery for Sustainable Production of Fuels, Chemicals, and Polymers*; Yang, S.T., El-Enshasy, H.A., Thongchul, N., Eds.; John Wiley & Sons, Inc.: Hoboken, NJ, USA, 2013; pp. 91–110.
22. Jönsson, L.J.; Martin, C. Pretreatment of lignocellulose: Formation of inhibitory by-products and strategies for minimizing their effects. *Bioresour. Technol.* **2016**, *199*, 103–112. [CrossRef] [PubMed]
23. Kim, T.H.; Lee, Y.Y. Pretreatment of corn stover by soaking in aqueous ammonia. *Appl. Biochem. Biotechnol.* **2005**, *121*, 1119–1132. [CrossRef]
24. Li, X.; Kim, T.H. Low-liquid pretreatment of corn stover with aqueous ammonia. *Bioresour. Technol.* **2011**, *102*, 4779–4786. [CrossRef] [PubMed]
25. Kumar, R.; Wyman, C.E. Effect of cellulase and xylanase enzymes on the deconstruction of solids from pretreatment of popular by leading technologies. *Biotechnol. Prog.* **2009**, *25*, 302–314. [CrossRef] [PubMed]
26. Kim, T.H.; Kim, J.S.; Sunwoo, C.; Lee, Y.Y. Pretreatment of corn stover by aqueous ammonia. *Bioresour. Technol.* **2003**, *90*, 39–47. [CrossRef]
27. Kim, T.H.; Lee, Y.Y.; Sunwoo, C.; Kim, J.S. Pretreatment of corn stover by low-liquid ammonia recycle percolation process. *Appl. Biochem. Biotechnol.* **2006**, *133*, 41–57. [CrossRef]
28. Kumar, A.K.; Sharma, S. Recent updates on different methods of pretreatment of lignocellulosic feedstocks: A review. *Bioresour. Bioprocess.* **2017**, *4*, 7. [CrossRef] [PubMed]
29. Sluiter, A.; Hames, B.; Ruiz, R.; Scarlata, C.; Sluiter, J.; Templeton, D. *Determination of Structural Carbohydrates and Lignin in Biomass*; Report No. TP-510-42618; National Renewable Energy Laboratory: Golden, CO, USA, 2012.
30. Resch, M.G.; Baker, J.O.; Decker, S.R. *Low Solid Enzymatic Saccharification of Lignocellulosic Biomass*; Report No. TP-5100-63351; National Renewable Energy Laboratory: Golden, CO, USA, 2015.

Sample Availability: Samples of the compounds are not available from the authors.

molecules

MDPI

Article

Combination of Superheated Steam with Laccase Pretreatment Together with Size Reduction to Enhance Enzymatic Hydrolysis of Oil Palm Biomass

Nur Fatin Athirah Ahmad Rizal [1], Mohamad Faizal Ibrahim [1,2,*], Mohd Rafein Zakaria [1,2], Ezyana Kamal Bahrin [1,2], Suraini Abd-Aziz [1] and Mohd Ali Hassan [1]

1 Department of Bioprocess Technology, Faculty of Biotechnology and Biomolecular Sciences, Universiti Putra Malaysia, 43400 UPM Serdang, Malaysia; nurfatinrizal@gmail.com (N.F.A.A.R.); mohdrafein@upm.edu.my (M.R.M.Z.); ezyana@upm.edu.my (E.K.B.); suraini@upm.edu.my (S.A.-A.); alihas@upm.edu.my (M.A.H.)

2 Laboratory of Biopolymer and Derivatives, Institute of Tropical Forestry and Forest Products, Universiti Putra Malaysia, 43400 UPM Serdang, Malaysia

* Correspondence: faizal_ibrahim@upm.edu.my; Tel.: +603-8947-1936

Received: 22 February 2018; Accepted: 23 March 2018; Published: 2 April 2018

check for updates

Abstract: The combination of superheated steam (SHS) with ligninolytic enzyme laccase pretreatment together with size reduction was conducted in order to enhance the enzymatic hydrolysis of oil palm biomass into glucose. The oil palm empty fruit bunch (OPEFB) and oil palm mesocarp fiber (OPMF) were pretreated with SHS and ground using a hammer mill to sizes of 2, 1, 0.5 and 0.25 mm before pretreatment using laccase to remove lignin. This study showed that reduction of size from raw to 0.25 mm plays important role in lignin degradation by laccase that removed 38.7% and 39.6% of the lignin from OPEFB and OPMF, respectively. The subsequent saccharification process of these pretreated OPEFB and OPMF generates glucose yields of 71.5% and 63.0%, which represent a 4.6 and 4.8-fold increase, respectively, as compared to untreated samples. This study showed that the combination of SHS with laccase pretreatment together with size reduction could enhance the glucose yield.

Keywords: oil palm biomass; physical pretreatment; biological pretreatment; lignin removal; lignocellulosic biomass

---

1. Introduction

Lignocellulosic biomass, produced mainly from agricultural industry and forestry wastes, is the most abundant plant material on Earth. There has been increasing interest in utilizing this lignocellulosic biomass in recent years due to its potential to be used as fermentation substrate for various valuable products, including biofuels and bio-based chemicals [1]. In Malaysia, oil palm biomass is the most abundant plant material generated every year, since palm oil is the biggest Malaysian agricultural commodity. In 2016, this crop occupied a total of 5.74 million hectares of Malaysia's land with the production of 17.32 million tonnes of crude palm oil [2]. Processing of palm oil from fresh fruit bunch (FFB) at the mills generates 7.34 million tonnes of oil palm empty fruit bunch (OPEFB), 7.72 million tonnes of oil palm mesocarp fiber (OPMF), 4.46 million tonnes of oil palm kernel shell (OPKS) and 64 million tonnes of palm oil mill effluent (POME) per year [3]. The OPEFB and OPMF, which are the most abundant oil palm biomass forms generated at the mills, has not yet been fully utilized. It is currently either being used as mulching at plantations or dumped at the nearby factories for natural degradation. Recently, both materials have been commercialized for biocompost [4], biochar and activated carbon production [5]. These materials also have been tested for various fermentation processes including biobutanol [1], bioethanol [6], biohydrogen [7] and many more. However, the major concern while utilizing these biomasses as feedstock for fermentation is the effectiveness of the conversion into fermentable sugars.

OPEFB and OPMF are composed of 60–75% [8,9] and 50–55% [10,11] of cellulose and hemicellulose, respectively. These sugar polymers can be hydrolyzed into sugar monomers which subsequently can be used as substrates for fermentation. Like other lignocellulosic biomasses, OPEFB and OPMF are also composed of lignin that protects cellulose and hemicellulose and hinders enzymatic hydrolysis into sugars by cellulase. Generally, lignin is the most complex structure and represents about 10–25% of the biomass weight [8–11]. It is a long chain and heterogenous polymer, composed of mostly phenylpropane units, linked by ether bonds [12]. It has aromatic and rigid biopolymer properties linked via covalent bonds to xylans. The lignin structure inside the plant cell wall makes lignocellulosic biomass more rigid and highly compact.

In order to utilize oil palm biomass as a fermentation substrate, suitable and effective pretreatments are required to reduce the recalcitrance of lignocellulosic biomass by extensive modification of its lignocellulosic structure, especially of lignin [13]. The modification process can be carried out using physical, physico-chemical, chemical and/or biological pretreatments [12]. Although chemical pretreatment using either alkali or acid has been reported as the most effective pretreatment to generate high sugar yields in a short time [14], this approach may cause negative impacts on the environment, especially water pollution when it is released into the water stream. Therefore, in order to make sure that lignocellulosic biomass is reliable as a fermentation substrate, combinations of chemical-free pretreatments should be explored and proved as being effective, clean and feasible on an industrial scale.

Superheated steam (SHS) is a type of steam pretreatment that has been reported as a good pretreatment to breakdown and loosening the structural arrangement of lignocellulosic components in biomass [15–17]. SHS is a dry steam that is produced by adding heat to wet steam. The additional heat aids in raising the saturated steam temperature to exceed the boiling point of the liquid at a certain pressure [15]. The lignocellulosic material exposed to a high steam temperature of more than 180 °C can degrade the hemicellulose components since hemicellulose is less thermally stable than lignin and cellulose. Degradation of hemicellulose reduces the recalcitrance of the lignocellulosic material. It should also be noted that SHS is safe to be used since it can be conducted at atmospheric pressure with low energy consumption of 3.30 kW, and could cause very little environmental impact if collected condensate is reused [16]. However, pretreating the lignocellulosic biomass using SHS resulted in a low sugar yield after the saccharification process [17].

Combining biological pretreatment after SHS could improve the whole pretreatment process to produce sugars. Biological pretreatment of lignocellulosic biomass can be carried out by applying microorganisms (microbial pretreatment) or ligninolytic enzymes (enzymatic pretreatment) to digest lignin components. Enzymatic pretreatment is faster than microbial pretreatment, hence the process is also easier to control. In addition, it requires only mild conditions and the process specifically only attacks the lignin [18,19]. Laccase (EC 1.10.3.2; benzenediol:oxygen oxidoreductase) is an oxidizing enzyme that was extensively studied for lignocellulosic biomass pretreatment [14,16]. It is a multicopper oxidase produced by fungi, plants and bacteria to specifically degrade lignin components. The oxidation of a laccase substrate leads to the formation of free radicals and reduction of molecular oxygen into water molecules [18,20]. However, laccase pretreatment alone does not produce a high yield of hydrolyzed sugars [21,22]. Therefore, combining this enzymatic pretreatment using laccase with SHS could enhance the saccharification performance of oil palm biomass into sugars. In addition, the effect of size reduction prior to laccase pretreatment was also conducted since the enzyme action is highly affected by the exposed surface area of the substrate.

2. Results and Discussion

2.1. Chemical Composition Analysis

The chemical compositions of raw OPEFB and OPMF in a dry basis are shown in Table 1. All the chemical components of OPEFB analyzed in this study are comparable with the results previously reported by Zakaria et al. [8]. However, the hemicellulose component was slightly lower as compared to the report of Kong et al. [9]. The value of cellulose, hemicellulose and acid insoluble lignin of OPMF are comparable with Zakaria et al. [10], but lower as compared with Iberahim et al. [11]. The variations of the chemical compositions might be due to the different factors affecting the collected samples, such as plantation area, planting batch, maturity level and year [23]. Besides, the compositional methods that have been employed might also contribute to the variation of chemical composition obtained [10]. Therefore, it is very important to tabulate the chemical compositional analysis for every experiment that was conducted since the total carbohydrates and lignin contents make up a major portion in biomass, and these constituents must be determined as a part of a comprehensive biomass analysis. Comparing between OPEFB and OPMF, results of this study showed that the total carbohydrates in OPEFB were 59.4%, which is higher than OPMF (51.8%). Lignin composition of OPMF was 31.3%, which is higher than OPEFB (25.6%), contributing to a tougher structural arrangement than OPEFB.

Table 1. Compositional analysis of oil palm empty fruit bunch (OPEFB) and oil palm mesocarp fiber (OPMF) in comparison with previous studies.

Samples	Components (%)							References
	Cellulose	Hemicellulose	Acid Insoluble Lignin	Acid Soluble Lignin	Water Extractives	Solvent Extractives	Ash	
OPEFB	38.1 ± 0.9	21.3 ± 1.1	22.6 ± 0.1	3.0 ± 0.0	8.3 ± 1.0	1.1 ± 0.1 [a]	3.5 ± 0.5	This study
	40.4 ± 2.4	20.2 ± 2.3	23.1 ± 0.5	-	-	2.5 ± 1.9 [b]	5.9 ± 0.3	[8]
	38.3 ± 0.1	35.3 ± 0.1	22.1 ± 1.6	-	-	2.7 ± 1.3 [a]	1.0 ± 0.1	[9]
OPMF	27.8 ± 0.5	24.0 ± 0.4	27.9 ± 0.1	3.4 ± 0.2	4.9 ± 0.3	7.4 ± 0.6 [a]	3.3 ± 0.3	This study
	25.5 ± 1.7	25.7 ± 3.3	25.5 ± 0.5	-	-	11.4 ± 0.4 [b]	5.8 ± 0.2	[10]
	28.8 ± 0.5	25.3 ± 0.7	28.9 ± 2.1	-	-	6.3 ± 0.5 [a]	2.6 ± 0.3	[11]

'-' not determined; [a] ethanol extractives; [b] acetone extractives.

2.2. Effect of Pretreatments on Chemical Compositions

2.2.1. Superheated Steam Pretreatment

The superheated steam (SHS) pretreatment acts as an initial pretreatment to open up the structure of biomass. The recalcitrance of the lignocellulosic material becomes loosened as the hemicellulose is solubilized when OPEFB and OPMF are exposed to a temperature of 180 °C for 60 min and 190 °C for 60 min, respectively. Degradation of hemicellulose can be observed in both samples of OPEFB and OPMF, with an 18.7% and 21.3% reduction of hemicellulose percentage, respectively (Table 2). Hemicellulose has a side chain (branched) and backbone that are sensitive to thermal processes [24]. High temperature causes the degradation of acetyl groups in hemicellulose in the form of acetic acid, while pentose and hexose sugars degrade into furfural and 5-HMF. Therefore, hemicellulose has a higher degree of depolymerization than cellulose and lignin. High reduction of hemicellulose after pretreatment with SHS caused an increment of the lignin and cellulose compositional percentage, but the lignin that interact while the hemicellulose was loosened up and this makes the structural arrangement weaker. This observation was supported by the increased of glucose yield after the saccharification of the SHS-pretreated sample. A slight increment of glucose yield equivalent to 18.4% for OPEFB and 15.6% for OPMF might be a result of the formation of pseudo-lignin from carbohydrate (hemicellulose) degradation that migrated to the surface of biomass and is deposited as lignin droplets [25,26]. This formation also resulted in an increment of total lignin composition in both SHS-pretreated OPEFB and OPMF.

Table 2. Chemical composition of untreated and pretreated oil palm empty fruit bunch (OPEFB) and oil palm mesocarp fiber (OPMF).

Samples	Chemical Components (%)					
	Cellulose	Hemicellulose	Acid Insoluble Lignin	Acid Soluble Lignin	Total Lignin	Lignin Removal
Untreated OPEFB	38.1 ± 0.9	21.3 ± 1.1	22.6 ± 0.1	3.0 ± 0.0	25.6	-
Raw size, SHS 180 °C 60 min	43.2 ± 0.9	17.3 ± 0.4	28.0 ± 0.1	4.0 ± 0.3	32.0	-
Raw size, laccase 100 U/g	45.5 ± 0.2	20.5 ± 0.3	19.6 ± 0.5	3.2 ± 0.0	22.8	10.9
Raw size, SHS 180 °C 60 min, 100 U/g laccase	47.4 ± 1.3	16.0 ± 0.4	17.8 ± 0.8	3.3 ± 0.2	21.1	17.6
2 mm, SHS 180 °C 60 min, 100 U/g laccase	49.7 ± 0.9	16.2 ± 0.7	15.4 ± 0.6	3.2 ± 0.0	18.8	27.4
1 mm, SHS 180 °C 60 min, 100 U/g laccase	52.5 ± 0.8	15.7 ± 0.9	14.5 ± 0.8	3.2 ± 0.3	17.7	30.9
0.5 mm, SHS 180 °C 60 min, 100 U/g laccase	54.0 ± 0.4	14.9 ± 1.3	14.0 ± 0.7	3.0 ± 0.3	17.0	33.6
0.25 mm, SHS 180 °C 60 min, 100 U/g laccase	57.3 ± 1.3	14.2 ± 1.8	12.7 ± 0.6	3.0 ± 0.0	15.7	38.7
Untreated OPMF	27.8 ± 0.5	24.0 ± 0.4	27.9 ± 0.1	3.4 ± 0.2	31.3	-
Raw size, SHS 190 °C 60 min	33.2 ± 0.6	18.9 ± 1.8	35.9 ± 0.6	3.8 ± 0.0	39.7	-
Raw size, laccase, 400 U/g	36.8 ± 0.4	22.7 ± 0.5	25.6 ± 0.1	3.1 ± 0.1	28.7	8.3
Raw size, SHS 190 °C 60 min, 400 U/g laccase	39.3 ± 0.3	18.6 ± 0.1	23.1 ± 0.5	3.4 ± 0.0	26.5	15.3
2 mm, SHS 190 °C 60 min, 400 U/g laccase	43.0 ± 0.1	17.4 ± 0.4	21.5 ± 0.2	3.2 ± 0.5	24.7	21.1
1 mm, SHS 190 °C 60 min, 400 U/g laccase	46.9 ± 1.2	16.5 ± 0.6	19.4 ± 1.2	3.1 ± 0.0	22.5	28.1
0.5 mm, SHS 190 °C 60 min, 400 U/g laccase	48.5 ± 0.0	15.7 ± 1.1	16.9 ± 1.3	3.1 ± 0.0	20.0	36.1
0.25 mm, SHS 190 °C 60 min, 400 U/g laccase	49.3 ± 1.5	15.0 ± 1.2	15.8 ± 0.4	3.1 ± 0.0	18.9	39.6

2.2.2. Effect of Laccase Loadings on Lignin Removal

Several experiments were conducted to determine the most suitable laccase loading for lignin degradation in OPEFB and OPMF. Both samples were treated with laccase loadings ranging between 20–100 U/g-substrate. However, the results show that OPMF pretreated with these laccase loadings does not lose any lignin. This situation might be due to a higher lignin composition in OPMF, and because its structural arrangement is tougher and more rigid than that of OPEFB. Therefore, a higher range of laccase loading (100–800 U/g-substrate) was applied to delignify OPMF and the results are tabulated in Table 3. These results showed that laccase loading had a significant effect on lignin removal for both OPEFB and OPMF. The lignin removal for OPEFB was improved from 3.5% to 10.9% when the laccase loading increased from 20 to 100 U/g-substrate. There is no further lignin removal observed when a laccase loading of more than 100 U/g-substrate was added. Meanwhile, OPMF has maximum lignin removal of 8.3% at a laccase loading of 400 U/g-substrate. It can be observed that there was a gradual decrease in the lignin removal percentage with the increase of laccase concentration until it reached the enzyme saturation point. Delignification by laccase occurs when the substrates are oxidized with the reduction of oxygen to water, which generates free radical electrons [18]. This experiment showed that the lignin composition and structural arrangement of lignocellulosic biomass could affect the amount of laccase needed for the pretreatment.

Table 3. Lignin removal at different laccase loadings on oil palm empty fruit bunch (OPEFB) and oil palm mesocarp fiber (OPMF).

Samples	Components (%)			Lignin Removal (%)
	Insoluble Lignin	Soluble Lignin	Total Lignin	
Untreated OPEFB	22.6 ± 0.1	3.0 ± 0.0	25.6	-
20 U/g-substrate	21.5 ± 0.2	3.2 ± 0.0	24.7	3.5
40 U/g-substrate	21.0 ± 0.4	3.3 ± 0.4	24.3	5.0
60 U/g-substrate	20.4 ± 1.1	3.3 ± 0.1	23.7	7.4
80 U/g-substrate	20.1 ± 0.0	3.4 ± 0.0	23.5	8.2
100 U/g-substrate	19.6 ± 0.5	3.2 ± 0.2	22.8	10.9
200 U/g-substrate	19.6 ± 0.3	3.3 ± 0.0	22.9	10.5
Untreated OPMF	27.9 ± 0.1	3.4 ± 0.2	31.3	-
100 U/g-substrate	27.4 ± 0.4	3.3 ± 0.1	30.7	1.9
200 U/g-substrate	26.8 ± 1.1	3.1 ± 0.4	29.9	4.5
400 U/g-substrate	25.6 ± 0.1	3.1 ± 0.1	28.7	8.3
600 U/g-substrate	25.3 ± 0.1	3.3 ± 0.0	28.6	8.6
800 U/g-substrate	25.5 ± 0.3	3.2 ± 0.2	28.7	8.3

In comparison with other studies, this experiment showed an improved delignification of lignocellulosic biomass as shown in Table 4. The previous study by Zanirun et al. [22] reported that OPEFB pretreatment with 50 U/g-substrate of laccase by *Pycnoporus sanguineus* UPM4 removed 3.1% of the lignin, which is lower than the lignin removal presented in this study. A sufficient amount of laccase loading is important to improve the removal of lignin components in lignocellulosic biomass. Compared with other types of biomass, wheat straw fiber pretreatment using 65 U/g-substrate of laccase loading from *Pycnoporus cinnabarinus* removed only 5.0% of the lignin [27]. A small reduction of lignin content (1.3%) has also been observed when furfural residues were pretreated using 100 U/g-substrate of laccase loading from *T. versicolor* [28]. Pretreatment of an *Eucalyptus globules* kraft pulp with 17.5 U/g-substrate of laccase from *T. vilosa* resulted in 23% lignin removal [29].

Table 4. Biological pretreatment using laccase on various lignocellulosic biomasses.

Substrates	Laccase Treatment	Laccase Loading (U/g-Substrate)	Lignin Removal (%)	References
Oil palm empty fruit bunch (OPEFB)	*P. sanguineus* UPM4	50	3.06	[22]
Wheat straw fiber	*P. cinnabarinus*	65	5.0	[27]
Furfural residue	*T. versicolor*	100	1.3	[28]
Eucalyptus globules kraft pulp	*T. vilosa*	17.5	23	[29]
Oil palm empty fruit bunch (OPEFB)	*T. versicolor*	100	10.9	This study
Oil palm mesocarp fiber (OPMF)	*T. versicolor*	400	8.3	This study

In addition, a higher percentage of lignin removal was observed when the OPEFB and OPMF samples were ground to a size of 0.25 mm. It should be noted that there are limited reports on delignification of OPMF through biological pretreatment by either microbial or enzymatic pretreatment. Beside the lignin removal, the polysaccharide compositions were also evaluated as shown in Table 2.

The cellulose composition of OPEFB and OPMF increased to 45.5% and 36.8%, respectively, as compared to the untreated biomass. On the other hand, the percentage of hemicellulose was reduced by 3.8% for OPEFB, and 5.4% for OPMF. The recalcitrance of the biomass was further reduced with the greater losses of lignin.

2.2.3. Total Phenolic Compounds after Laccase Pretreatment

To evaluate the effect of different laccase loadings on the removal of phenolic compounds from lignin components, the concentration of total phenolic compounds was measured and the degradation products were recovered in the liquid fraction after the pretreatment. Phenols are released due to partial solubilization and degradation of the lignin during the pretreatment [18,19]. The concentration of total phenolic content obtained after laccase pretreatment of OPEFB and OPMF with different laccase loading is shown in Figure 1.

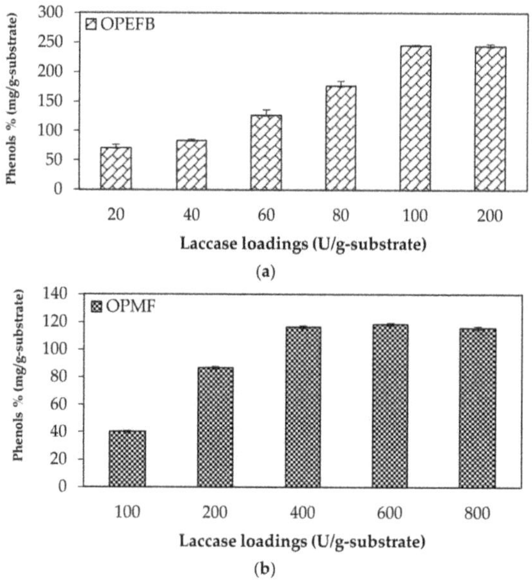

Figure 1. Total phenolic compounds after laccase pretreatment on (**a**) oil palm empty fruit bunch (OPEFB) and (**b**) oil palm mesocarp fiber (OPMF).

Phenols have inhibitory effects on saccharification and fermentation processes by inhibiting the activities of cellulolytic enzymes and microbes, thus, decreasing yields and lowering the productivity. Phenols also can alter the growth of fermenting microorganisms [19]. Evaluation of the laccase loading for OPEFB showed that the concentration of total phenolic content increased gradually until the concentration become constant at a laccase loading of more than 100 U/g-substrate, with a phenol concentration of 245.17 mg/g. Meanwhile, in OPMF, the highest phenolic compound content was 116.46 mg/g after pretreatment using 400 U/g-substrate and become constant at higher laccase loadings. These results were compatible with the total lignin removal obtained as presented in Table 3, where the lignin degradation reflects the total quantified phenolic compounds. Total phenolic compound is an indicator to verify the concentration of phenolic compounds present, in which it is also related to the structure, reactivity and mechanism of lignin degradation [18,20]. Laccase catalyzes the oxidation of phenols and form unstable phenoxy radicals. These radicals can interact with each other and contribute to destroying aromatic compounds [20].

2.2.4. Combination of Pretreatments with Size Reduction

OPEFB and OPMF pretreatment using SHS + laccase has been conducted to examine the suitability of this combination. Results showed that OPEFB (raw size) pretreated using SHS + laccase had an increased cellulose percentage from 38.1% (untreated) to 47.4%, with lignin removal of 17.6%. It can be clearly observed that the cellulose composition was increased when the OPEFB was pretreated using SHS only, followed by laccase only, and SHS + laccase as shown in Table 2. Similar situations can be observed for OPMF that followed the same trend as OPEFB. However, a lower lignin removal percentage was observed might be due to a tougher structural arrangement than in case of OPEFB.

In order to improve the enzymatic pretreatment by laccase, the SHS pretreated OPEFB and OPMF were ground to 2, 1, 0.5 and 0.25 mm using a hammer mill and delignified by laccase at 100 U/g-substrate for OPEFB, and 400 U/g-substrate for OPMF. Reduction of size from raw to 0.25 mm had significantly increased the lignin removal of OPEFB up to 38.7% and increased the cellulose composition to 57.3%. A similar trend was observed for OPMF, where the cellulose composition increased from 39.3% (raw size) to 49.3% (0.25 mm) with 39.6% of lignin removal. Both OPEFB and OPMF showed greater lignin removal when the substrate size was reduced from raw to 0.25 mm. Small particle size increases the total surface area, homogeneity and heat transfer efficiency [30–32]. Therefore, enzymatic digestibility by laccase has been improved by increasing the surface area of the substrate.

Although the enzyme action could be enhanced by reducing the substrates' size to less than 0.25 mm, the milling process using a hammer mill consumes more energy to generate smaller particle sizes. According to Ndukwu et al. [32], the specific energy requirement ($kWht^{-1}$) to grind palm kernel using a hammer mill to a size of 5–0.8 mm consumes 0.2–2.3 kWh of energy. However, the hammer mill has been reported as a convenient and probably the most commonly used method in order to obtain a suitable substrate size for subsequent processing [31,32]. In addition, it should be noted that particle size of less than 0.25 mm is not suitable for the pretreatment process because it may result in a low bias for carbohydrate and high bias for lignin content due to excessive carbohydrate degradation [33].

2.3. *Structural Analysis Using SEM*

Scanning electron microscope (SEM) images were taken to investigate the morphological changes of OPEFB and OPMF after pretreatment, as shown in Figure 2. SEM images showed similar fiber-like structures containing silica bodies for both OPEFB and OPMF. The untreated OPEFB and OPMF had a rough surface on the whole area of the fiber (Figure 2a,e). Therefore, the structure of untreated OPEFB and OPMF displayed a rigid and highly ordered fibrils arrangement. A great amount of silica bodies that attached to circular craters over the strand surface of the fibers could also be observed in both the OPEFB and OPMF SEM images.

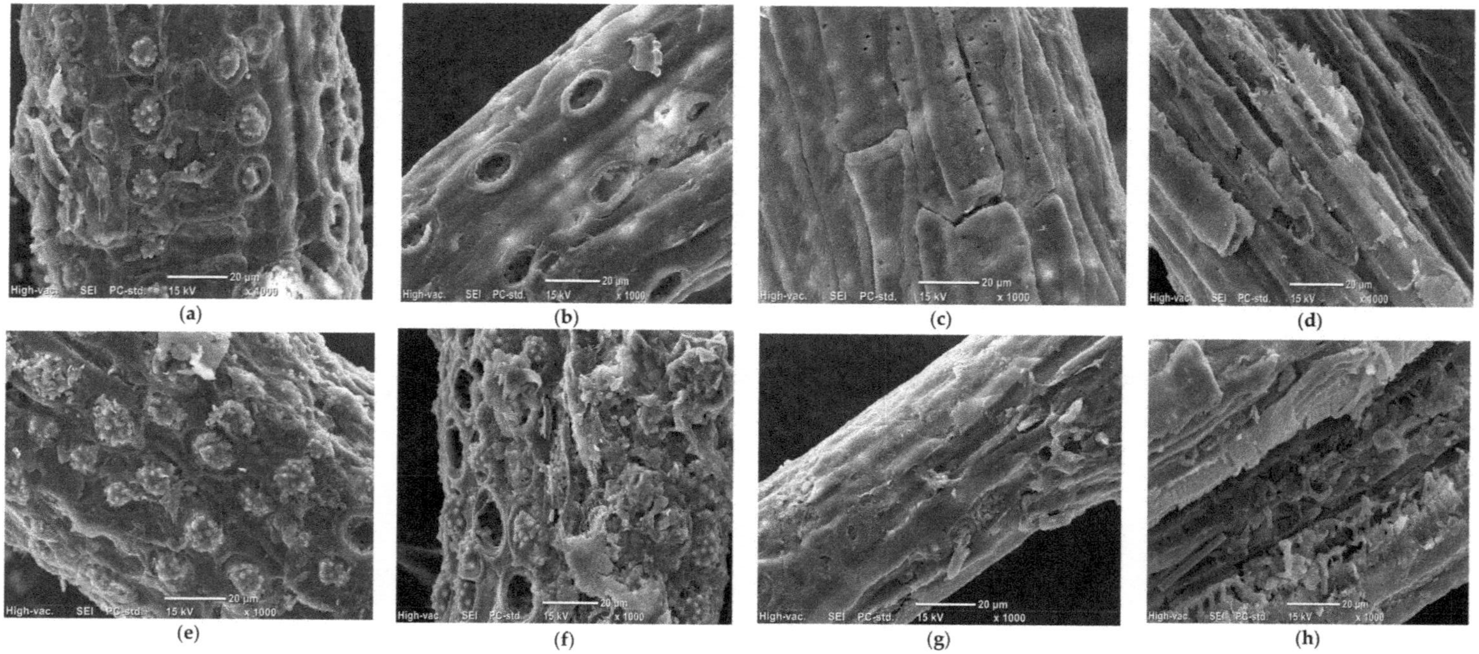

Figure 2. Scanning electron micrographs of oil palm empty fruit bunch (OPEFB) (**a**) untreated (**b**) SHS (180 °C, 60 min) (**c**) laccase (100 U/g-substrate) (**d**) SHS + laccase at 0.25 mm size and oil palm mesocarp fiber (OPMF) (**e**) untreated (**f**) SHS (190 °C, 60 min) (**g**) laccase (400 U/g-substrate) and (**h**) SHS + laccase at 0.25 mm size.

Based on the SEM images, the SHS pretreatment was able to remove the silica bodies from the structure and empty craters can be clearly observed in both substrates (Figure 2b,f). This observation indicates that sufficient energy from SHS was able to remove the silica bodies, hence the steam appeared can interrupt lignocellulosic materials beside its ability to loosen up the recalcitrance of the structure. The SEM images also showed that the surface of laccase-pretreated OPEFB and OPMF appeared to be more uniform and smooth, and all the silica bodies were removed. There were some cracks and formation of micropores that can be observed on the strands of the OPEFB and OPMF structure (Figure 2c,g). The microscopic alterations in the fiber have been generally considered as a result of lignin removal [34]. In Figure 2d,h, SHS and laccase-pretreated OPEFB and OPMF with particle size of 0.25 mm showed that the structural arrangements of the substrates have been altered and the outer layer of the fiber was 'peeled off'. The degree of defibrillation and particle size reduction played a very important role to enhance the conversion of lignocellulosic material into hydrolyzed sugars.

2.4. Saccharification of Pretreated OPEFB and OPMF

An efficient saccharification process is highly dependent on an effective pretreatment being applied to lignocellulosic biomass [12]. To investigate the efficiency of various pretreatments, the pretreated OPEFB and OPMF were subsequently submitted to a saccharification process using cellulase (Celluclast 1.5 L). This experiment showed that the glucose yield obtained for untreated OPEFB was only 15.5%, and 13.1% for untreated OPMF. The glucose yield was increased to 18.4% for OPEFB, and 15.6% for OPMF when treated with SHS alone. Meanwhile, substrates pretreated by laccase only generated up to 29.5% of glucose yield for OPEFB, and 27.5% of glucose yield for OPMF, which was higher than the SHS pretreatment. Higher degradation of lignin after laccase pretreatment than SHS pretreatment contributed to a higher digestibility of the cellulose structure into glucose. Several studies reported that lignin removal enhanced enzyme digestibility in the saccharification of lignocellulosic materials [8,18,27]. Besides, there was a little increment of lignin percentage observed after SHS pretreatment, which was due to attribution of pseudo-lignin that still adhered to the surface of the substrates, which constrains the saccharification process [25,26]. In this study, the glucose yield was further improved by reducing the substrate particle size from raw to 0.25 mm. The saccharification of OPEFB and OPMF pretreated with SHS + laccase at 0.25 mm size reduction as shown in Figure 3 resulted in the highest glucose yield of 71.5% and 63% for OPEFB and OPMF, respectively.

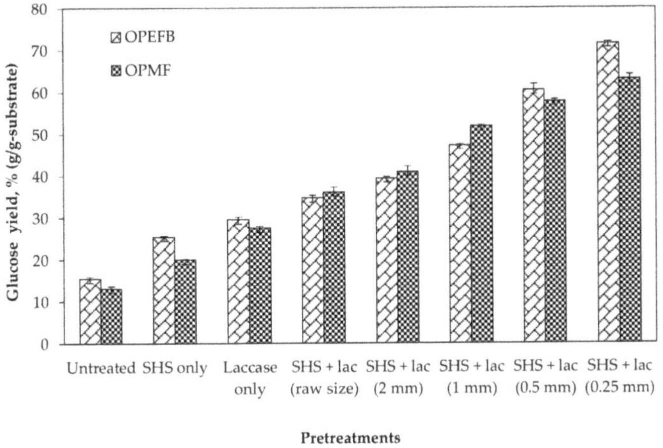

Figure 3. Hydrolysis performance of untreated and pretreated oil palm empty fruit bunch (OPEFB) and oil palm mesocarp fiber (OPMF). 'SHS'—superheated steam, 'lac'—laccase.

These values are equivalent to a 4.6-fold increment of the glucose yield for OPEFB, and a 4.8-fold increment for OPMF as compared with the untreated substrates. The efficiency of enzymatic hydrolysis was greatly improved due to the structural modification and lignin degradation of OPEFB and OPMF by combining SHS + laccase pretreatments, which made the cellulose more accessible to the cellulase. On top of that, reduction in particle size increased the surface area, and provided more accessible lignin components to be further degraded by laccase, and exposed more cellulose for enzymatic hydrolysis by cellulase. Based on these results, a combination of SHS + laccase pretreatment with size reduction to 0.25 mm enhanced the saccharification and increased the glucose recovery yield of both OPEFB and OPMF.

After 48 h of saccharification, the liquid fraction from OPEFB and OPMF were taken to quantify the inhibitory compounds using HPLC. The presence of inhibitory compounds in sugars could negatively affect the subsequent fermentation process. The main inhibitory components include furan derivatives, aliphatic acids, phenolic and other aromatic compounds [18,20]. Furfural and 5-HMF are generated from the furan derivatives derived from cellulose and hemicellulose and can be further degraded to form levulinic acid and formic acid. Hydrolysis of the acetyl groups in the hemicellulose generates acetic acid. Meanwhile, phenolic compounds like 4-hydroxybenzoic acid, 4-hydroxybenzaldehyde, vanillic acid, syringic acid, vanillin, syringaldehyde, *p*-coumaric acid, ferulic acid and coniferyl aldehyde are derived from the degradation of lignin [18,20]. However, these inhibitory compounds were not detected in the sugar produced in this study. This is because after the combination pretreatment of SHS + laccase, the whole slurry was filtered and washed. The purpose of washing the substrates after the pretreatment is to prevent the inhibitory compounds from affecting the saccharification.

3. Materials and Methods

3.1. *Raw Materials*

Pressed and shredded OPEFB and OPMF were obtained from Seri Ulu Langat Palm Oil Mill, Dengkil (Selangor, Malaysia). The OPMF was manually separated from the crushed kernels and shells to prevent errors in the experiments. The pressed and shredded OPEFB (10–50 mm) and OPMF (10–30 mm) were sun dried and stored in sealed plastic bags at room temperature prior to further use.

3.2. *Characterization*

3.2.1. Determination of Extractives

The determination of extractives in OPEFB and OPMF were carried out according to the NREL laboratory analytical procedure [35]. The analysis was carried out using a two-step Soxhlet extraction. The cellulose thimble was weighed, and the samples were added to a cellulose thimble. The cellulose thimble was inserted into the Soxhlet tube and the round bottom flask containing 200 mL of deionized water. First, hot water extraction was carried out for 8 h to remove water-soluble compounds and nitrogenous material. After this process completed, the thimble was carefully removed and dried in an oven at 60 °C for 24 h and the thimble weight was measured. Second, the process was continued using another Soxhlet extraction with 200 mL of 95% ethanol for 8 h and the thimble was carefully removed and dried in an oven at 60 °C for 24 h. Triplicate samples were used, and average values were calculated.

3.2.2. Determination of Lignocellulosic Compositions

The composition of cellulose, hemicellulose, acid insoluble lignin and acid soluble lignin in OPEFB and OPMF were determined according to NREL laboratory analytical procedures as described by Sluiter et al. [36]. Approximately 0.3 g of dried OPEFB and OPMF was placed into a glass vial and hydrolyzed in 72% (*w/w*) of H_2SO_4 at 30 °C for 60 min, and the slurry was further hydrolyzed in diluted

4% (*w/w*) of H_2SO_4 followed by autoclaved at 121 °C for 60 min. The samples were vacuum filtered, and the liquid sugars were filtered again using 0.22 μm of nylon membrane filter and analyzed using an HPLC instrument equipped with a refractive index detector (Shimadzu, Kyoto, Japan). The residue left on the filter paper was dried overnight in an oven at 105 °C. Final weight of the residual after acid hydrolysis was measured as acid insoluble lignin while its filtrate was measured as acid soluble lignin. Acid soluble lignin was determined using a UV-Vis spectrophotometer (Shimadzu) at the wavelength of 205 nm. Triplicate samples were used, and average values were calculated.

3.2.3. Determination of Ash Content

The ash content in OPEFB and OPMF was determined based on the NREL laboratory analytical procedures reported by Sluiter et al. [37]. Triplicate samples of OPEFB and OPMF in porcelain crucibles were placed in a muffle furnace and heated at 575 °C for 4 h. After the heating process, the crucibles were removed from the furnace and cooled down to room temperature in a desiccator before weighing the crucibles and ash. The procedures were repeated until a constant weight of samples were obtained.

3.3. *Superheated Steam Pretreatment*

The OPEFB and OPMF were pretreated using a lab scale SHS oven (DC Quto, QF-5200C, Naomoto Corporation, Osaka, Japan) with treatment chamber dimension of 300 mm × 265 mm × 100 mm. The SHS oven consists of a stainless-steel heating chamber and a boiler. The heater power of the SHS oven and the steam flow rate were conducted at 6.6 kW and 4.95 kg/h, respectively. The selection of pretreatment condition using SHS was based on the best pretreatment condition reported. For OPEFB, the pretreatment was conducted at 180 °C for 60 min [15], and 190 °C for 60 min for OPMF [17]. The SHS pretreated samples were ground using a hammer mill (Hsiangtai CW-1, Taipei, Taiwan) to 2, 1, 0.5 and 0.25 mm for subsequent use.

3.4. *Laccase Pretreatment*

The OPEFB and OPMF were pretreated using enzyme laccase produced by *T. versicolor* (Sigma-Aldrich, St. Louis, MO, USA) with enzyme loading of 20–200 U/g-substrate for OPEFB and 100–800 U/g-substrate for OPMF. The pretreatment was carried out using 2.5% of substrate concentration mixed with 0.05 M of sodium acetate buffer (pH 4.8) and incubated in a rotary incubator operated at 150 rpm, 50 °C for 24 h [28]. All experiments were performed in triplicates. After incubation, the pretreated sample mixtures were filtered using filter papers (No. 1, Whatman, Maidstone, UK) and then washed with deionized water until a neutral pH was obtained. Then, the pretreated samples were oven dried at 60 °C for 24 h.

3.5. *Saccharification*

The saccharification was conducted using commercial cellulase (Celluclast 1.5 L) purchased from Novozymes (Bassvaerd, Denmark). The experiment was performed by adding 5% of substrate concentration in 0.05 M of sodium acetate buffer (pH 4.8) with 30 FPU/g-substrate of cellulase activity. The mixtures were incubated at 50 °C in a rotary incubator shaker at 200 rpm for 48 h [15]. Samples were taken from the mixture and centrifuged for 10 min at 10,000 rpm for sugar determination. All experiments were performed in triplicates and results were presented as an average value.

3.6. *Analytical Procedures*

Monomeric sugars from saccharification were analyzed using a HPLC equipped with a refractive index detector (RID-10A, Shimadzu) and a Rezex RCM-monosaccharide column (Phenomenex, Torrance, CA, USA) equipped with a Carbo-Pb micro-guard cartridge. The column oven was set at 80 °C and samples were eluted at 0.60 mL/min using deionized water as a mobile phase [38]. The enzymatic digestibility was represented by the sugar yield (%) calculated as of the formula below:

$$\text{Sugar yield (\%)} = \frac{\text{Weight of monomeric sugars after enzymatic hydrolysis (mg)}}{\text{Weight of potential total monomeric sugars after sulfuric acid hydrolysis of oil palm biomass (mg)}} \times 100 \quad (1)$$

Laccase activity was measured using 2,2′azinobis-(3-ethylbenzenthiazoline-6-sulfonic acid) (ABTS) as a substrate by a UV-Vis spectrophotometer (Shimadzu) at 420 nm with molar extinction coefficient, ε = 36,000 $M^{-1}\cdot cm^{-1}$ based on Bourbonnais et al. [39]. The kinetic of the graph slope values were calculated to determine the activities of enzyme in Unit per millilitre (U/mL). Total phenolic content of the liquid fraction was quantified according to the Folin–Ciocalteau method described by Makkar et al. [40]. The sample (0.5 mL) was put into a test tube with 0.25 mL of Folin-Ciocalteu reagent and 1.25 mL of sodium carbonate solution. All the samples were vortexed and the absorbance was measured after 40 min at wavelength of 725 nm. Furfural, 5-hydroxymethylfurfural (5-HMF), acetic acid and formic acid were analyzed using a HPLC equipped with a refractive index detector (RID-10A, Shimadzu) and a BioRad Aminex HPX-87H column (Bio-Rad, Hercules, CA, USA) with a Carbo-H micro-guard cartridge. The column oven was set at 65 °C and samples were eluted at 0.60 mL/min using 0.005 M of H_2SO_4 as a mobile phase [41]. The surfaces morphological images of raw and pretreated fibers were examined using a scanning electron microscopy (SEM) with JCM-6000 PLUS Neo Scope Bench top SEM (Jeol, Tokyo, Japan). The fiber was coated with Pt for 30 s using an ion sputtering system (Hitachi, Tokyo, Japan) prior to images observation. The instrument was operated at a beam voltage of 15 kV and with 1000× magnification.

4. Conclusions

The combination of chemical-free pretreatments using SHS followed by laccase was successfully performed. Reduction of substrate size from raw to 0.25 mm improved lignin removal of OPEFB and OPMF by 38.7% and 39.6%, respectively. This pretreatment increased the glucose yield by 71.5% and 63.0%, respectively, as compared to the untreated substrates. This present study revealed the suitability of combining SHS with laccase pretreatment together with the positive effect of particle size reduction of OPEFB and OPMF.

Acknowledgments: This work was partly supported by the Science and Technology Research Partnership for Sustainable Development (SATREPS), organized by Japan Science and Technology Agency (JST) and Japan International Cooperation Agency (JICA). We are grateful to Seri Ulu Langat Palm Oil Mill, Malaysia for providing us the raw materials. The authors gratefully acknowledge the financial support from Ministry of Higher Education (MOHE), Malaysia and Universiti Putra Malaysia (UPM). The authors would also like to thank Shazrin Ahmad Kamaruzzaman for her help in proof reading this manuscript.

Author Contributions: N.F.A.A.R., M.F.I., M.R.M.Z., E.K.B., S.A.-A. and M.A.H. conceived and designed the experiments; N.F.A.A.R. performed the experiments and analyzed the data; All the authors had contributed in writing the article.

Conflicts of Interest: The authors declare no conflict of interest.

References

1. Ibrahim, M.F.; Ramli, N.; Bahrin, E.K.; Abd-Aziz, S. Cellulosic biobutanol by Clostridia: Challenges and improvements. *Renew. Sustain. Energy Rev.* **2017**, *79*, 1241–1254. [CrossRef]
2. Malaysian Palm Oil Board, Oil Palm Estates, Economic & Industry Development Division. Oil Palm Planted Area. January—December 2016. Available online: http://bepi.mpob.gov.my/index.php/statistics/area.html (accessed on 7 October 2017).
3. Loh, S.K. The potential of the Malaysian oil palm biomass as a renewable energy source. *Energy Convers. Manag.* **2017**, *141*, 285–298. [CrossRef]
4. Siddiquee, S.; Shafawati, S.N.; Naher, L. Effective composting of empty fruit bunches using potential *Trichoderma* strains. *Biotechnol. Rep.* **2017**, *13*, 1–7. [CrossRef] [PubMed]

5. Zainal, N.H.; Aziz, A.A.; Idris, J.; Mamat, R.; Hassan, M.A.; Bahrin, E.K.; Abd-Aziz, S. Microwave-assisted pre-carbonisation of palm kernel shell produced charcoal with high heating value and low gaseous emission. *J. Clean. Prod.* **2017**, *142*, 2945–2949. [CrossRef]
6. Limayem, A.; Ricke, S.C. Lignocellulosic biomass for bioethanol production: Current perspectives, potential issues and future prospects. *Prog. Energy Combust. Sci.* **2012**, *38*, 449–467. [CrossRef]
7. Taifor, A.F.; Zakaria, M.R.; Mohd Yusoff, M.Z.; Toshinari, M.; Hassan, M.A.; Shirai, Y. Elucidating substrate utilization in biohydrogen production from palm oil mill effluent by *Escherichia coli*. *Int. J. Hydrog. Energy* **2017**, *42*, 5812–5819. [CrossRef]
8. Zakaria, M.R.; Fujimoto, S.; Hirata, S.; Hassan, M.A. Ball milling pretreatment of oil palm biomass for enhancing enzymatic hydrolysis. *Appl. Biochem. Biotechnol.* **2014**, *173*, 1778–1789. [CrossRef] [PubMed]
9. Kong, S.H.; Loh, S.K.; Bachmann, R.T.; Rahim, S.A.; Salimon, J. Biochar from oil palm biomass: A review of its potential and challenges. *Renew. Sustain. Energy Rev.* **2014**, *39*, 729–739. [CrossRef]
10. Zakaria, M.R.; Hirata, S.; Hassan, M.A. Combined pretreatment using alkaline hydrothermal and ball milling to enhance enzymatic hydrolysis of oil palm mesocarp fiber. *Bioresour. Technol.* **2014**, *169*, 236–243. [CrossRef] [PubMed]
11. Iberahim, N.I.; Jahim, J.M.; Harun, S.; Nor, M.T.M.; Hassan, O. Sodium hydroxide pretreatment and enzymatic hydrolysis of oil palm mesocarp fiber. *Int. J. Chem. Eng. Appl.* **2013**, *4*, 101–105. [CrossRef]
12. Sun, S.; Sun, S.; Cao, X.; Sun, R. The role of pretreatment in improving the enzymatic hydrolysis of lignocellulosic materials. *Bioresour. Technol.* **2016**, *199*, 49–58. [CrossRef] [PubMed]
13. Isroi; Ishola, M.M.; Millati, R.; Syamsiah, S.; Cahyanto, M.N.; Niklasson, C.; Taherzadeh, M.J. Structural changes of oil palm empty fruit bunch (OPEFB) after fungal and phosphoric acid pretreatment. *Molecules* **2012**, *17*, 14995–15002. [CrossRef] [PubMed]
14. Kshirsagar, S.D.; Waghmare, P.R.; Loni, P.C.; Patil, S.A.; Govindwar, S.P. RSC Advances characterization and optimization of enzymatic hydrolysis conditions by response surface methodology. *RSC Adv.* **2015**, *5*, 46525–46533. [CrossRef]
15. Bahrin, E.K.; Baharuddin, A.S.; Ibrahim, M.F.; Abdul Razak, M.N.; Sulaiman, A.; Abd-Aziz, S.; Hassan, M.A.; Shirai, Y.; Nishida, H. Physicochemical property changes and enzymatic hydrolysis enhancement of oil palm empty fruit bunches treated with superheated steam. *BioResources* **2012**, *7*, 1784–1801.
16. Warid, M.N.M.; Ariffin, H.; Hassan, M.; Shirai, Y. Optimization of superheated steam treatment to improve surface modification of oil palm biomass fiber. *BioResources* **2016**, *11*, 5780–5796.
17. Zakaria, M.R.; Norrrahim, M.N.F.; Hirata, S.; Hassan, M.A. Hydrothermal and wet disk milling pretreatment for high conversion of biosugars from oil palm mesocarp fiber. *Bioresour. Technol.* **2015**, *181*, 263–269. [CrossRef] [PubMed]
18. Catherine, H.; Frédéric, D.; Penninckx, M. Product formation from phenolic compounds removal by laccases: A review. *Environ. Technol. Innov.* **2016**, *5*, 250–266. [CrossRef]
19. Moreno, A.; Ibarra, D.; Mialon, A.; Ballesteros, M. A bacterial laccase for enhancing saccharification and ethanol fermentation of steam-pretreated biomass. *Fermentation* **2016**, *2*, 11. [CrossRef]
20. Fillat, Ú.; Ibarra, D.; Eugenio, M.E.; Moreno, A.D.; Tomás-Pejó, E.; Martín-Sampedro, R. Laccases as a potential tool for the efficient conversion of lignocellulosic biomass: A Review. *Fermentation* **2017**, *3*, 17. [CrossRef]
21. Saha, B.C.; Qureshi, N.; Kennedy, G.J.; Cotta, M.A. Biological pretreatment of corn stover with white-rot fungus for improved enzymatic hydrolysis. *Int. Biodeterior. Biodegrad.* **2016**, *109*, 29–35. [CrossRef]
22. Zanirun, Z.; Bahrin, E.K.; Lai-Yee, P.; Hassan, M.A.; Abd-Aziz, S. Enhancement of fermentable sugars production from oil palm empty fruit bunch by ligninolytic enzymes mediator system. *Int. Biodeterior. Biodegrad.* **2015**, *105*, 13–20. [CrossRef]
23. Sukri, S.S.M.; Rahman, R.A.; Md Illias, R.; Yaakob, H. Optimization of alkaline pretreatment conditions of oil palm fronds in improving the lignocelluloses contents for reducing sugar production. *Rom. Biotechnol. Lett.* **2014**, *19*, 9006–9018.
24. Agbor, V.B.; Cicek, N.; Sparling, R.; Berlin, A.; Levin, D.B. Biomass pretreatment: Fundamentals toward application. *Biotechnol. Adv.* **2011**, *29*, 675–685. [CrossRef] [PubMed]
25. Donohoe, B.S.; Decker, S.R.; Tucker, M.P.; Himmel, M.E.; Vinzant, T.B. Visualizing lignin coalescence and migration through maize cell walls following thermochemical pretreatment. *Biotechnol. Bioeng.* **2008**, *101*, 913–925. [CrossRef] [PubMed]

26. Pu, Y.; Hu, F.; Huang, F.; Davison, B.H.; Ragauskas, A.J. Assessing the molecular structure basis for biomass recalcitrance during dilute acid and hydrothermal pretreatments. *Biotechnol. Biofuels* **2013**, *6*, 15. [CrossRef] [PubMed]
27. Rencoret, J.; Pereira, A.; Río, J.C.D.; Martínez, A.T.; Gutiérrez, A. Laccase-mediator pretreatment of wheat straw degrades lignin and improves saccharification. *BioEnergy Res.* **2016**, *9*, 917–930. [CrossRef]
28. Yu, H.; Li, X.; Xing, Y.; Liu, Z.; Jiang, J. A sequential combination of laccase pretreatment and enzymatic hydrolysis for glucose production from furfural residues. *BioResources* **2014**, *9*, 4581–4595. [CrossRef]
29. Moldes, D.; Díaz, M.; Tzanov, T.; Vidal, T. Comparative study of the efficiency of synthetic and natural mediators in laccase-assisted bleaching of eucalyptus kraft pulp. *Bioresour. Technol.* **2008**, *99*, 7959–7965. [CrossRef] [PubMed]
30. Alvira, P.; Moreno, A.D.; Ibarra, D.; Sáez, F.; Ballesteros, M. Improving the fermentation performance of saccharomyces cerevisiae by laccase during ethanol production from steam-exploded wheat straw at high-substrate loadings. *Biotechnol. Prog.* **2013**, *29*, 74–82. [CrossRef] [PubMed]
31. Bitra, V.S.P.; Womac, A.R.; Chevanan, N.; Sokhansanj, S. Comminution properties of biomass in hammer mill and its particle size characterization. In Proceedings of the ASABE Annual Internationnal Meeting, Detroit, MI, USA, 29 June–2 July 2008. [CrossRef]
32. Ndukwu, M.; Nwakuba, N.; Henry, O. Measurement of energy requirements for size reduction of palm kernel and groundnut shells for downstream bioenergy generation. *J. Eng. Technol. Res.* **2016**, *8*, 47–57. [CrossRef]
33. Hames, B.; Ruiz, R.; Scarlata, C.; Sluiter, A.; Sluiter, J.; Templeton, D. *Preparation of Samples for Compositional Analysis*; Technical Report No. NREL/TP 510-42620; National Renewable Energy Laboratory: Golden, CO, USA, 2008; pp. 1–12.
34. Castoldi, R.; Bracht, A.; de Morais, G.R.; Baesso, M.L.; Correa, R.C.G.; Peralta, R.A.; Moreira, R.d.F.P.M.; Polizeli, M.d.L.T.d.M.; de Souza, C.G.M.; Peralta, R.M. Biological pretreatment of *Eucalyptus grandis* sawdust with white-rot fungi: Study of degradation patterns and saccharification kinetics. *Chem. Eng. J.* **2014**, *258*, 240–246. [CrossRef]
35. Sluiter, A.; Ruiz, R.; Scarlata, C.; Sluiter, J.A.; Templeton, D. *Determination of Extractives in Biomass*; Technical Report No. NREL/TP-510-42619; National Renewable Energy Laboratory: Golden, CO, USA, 2008; pp. 1–12.
36. Sluiter, A.; Hames, B.; Ruiz, R.; Scarlata, C.; Sluiter, J.; Templeton, D.; Crocker, D.C. *Determination of Structural Carbohydrates and Lignin in Biomass*; Technical Report No. NREL/TP-510-42618; National Renewable Energy Laboratory: Golden, CO, USA, 2012; pp. 1–18.
37. Sluiter, A.; Hames, B.; Ruiz, R.; Scarlata, C.; Sluiter, J.; Templeton, D. *Determination of Ash in Biomass*; Technical Report No. NREL/TP-510-42622; National Renewable Energy Laboratory: Golden, CO, USA, 2008; pp. 1–8.
38. Wi, S.G.; Cho, E.J.; Lee, D.-S.; Lee, S.J.; Lee, Y.J.; Bae, H.-J. Lignocellulose conversion for biofuel: A new pretreatment greatly improves downstream biocatalytic hydrolysis of various lignocellulosic materials. *Biotechnol. Biofuels.* **2015**, *8*, 228. [CrossRef] [PubMed]
39. Bourbonnais, R.; Paice, M.G. Veratryl alcohol oxidases from the lignin-degrading basidiomycete *Pleurotus sajor-caju*. *Biochem. J.* **1988**, *255*, 445–450. [CrossRef] [PubMed]
40. Makkar, H.P.S. Measurement of total phenolics and tannins using Folin-Ciocalteu Method. In *Quantification of Tannins in and Tree Shrub Foliage*; Springer: Dordrecht, The Netherlands, 2003; pp. 49–51, ISBN 978-94-017-0273-7.
41. Sluiter, A.; Hames, B.; Ruiz, R.; Scarlata, C.; Sluiter, J.; Templeton, D. *Determination of Sugars, Byproducts, and Degradation Products in Liquid Fraction Process Samples*; Technical Report No. NREL/TP-510-42623; National Renewable Energy Laboratory: Golden, CO, USA, 2008; pp. 1–14.

Sample Availability: Samples of the raw OPEFB and OPMF are available from the authors.

molecules

MDPI

Article

Mechano-Enzymatic Deconstruction with a New Enzymatic Cocktail to Enhance Enzymatic Hydrolysis and Bioethanol Fermentation of Two Macroalgae Species

Sameh Amamou [1], Cecilia Sambusiti [1], Florian Monlau [1,2], Eric Dubreucq [1] and Abdellatif Barakat [1,3,*]

1 UMR, Ingénierie des Agropolymères et des Technologies Emergentes (IATE), CIRAD, Montpellier SupAgro, INRA, Université de Montpellier, 34060 Montpellier, France; sameh.amamou@supagro.fr (S.A.); cecilia.sambusiti@gmail.com (C.S.); Florian.Monlau@Apesa.Fr (F.M.); Eric.Dubreucq@supagro.fr (E.D.)

2 APESA, Plateau Technique, Cap Ecologia, Avenue Fréderic Joliot Curie, 64230 Lescar, France

3 AgroBioSciences, Mohammed VI Polytechnic University, Lot 660-Hay Moulay Rachid, Ben Guerir 43150, Morocco

* Correspondence: abdellatif.barakat@inra.fr; Tel.: +33-(0)4-99-61-25-81; Fax: +33-(0)4-99-61-30-76

Received: 14 November 2017; Accepted: 7 January 2018; Published: 17 January 2018

Abstract: The aim of this study was to explore the efficiency of a mechano-enzymatic deconstruction of two macroalgae species for sugars and bioethanol production, by using a new enzymatic cocktail (Haliatase) and two types of milling modes (vibro-ball: VBM and centrifugal milling: CM). By increasing the enzymatic concentration from 3.4 to 30 g/L, the total sugars released after 72 h of hydrolysis increased (from 6.7 to 13.1 g/100 g TS and from 7.95 to 10.8 g/100 g TS for the green algae *U. lactuca* and the red algae *G. sesquipedale*, respectively). Conversely, total sugars released from *G. sesquipedale* increased (up to 126% and 129% after VBM and CM, respectively). The best bioethanol yield (6 g_{eth}/100 g TS) was reached after 72 h of fermentation of *U. lactuca* and no increase was obtained after centrifugal milling. The latter led to an enhancement of the ethanol yield of *G. sesquipedale* (from 2 to 4 g/100 g TS).

Keywords: bioethanol; enzymatic hydrolysis; macroalgae; mechanical pretreatment

1. Introduction

Over the last decades, the world has been facing critical economic and environmental issues, such as the exhaustion of fuels, environmental pollution and climate change, combined with the increase of the world population. These issues led to the expansion of research and development on renewable and sustainable biofuels [1,2]. Both lignocellulosic biomasses and algae constitute sustainable sources of bioenergy and biomolecules (i.e., surfactant, bioethanol, biogas and biodiesel) and they represent promising alternative sources to petroleum-based fuels and chemicals.

In particular, macro- and/or micro-algae permit us to overcome the major limitations associated with lignocellulosic plants [3]. Macroalgae, also called seaweeds, represent renewable abundant biomasses, which could be easily cultivated in aquatic environment. Thus, they do not compete with land use, and water consumption necessary for terrestrial plants.

Furthermore, macroalgae are characterized by a higher growth rate than lignocellulosic biomasses and higher hydrolysable sugar contents than crops with almost no lignin [4–6]. Macroalgae are multicellular photosynthetic organisms divided into three major groups: green, red and brown algae, according to the thallus color derived from natural pigments and chlorophylls [4,7]. Generally, the amounts of carbohydrates vary between 25–60%, 30–60% and 30–50% dry wt. for green, red, and brown algae, respectively. In particular, green algae are mainly composed of mannan, ulvan, starch and cellulose, while red algae are mainly composed of carrageenan (up to 75% dry wt.) and agar (up to 52% dry wt.) as polysaccharides [6].

Several researches have reported the use of marine alga biomass as bioethanol feedstock. Different macroalgae groups such as *Gelidium amansii* [8], *Gracilaria salicornia* [9] and *Kappaphycus alvarezii* belonging to red seaweed and green algal species such as *Ulva* spp. have been considered as potential sources for conversion to bioethanol. As the interests in seaweeds were expanding, intense research was required for an efficient use of this biomass. However, it still faces technical and economic challenges and still depends on the development of eco-friendly pretreatment and conversion methods [10]. Since this step is often required to facilitate the enzymatic hydrolysis of macroalgae and their further sugars and bioethanol conversion. So far, the most common pretreatments used to enhance the hydrolysis and thus bioethanol production of macroalgae are physical (wet oxidation, thermal, milling and oven drying), chemical (acidic and alkaline), and thermo-chemical pretreatments [10]. However, one of the major drawbacks of using thermal and thermo-chemical pretreatments is the possible formation of organic acids and furan derivatives, which can inhibit bioethanol fermentation [11,12]. In addition, to find the most sustainable and cost-effective pretreatment, another challenge of producing bioethanol and interesting molecules from *G. sesquipedale* (red) and green *U. lactuca* (green) macroalgae is to find the specific enzymes able to efficiently hydrolyze their polysaccharides (Figure 1). Mechanical fractionation of biomass is one promising route that can contribute to a future sustainable dry biorefinery without water consumption and without waste production. Grinding or dry fractionation can be easily introduced in a biorefinery scheme improving the overall sustainability process [13]. Thus, coupling mechanical fractionation with enzymes is a promising biorefinery scheme of algae biomass valorization. In addition, the use of this natural enzymatic cocktail (i.e., Haliatase) coupling to mechanical fractionation has never yet been investigated in algae biomass biorefineries. Thus, the main objectives of this study were the following:

(i) Explore the efficiency of a new enzymatic cocktail to hydrolyze polysaccharides of two macroalgae species (red and green sp.)
(ii) Study the effect of two mechanical pretreatments, centrifugal milling (CM) and vibro ball milling (VBM) on enzymatic hydrolysis and bioethanol fermentation of the two-macroalgae species (red *G. sesquipedale* and green *U. lactuca*) (Figure 1).

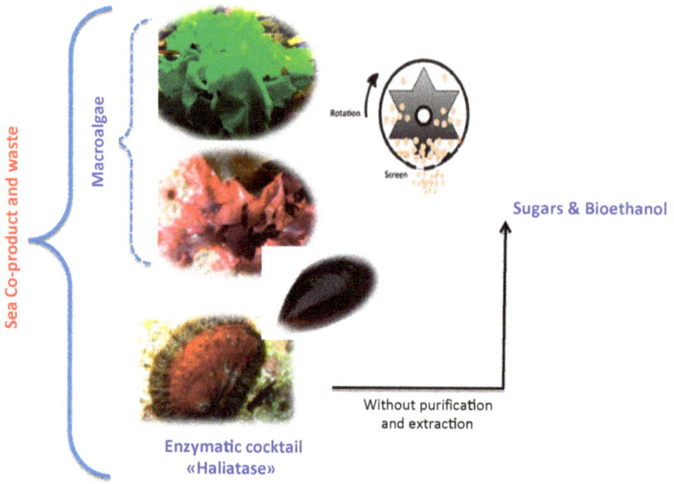

Figure 1. Mechano-enzymatic pretreatment and deconstruction of macroalgae developed in this study.

2. Results and Discussion

2.1. Chemical Composition

The chemical compositions of both red *G. sesquipedale* and green *U. lactuca* macroalgae are shown in Table 1. *U. lactuca* had higher ash content (32 g/100 g TS) than *G. sesquipedale* (11 g/100 g TS). These values are in agreement with literature studies that reported ash values ranging from 11 to 34 g/TS and from 9 to 20 g/100 g TS, for green and red macroalgae, respectively [8,14,15]. Similar protein content (13 and 16 g/100 g TS) was observed for *U. lactuca* and *G.sesquipedale*, in accordance with literature data for red (10–16 g/100 g TS) and green (12–21 g/100 g TS) macroalgae [14–16].

Table 1. Chemical composition of red *G. sesquipedale* and green *U. lactuca* Values correspond to mean ± SD (standard deviation) of measurement performed in duplicate.

Composition	*G. sesquipedale*	*U. lactuca*
TS (g/100 g FM)	91 ± 0	90 ± 0
VS (g/100 g TS)	78 ± 0.1	62 ± 0.7
Ash (g/100 g TS)	11 ± 0.3	32 ± 0.7
C (g/100 g TS)	34.7 ± 0.5	26.8 ± 1.5
N (g/100 g TS)	2.5 ± 0.1	2.3 ± 0.4
H (g/100 g TS)	5.7 ± 0.05	4.5 ± 0.6
S (g/100 g TS)	2.1 ± 0.02	3.7 ± 0.12
Proteins (g/100 g TS)	19.9 ± 0.82	15.9 ± 1.91
Total Sugars (g/100 g TS)	30.9	25.8
Monomeric sugars *		
Glucose (g/100 g TS)	9.6 ± 0.06	15.2 ± 1.01
Galactose (g/100 g TS)	20.3 ± 0.78	n.d.
Arabinose (g/100 g TS)	0.9 ± 0.06	n.d.
Xylose (g/100 g TS)	n.d.	3.1 ± 0.18
Rhamnose (g/100 g TS)	n.d.	7.5 ± 0.13
Fucose (g/100 g TS)	n.d.	0.5 ± 0.04
Glucuronic acid (g/100 g TS)	0.3 ± 0.03	3.86 ± 0.01
Galacturonic acid (g/100 g TS)	3.0 ± 0.06	1.15 ± 0.00

n.d.: Not detected; * Monosaccharide profile of polymeric carbohydrates determined after acid hydrolysis and HPLC quantification; SD: standard deviation.

Total sugar content of *U. lactuca* and *G. sesquipedale* were 25.8 and 30.9 g/100 g TS, respectively. Similar sugar content was reported by Jard et al. [16] for red and green macroalgae. As reported by Jung et al. [7], the red alga *G. sesquipedale* was mainly composed of glucose, galactose, and agar, while *U. lactuca* consisted of glucose, xylose, and rhamnose. Glucuronic and galacturonic acids were also detected in both algae with values of 5.01 and 3.32 g/100 g TS for the green and red algae, respectively. However, it is noteworthy that the chemical composition of macroalgae presents a great variability in the literature, which is related to several factors, such as species, geographical origin, season, environmental, and physiological variations, but also to the analytical method used for their characterization.

2.2. Particle Size of Macroalgae

Particle size of the untreated and milled macroalgae is reported in Figure 2. After milling, a lower mean particle size was obtained for the green alga *U. lactuca* (147–161 µm) than the red alga *G. sesquipedale* (201–355 µm), which can be explained by the lower particle size distribution of the untreated green algae biomass (289 µm) than the red one (472 µm). Furthermore, it was observed that for the red alga, CM was more effective than VBM in particle size reduction (Anova p-values < 0.05) (Figure 2); while for the green alga similar effect of VBM and CM was noticed with a slightly higher effect of VBM but not significant with an Anova p-values of 0.59. This could be explained by the high ash content (around 30%) and a possible synergistic impact between the mineral ash and the mechanical process (i.e., VBM) as previously mentioned by Motte et al. [13]. Indeed, they highlighted in their study that a mineral-vegetal co-milling in a VBM could significantly reduce the final particle size of the lignocellulosic biomass compared to a simple milling of lignocellulosic biomass.

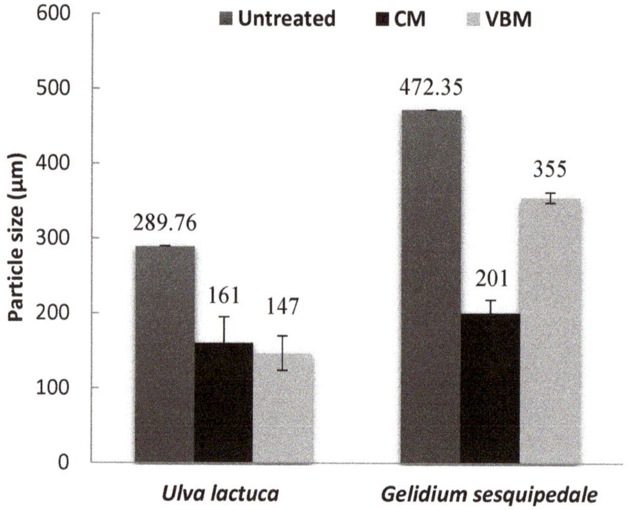

Figure 2. Mean particle sizes of untreated and milled algae biomass. Values correspond to mean ± SD (standard deviation) of measurement performed in duplicate.

2.3. *Effect of Haliatase Cocktail Activity on Sugars Yield of Untreated Macroalgae*

The effect of Haliatase dosage on total sugars released during enzymatic hydrolysis of untreated *U. lactuca* and *G. sesquipedale* was investigated (Figure 3). For both macroalgae, the increase of the enzymatic concentration led to higher total sugars released. By increasing the enzymatic concentration from 3.4 to 30 g/L, the total sugars released after 72 h of hydrolysis varied from 6.7 to 13.1 g/100 g TS and from 7.9 to 10.8 g/100 g TS for *U. lactuca* and *G. sesquipedale*, respectively (significant difference with Anova *p*-values < 0.05) (Table 2).

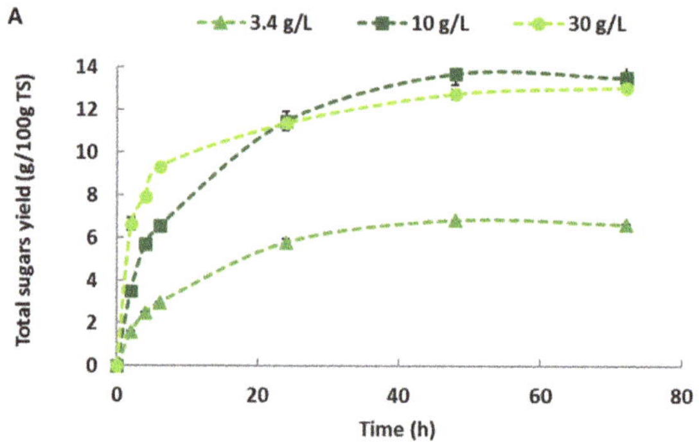

Figure 3. *Cont.*

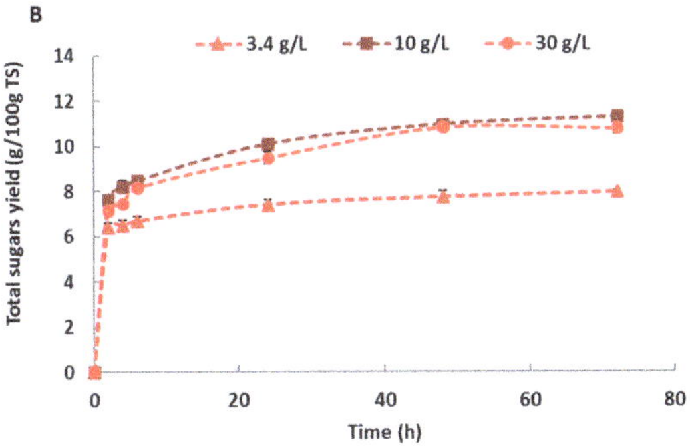

Figure 3. Total sugars released after 72 h of enzymatic hydrolysis for various enzyme dosages (3.4, 10 and 30 g/L) for both untreated (**A**) *Ulva lactuca* and (**B**) *Gelidium sesquipedale*.

Table 2. Total sugar yields (g/100 g TS) obtained after 72 h of enzymatic hydrolysis of untreated and milled algae with an enzyme dosage of 3.4, 10 and 30 g/L.

Samples	Enzyme Loading	3.4 g/L	10 g/L	30 g/L
Green alga: *Ulva lactuca*	Untreated	6.66 ± 0.04	13.52 ± 0.51	13.05 ± 0.16
	Centrifugal milling	6.48 ± 0.40	13.46 ± 0.61	13.24 ± 0.37
	Vibro ball milling	6.70 ± 0.35	13.33 ± 0.37	12.49 ± 0.20
Red alga: *Gelidium sesquipedale*	Untreated	7.96 ± 0.09	11.30 ± 0.13	10.79 ± 0.26
	Centrifugal milling	10.28 ± 0.06	13.28 ± 0.19	13.09 ± 0.48
	Vibro ball milling	10.03 ± 0.20	12.70 ± 0.11	13.59 ± 0.34

However, the increase of enzyme concentration from 10 to 30 g/L did not lead to a further enhancement of total sugar yield of both red and green algae witch is assumed with Anova *p*-values of 0.13 and 1.55, respectively, for green and red algae. Finally, for both algae species, the major soluble sugar released during the enzymatic hydrolysis was glucose which is very pertinent in the case of ethanol production using the *Saccharomyces cerevisae* strain. Glucose released after 72 h of enzymatic hydrolysis varied from 5.8 to 10.5 g/100 g TS for *U. lactuca* and from 1.5 to 4.1 g/100 g TS for *G. sesquipedale* after increasing the enzymatic dosage from 3.4 to 30 g/L (significant difference with Anova *p*-values < 0.05) (Table 3). If the results of glucose yields were not significantly different from 10 g/L to 30 g/L for red algae (Anova *p*-values of 0.95), they were for green algae (Anova *p*-values < 0.05). Cho et al. [17] reported an enzymatic saccharification of *Gelidium amansii* using Celluclast 1.5 L (endoglucanase: (8.4 U/mL), Viscozymes L (β-glucanase: 1.2 U/mL) and a mixture of both enzymes. They obtained a glucose concentration of 5.5 g/L after an enzymatic hydrolysis using Celluclast 1.5 L. Interestingly, the glucose concentration released was improved to 7.6 g/L by using the mixture of "Celluclast + Viscozymes" enzymes after 48 h of hydrolysis. Nonetheless, in both assays, only the fibers rich in cellulose were hydrolyzed whereas agar was not solubilized [17].

Table 3. Glucose yield (g/100 g TS) obtained after 72 h of enzymatic hydrolysis of untreated and milled algae with an enzyme dosage of 3.4, 10 and 30 g/L.

Samples	Enzyme Loading	3.4 g/L	10 g/L	30 g/L
Green alga: *Ulva lactuca*	Untreated	5.78 ± 0.00	12.45 ± 0.47	10.49 ± 0.10
	Centrifugal milling	5.79 ± 0.52	12.59 ± 0.50	10.71 ± 0.13
	Vibro ball milling	6.01 ± 0.60	12.63 ± 0.46	10.21 ± 0.08
Red alga: *Gelidium sesquipedale*	Untreated	1.48 ± 0.04	4.07 ± 0.22	4.12 ± 0.13
	Centrifugal milling	3.87 ± 0.17	7.09 ± 0.22	6.68 ± 0.70
	Vibro ball milling	3.17 ± 0.07	5.45 ± 0.10	6.35 ± 0.58

2.4. Effect of Mechanical Pretreatments on Enzymatic Hydrolysis of Macroalgae

The effect of mechanical pretreatments (i.e., centrifugal milling, vibro-ball milling) on total sugars released during enzymatic hydrolysis of untreated *U. lactuca* and *G. sesquipedale* was investigated (Figure 4). The results revealed that, whatever the enzymatic dosage, mechanical pretreatments did not have any effect on the total sugars released from green alga *U. lactuca* (Anova p-values > 0.05) Conversely, total sugars released from *G. sesquipedale* increased after mechanical pretreatments (up to 126% and 129% after vibro-ball and centrifugal milling fractionation, respectively, Anova p-values < 0.05). Otherwise, fractionation was more effective in glucose releasing (up to 214% and 261% after vibro-ball and centrifugal milling, respectively). It is important to note also that mechanical fractionation seems to be less effective after increasing the enzymatic dosage from 3.4 g/L to 30 g/L. Thus, glucose released from *G sesquipedale* increased after CM by 161% and 62% for 3.4 and 30 g/L of enzymatic dosage, respectively. Furthermore, whatever the enzymatic dosage, CM was more effective than VBM in improving the total sugars and glucose released. Moreover, it is noteworthy that the high-energy requirement is one of the drawbacks of mechanical treatments. In a previous study, [18–20] reported that the energy requirement for CM (100 kWh t^{-1} TS) was lower than that of VBM (2000 kWh t^{-1} TS). Thus, CM was chosen for the performance of experimentation.

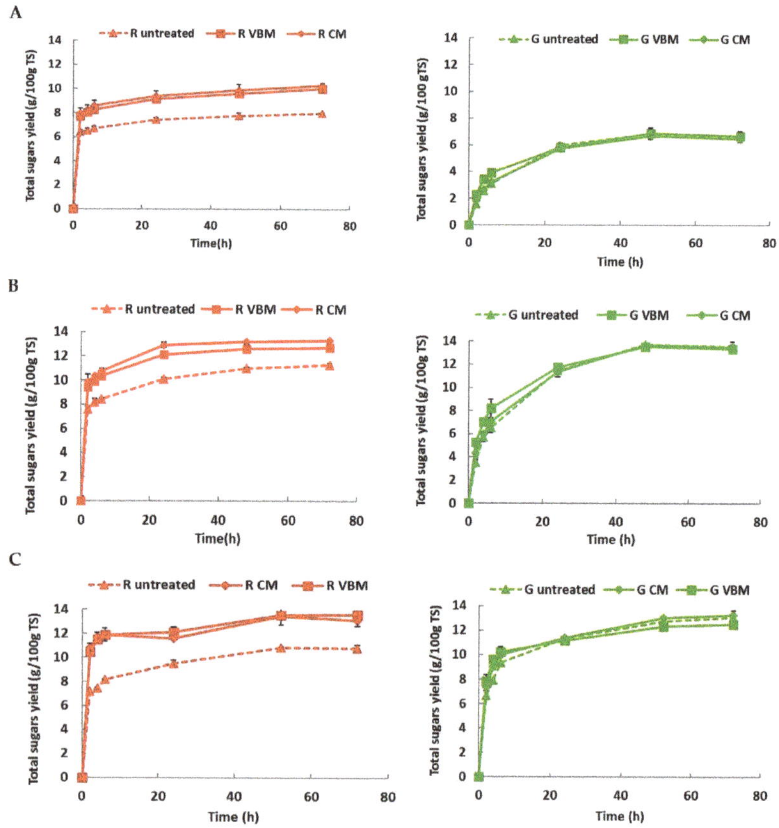

Figure 4. Total sugars released after 72 h of enzymatic hydrolysis for untreated and milled (G: Green, R: Red) macroalgae with an enzyme dosage of (**A**) 3.4; (**B**) 10 and (**C**) 30 g/L. Values correspond to mean ± SD of measurement performed in duplicate.

2.5. Bioethanol Fermentation of *U. lactuca* and *G. sesquipedale*

Taking into account the previous results, CM was chosen as mechanical treatment prior to a simultaneous saccharification and fermentation (SSF). Also, 10 g/L was selected as the optimal enzymatic dosage of Haliatase and, thus, used in saccharification process. Bioethanol yields of untreated and centrifugal milled macroalgae were evaluated and compared through SSF experiments (Figure 5). CM treatment did not affect the ethanol production of *U. lactuca* with values around 6 g_{eth}/100 g TS (Anova *p*-values of 0.35), confirming the enzymatic hydrolysis data (Figure 4). Conversely, the CM led to an enhancement of the ethanol yield of the red algae (from 1.95 to 3.51 g/100 g TS, significantly different, with Anova p-values < 0.05). However, although bioethanol yield obtained with *U. lactuca* was higher than that of *G. sesquipedale*, the bioethanol conversion efficiency of the red one was higher, because the glucose content of red algae (9.62%) is lower than that (15.2%) of green algae (Table 1). Thus, after CM fractionation, 64% and 69% of bioethanol conversion efficiency (expressed in % of the theoretical yield) (Table 4) was obtained for green and red algae, respectively. Furthermore, for algae strains, the galactose and xylose were not consumed. Such observation could be attributed to a diauxic effect commonly observed but it is not totally satisfactory, as the galactose consumption of the red algae did not start even after a total depletion of the glucose, so probably 72 h would be too short to initiate diauxic effect since Berlowska et al. (2017) [21] Berlowska et al., 2017 have demonstrated that actually, *S. cerevisae* ethanol red was capable to metabolize galactose but in absence of glucose. Nevertheless, despite the fact that a large number of yeast species can metabolize xylose, only 1% of strains convert xylose to ethanol [22]. Thus, it is important to find the most active yeast species for bioethanol fermentation of hexoses (other than glucose) and pentose sugars, in order to achieve higher ethanol yield.

(A)

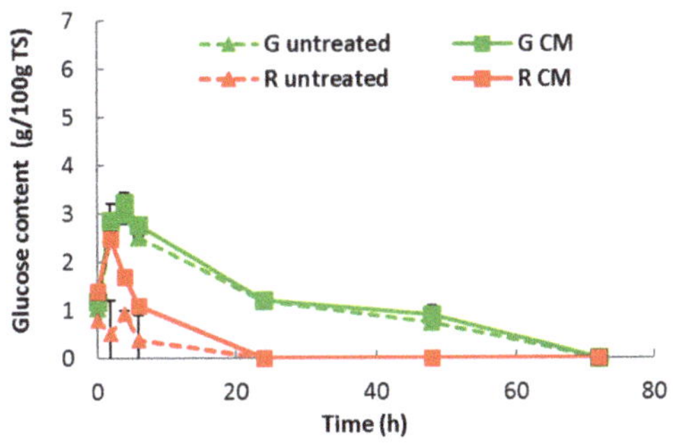

(B)

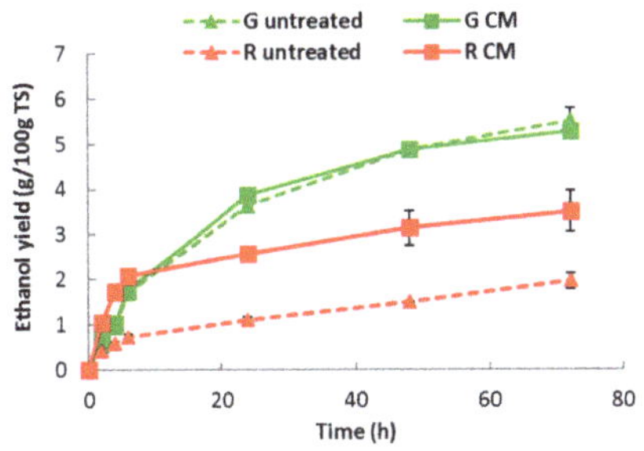

Figure 5. Simultaneous saccharification and fermentation of untreated and milled macroalgae (G: Green, R: Red) at an enzyme dosage of 10 g/L. (**A**) Glucose content (g/100 g TS); (**B**) Ethanol yield (g/100 g TS). Values correspond to mean ± SD of measurement performed in duplicate.

Table 4. Ethanol yield (g/100 g TS) obtained after 72 h of SSF of untreated and milled algae with an enzyme dosage of 10 g/L.

	Ethanol Yield (g/100 g TS)		Ethanol Efficiency (% Theoretical Yield *)	
Samples	**Untreated**	**Centrifugal Milling**	**Untreated**	**Centrifugal Milling**
Ulva lactuca	5.51 ± 0.29	5.27 ± 0.02	67.2%	64.3%
Gelidium sesquipedale	1.95 ± 0.17	3.51 ± 0.46	38.2%	68.7%

* Theoretical ethanol yield: 8 g/100 g TS and 5 g/100 g TS for *U. lactuca* and *G. sesquipedale* respectively.

In this matter, Cho et al., 2014 [17], reported that glucose causes the repression of galactose uptake which decreased ethanol yield. The acclimation of galactose was then reported as the key of a fermentation process since it has allowed simultaneous utilization of glucose and galactose. In fact, ethanol yield doubled (from 0.21 to 0.44 g/100 g TS) after using *S. cerevisiae* acclimated to high concentration of galactose.

Regarding treatment effects on ethanol fermentation, Schultz-Jensen et al. [10] investigated the ethanol fermentation of the green macroalga *Chaetomorpha linum* after wet oxidation, hydrothermal treatment, plasma, and ball milling for 48 h at 40 °C. Interestingly, the best ethanol recovery was obtained after ball milling with an ethanol yield of 18 g/100 g TS, corresponding to 78% of the theoretical ethanol yield.

3. Materials and Methods

3.1. Macroalgae

Red alga *Gelidium sesquipedale* and green alga *Ulva lactuca* were obtained from the Morocco coast. Once collected, samples were washed with tap water and further air-dried (8% DM) and milled using a cutting mill to a particle size less than 2 mm (SM100 Retsch, Haan, Germany). Then, they were further milled using two equipments, characterized by different mechanical stresses, such as impact, compression, friction, and shear: (i) a centrifugal mill "CM" (Retsch ZM 200, Haan, Germany) with 0.25 mm screen size, operated at ambient temperature with a speed of 12000 rpm; (ii) a vibratory ball mill "VBM" (Retsch MM400, Haan, Germany) operated at ambient temperature, at a frequency of 15 s^{-1} for 5 min.

3.2. Enzymatic Cocktail

The enzymatic cocktail (i.e., Haliatase enzyme) has been obtained from KURA BIOTECH SPA, (Puerto Varas, Chile) and it is derived from the hepatopanchreas of cultured abalone (Haliotis rufescens). It is a multi-enzymatic cocktail capable of degrading the cell walls of macroalgae by hydrolyzing most of their polysaccharides components. It is composed of mainly β-glucanase (1875 U/g), carragenase (315 U/g) and agarase (440 U/g).

3.3. Enzymatic Hydrolysis

Enzymatic hydrolysis of untreated and milled samples was performed in 40 mL of closed flasks (working volume of 20 mL). An amount (1 g) of each sample (solid loading of 50 g/L), 2 mL of acetate buffer (500 mM) and 15.8 mL of ultra-pure water were added to each flask. The pH was then adjusted to 5.5 with NaOH (1 N) or HCl (2 N). Finally, 1.2 mL of sodium azide (final concentration 1 g/L) and 1 mL of concentrated Haliatase enzyme were added to have final enzymatic concentrations of 3.4, 10 and 30 g/L, respectively. Flasks were kept at 37 °C for 72 h with stirring 500 rpm. Samples were withdrawn at 0, 2, 4, 6, 24, 48 and 72 h and the corresponding supernatants were analyzed by HPLC (Waters corporation, Milford, CA, USA), equipped with a BioRad HPX-87H column (Biorad, Hercules, CA, USA) at 40 °C, a refractive index detector at 40 °C and a 0.005 M H_2SO_4 solvent at

0.3 mL/min. Tests were performed in duplicate, to evaluate the amount of C5-C6 sugars released during the enzymatic hydrolysis.

Sugar yields (g_i/100 g TS) were calculated according to Equation (1):

$$\text{Sugar “}i\text{” yield}_t = [C_t \text{ sugar “}i\text{”}/C \text{ solid}] \times 100 \quad (1)$$

where: C_t sugar "i" ($g_{\text{sugar “}i\text{”}}$/L) is the concentration of C5 and C6 sugars produced during hydrolysis, at time t; C solid (g TS/L) is the total solids concentration in the flask.

The analysis of variance (Anova) method was used to analyse the impact of the enzymatic dose and mechanical fractionation (CM and VBM) on both red and green algae, the confidence level considered was 95%.

3.4. Bioethanol Fermentation

Bioethanol yields of untreated and milled (CM) macroalgae were evaluated and compared through simultaneous saccharification and fermentation (SSF) experiments. Tests were performed by using unsterilized samples, into 40 mL flasks (working volume of 20 mL) closed with rubber septa and equipped with an air vent system, constituted of sterilized needle and filter, in order to evacuate the CO_2 produced during the bioconversion. A lyophilized *S. cerevisiae* yeast strain (Ethanol Red®, FERMENTIS, a division of S. I. LESAFFRE, Lille, France) was used as inoculum. For this purpose, lyophilized cells were previously washed and then suspended in sterilized distilled water to a concentration of 30 g TS/L. Each flask contained: 1 g TS of sample (solid loading of 50 g/L), 1 mL of concentrated Haliatase enzyme, to have an enzymatic concentration of 10 g/L in each flask, 1 mL of yeast (30 g TS/L), 2 mL of nutrients, containing: 50 g TS/L yeast extract (Difco®), 4 g TS/L urea, 0.5 g TS/L chloramphenicol and 50 mM acetate buffer (pH = 5). Flasks were incubated at 37 °C for 72 h under stirring. Samples were withdrawn at 0, 2, 4, 6, 24, 48 and 72 h and the cell free supernatants were evaluated for ethanol and C6 sugars (i.e., glucose, galactose, fucose and rhamnose) concentrations by HPLC as previously mentioned.

Ethanol yields ($g_{ethanol}$/100 g TS) were calculated according to Equation (2):

$$\text{Ethanol yield}_t = [C_t \text{ ethanol}/C \text{ solid}] \times 100 \quad (2)$$

where C_t ethanol ($g_{ethanol}$/L) is the concentration of ethanol produced during SSF, at time t; C solid (g TS/L) is the total solids concentration in the flask.

The analysis of variance (Anova) method was used to analyze the impact of mechanical fractionation (CM and VBM) on both red and green algae bioethanol fermentation, the confidence level considered was 95%.

3.5. Analytical Determinations

Particle size distribution of untreated macroalgae was determined by a vibratory sieving apparatus (Analytical Sieve Shaker AS 200, Retsch®, Haan, Germany) equipped with six sieves of different sizes (1, 0.8, 0.71, 0.56, 0.32 and 0.2 mm). Particle size distribution of milled macroalgae was analyzed by a laser granulometry (MASTERSIZER 2000, Malvern Instrument, Orsay, France). Total Solids (TS), Volatile Solids (VS) and ash contents were determined according to APHA methods [22]. Ultimate analysis (C, N, H and S) was accomplished with an elemental analyzer (Elementar "VarioMacroCube", Elementar group, Langenselbold, Germany). Proteins content was estimated by multiplying N by 6.25. Carbohydrates and uronic acids were determined according to a reduced scale hydrolysis procedure, based on the NREL Laboratory Analytical Procedure [23]. Briefly, 80 ± 1 mg of milled algae biomass was subjected to a two-stage sulfuric acid hydrolysis (1 h at 30 °C in 72 wt % H_2SO_4, followed by 1 h at 121 °C in 4 wt % H_2SO_4 for red algae and 3 h at 120 °C in 6 wt % H_2SO_4 for the green algae). Samples were withdrawn at 1 h, 2 h and 3 h and the cell free supernatants were evaluated for sugars (i.e., glucose, xylose, arabinose, galactose, fucose, rhamnose) and uronic acids (galacturonic acid and glucuronic acid) concentrations by high-performance liquid chromatography by HPLC system (Waters corporation), equipped with a BioRad HPX-87H column at 40 °C, a refractive index detector at 40 °C and a 0.005 M H_2SO_4 solvent at 0.3 mL/min. It is noteworthy that analytical determinations were performed in duplicate.

4. Conclusions

A comprehensive study was performed on the efficiency of a new natural enzymatic cocktail to hydrolyze polysaccharides of two types of seaweeds and produce bioethanol. The effects of two mechanical pretreatments were also tested for increasing bioethanol fermentation. The most effective enzymatic dosage for the saccharification process of green *U. lactuca* and red *G. sesquipedale* was 10 g/L and the highest values of glucose released were obtained with green algae after 72 h of enzymatic hydrolysis. Centrifugal milling was more effective in hydrolyzing red *G. sesquipedale* compared to vibro-ball milling, while the mechanical pretreatments applied did not show any effect on green *U. lactuca*. However, green *U. lactuca* showed the highest bioethanol yield compared to red *G. sesquipedale*.

Acknowledgments: The authors are grateful to the LipPolGreen Platform of UMR IATE and (INRA Montpellier) for her help in biochemical analysis and fermentation and INRA for the financial support. The authors are grateful to KURA BIOTECH SPA (Chile) for giving us the Haliatase enzyme cocktail.

Author Contributions: A.B. and F.M. conceived and designed the experiments; S.A. and C.S. performed the experiments; E.D., A.B. and F.M. analyzed the data; S.A. contributed analysis tools; all authors wrote the paper.

Conflicts of Interest: The authors declare no conflict of interest

Abbreviations

VBM	Vibro-Ball Milling
CM	Centrifugal Milling
TS	Total Solids
VS	Volatile Solids
SSF	Simultaneous Saccharification Fermentation

References

1. John, R.P.; Anisha, G.S.; Nampoothiri, K.M.; Pandey, A. Micro and macroalgal biomass: A renewable source for bioethanol. *Bioresour. Technol.* **2011**, *102*, 186–193. [CrossRef] [PubMed]
2. Borines, M.G.; de Leon, R.L.; Cuello, J.L. Bioethanol production from the macroalgae *Sargassum* spp. *Bioresour. Technol.* **2013**, *138*, 22–29. [CrossRef] [PubMed]
3. Pham, T.N.; Um, Y.; Yoon, H.H. Pretreatment of macroalgae for volatile fatty acid production. *Bioresour. Technol.* **2013**, *146*, 754–757. [CrossRef] [PubMed]
4. Lee, J.Y.; Li, P.; Lee, J.; Ryu, H.J.; Oh, K.K. Ethanol production from Saccharina japonica using an optimized extremely low acid pretreatment followed by simultaneous saccharification and fermentation. *Bioresour. Technol.* **2013**, *127*, 119–125. [CrossRef] [PubMed]
5. Park, J.-H.; Hong, J.-Y.; Jang, H.C.; Oh, S.G.; Kim, S.-H.; Yoon, J.-J.; Kim, Y.J. Use of *Gelidium amansii* as a promising resource for bioethanol: A practical approach for continuous dilute-acid hydrolysis and fermentation. *Bioresour. Technol.* **2012**, *108*, 83–88. [CrossRef] [PubMed]
6. Sambusiti, C.; Bellucci, M.; Zabaniotou, A.; Beneduce, L.; Monlau, F. Algae as promising feedstocks for fermentative biohydrogen production according to a biorefinery approach: A comprehensive review. *Renew. Sustain. Energy Rev.* **2015**, *44*, 20–36. [CrossRef]
7. Jung, K.A.; Lim, S.-R.; Kim, Y.; Park, J.M. Potentials of macroalgae as feedstocks for biorefinery. *Bioresour. Technol.* **2013**, *135*, 182–190. [CrossRef] [PubMed]
8. Meinita, M.D.N.; Hong, Y.-K.; Jeong, G.-T. Comparison of sulfuric and hydrochloric acids as catalysts in hydrolysis of *Kappaphycus alvarezii* (cottonii). *Bioprocess Biosyst. Eng.* **2012**, *35*, 123–128. [CrossRef] [PubMed]
9. Wang, X.; Liu, X.; Wang, G. Two-stage Hydrolysis of Invasive Algal Feedstock for Ethanol FermentationF. *J. Integr. Plant Biol.* **2011**, *53*, 246–252. [CrossRef] [PubMed]

10. Schultz-Jensen, N.; Thygesen, A.; Leipold, F.; Thomsen, S.T.; Roslander, C.; Lilholt, H.; Bjerre, A.B. Pretreatment of the macroalgae *Chaetomorpha linum* for the production of bioethanol—Comparison of five pretreatment technologies. *Bioresour. Technol.* **2013**, *140*, 36–42. [CrossRef] [PubMed]
11. Monlau, F.; Sambusiti, C.; Antoniou, N.; Zabaniotou, A.; Solhy, A.; Barakat, A. Pyrochars from bioenergy residue as novel bio-adsorbents for lignocellulosic hydrolysate detoxification. *Bioresour. Technol.* **2015**, *187*, 379–386. [CrossRef] [PubMed]
12. Delgenes, J.P.; Moletta, R.; Navarro, J.M. Effects of lignocellulose degradation products on ethanol fermentations of glucose and xylose by *Saccharomyces cerevisiae*, *Zymomonas mobilis*, *Pichia stipitis*, and *Candida shehatae*. *Enzym. Microb. Technol.* **1996**, *19*, 220–225. [CrossRef]
13. Motte, J.C.; Delenne, J.Y.; Rouau, X.; Mayer-Laigle, C. Mineral–vegetal co-milling: An effective process to improve lignocellulosic biomass fine milling and to increase interweaving between mixed particles. *Bioresour. Technol.* **2015**, *192*, 703–710. [CrossRef] [PubMed]
14. Kim, N.-J.; Li, H.; Jung, K.; Chang, H.N.; Lee, P.C. Ethanol production from marine algal hydrolysates using *Escherichia coli* KO11. *Bioresour. Technol.* **2011**, *102*, 7466–7469. [CrossRef] [PubMed]
15. Trivedi, N.; Gupta, V.; Reddy, C.R.K.; Jha, B. Enzymatic hydrolysis and production of bioethanol from common macrophytic green alga *Ulva fasciata* Delile. *Bioresour. Technol.* **2013**, *150*, 106–112. [CrossRef] [PubMed]
16. Jard, G.; Marfaing, H.; Carrère, H.; Delgenes, J.P.; Steyer, J.P.; Dumas, C. French Brittany macroalgae screening: Composition and methane potential for potential alternative sources of energy and products. *Bioresour. Technol.* **2013**, *144*, 492–498. [CrossRef] [PubMed]
17. Cho, H.; Ra, C.-H.; Kim, S.-K. Ethanol Production from the Seaweed *Gelidium Amansii*, Using Specific Sugar Acclimated Yeasts. *J. Microbiol. Biotechnol.* **2014**, *24*, 264–269. [CrossRef] [PubMed]
18. Motte, J.-C.; Sambusiti, C.; Dumas, C.; Barakat, A. Combination of dry dark fermentation and mechanical pretreatment for lignocellulosic deconstruction: An innovative strategy for biofuels and volatile fatty acids recovery. *Appl. Energy* **2015**, *147*, 67–73. [CrossRef]
19. Barakat, A.; De Vries, H.; Rouau, X. Dry fractionation process as an important step in current and future lignocellulose biorefineries: A review. *Bioresour. Technol.* **2013**, *134*, 362–373. [CrossRef] [PubMed]
20. Barakat, A.; Chuetor, S.; Monlau, F.; Solhy, A.; Rouau, X. Eco-friendly dry chemo-mechanical pretreatments of lignocellulosic biomass: Impact on energy and yield of the enzymatic hydrolysis. *Appl. Energy* **2014**, *113*, 97–105. [CrossRef]
21. Berlowska, J.; Pielech-Przybylska, K.; Balcerek, M.; Cieciura, W.; Borowski, S.; Kregiel, D. Integrated Bioethanol Fermentation/Anaerobic Digestion for Valorization of Sugar Beet Pulp. *Energies* **2017**, *10*, 1255. [CrossRef]
22. American Public Health Association (APHA). *Standard Methods for the Examination of Water and Wastewater*, 21st ed.; American Public Health Association: Washington, DC, USA, 2005.
23. Wychen, V.; Laurens, L.M.L. *Determination of Total Carbohydrates in Algal Biomass*; Laboratory Analytical Pro-Cedure (LAP): Golden, CO, USA, 2013.

Sample Availability: Samples of the compounds are not available from the authors.

MDPI
St. Alban-Anlage 66
4052 Basel
Switzerland
Tel. +41 61 683 77 34
Fax +41 61 302 89 18
www.mdpi.com

Molecules Editorial Office
E-mail: molecules@mdpi.com
www.mdpi.com/journal/molecules

www.ingramcontent.com/pod-product-compliance
Lightning Source LLC
LaVergne TN
LVHW070452061126
851858LV00011B/222